Air Transport System

Dieter Schmitt · Volker Gollnick

Air Transport System

Springer

Dieter Schmitt
ARTS-DS Aeronautical Research & Technology Service
Frankfurt/Main
Germany

Volker Gollnick
Institute for Air Transportation Systems
Technical University Hamburg-Harburg
Hamburg
Germany

ISBN 978-3-7091-1879-5 ISBN 978-3-7091-1880-1 (eBook)
DOI 10.1007/978-3-7091-1880-1

Library of Congress Control Number: 2015943840

Springer Wien Heidelberg New York Dordrecht London

Printed on acid-free paper

Springer-Verlag GmbH Wien is part of Springer Science+Business Media (www.springer.com)

Preface

This book intends to provide an overview and introduction into the physical and operational mechanism of the air transportation system. To think about new aircraft technologies or new airline business models, it is of paramount importance to understand the major interdependencies and interactions between the main stakeholders like airline, airport, air navigation services and aircraft manufacturer.

Compared to other publications on air transportation the focus is on the basic and major technical and operational characteristics of different technologies and procedures to show the functional principles. The functional and process-oriented perspective on air transportation seems to be a key for future developments and progress.

Therefore, the book starts with an introduction to the definition of the air transportation system and its main stakeholder.

A historical look back on the development of the air transportation system highlighting the big steps forward is given in Chap. 2.

Methods to predict the future of aviation, such as scenario technique and market forecasts of the various manufacturers, are presented in Chap. 3.

Chapter 4 gives an overview of governmental rules and organizations, which directly affect air transportation. The safety philosophy of aviation is presented with an introduction to the certification of aircraft and ATM-systems. Also, security as an upcoming major issue is addressed.

Chapter 5 presents an introduction to the physics of flight and the principles of aircraft design. Also, boundaries and limitations of aircraft operations are discussed. A discussion of various aircraft configurations including an outlook to unconventional future configurations closes this chapter.

Chapter 6 is dedicated to the aircraft manufacturer. A focus is put on the organization and development process in international companies. The cashflow and economical assessments of aircraft programmes are also part of this chapter. Finally, the actual supply chain and the role of the engine manufacturer is addressed.

Ways of how an aircraft is operated by an airline are discussed in Chap. 7. The development of global operation strategies is discussed including the different concepts of low-cost carrier and flag carrier. The relevance of alliances, fleet planning and network development is investigated as well. Also, pricing and ticketing are part of this chapter as well as the role of aircraft maintenance.

Chapter 8 addresses the airport as a major stakeholder. Principal airport concepts and layout are introduced and the various operations on an airport around the aircraft, especially during turn around and taxiing are presented.

The airspace structure and the principal air traffic management processes are part of Chap. 9. Also, the basics of navigation and guidance technologies including the modern satellite-based systems Gallileo and GPS are presented. The safety issues of aircraft separation and wake vortex are also part of this section.

Chapter 10 is dedicated to the environmental boundaries of air transportation. The principles of climate impact and atmospheric implications are presented. Also noise as one of the most significant environmental impacts is discussed. Within this context, emission trading concepts and fees are also presented.

Air transport and its competitors are highlighted in Chap. 11 discussing future challenges. The role of high-speed trains as automotives is investigated and also the impact of new communication technologies on the air transport market is described.

The book closes with an outlook to future challenges and perspectives of air transportation in Chap. 11.

To cover the deeper context of the entire air transportation system would not have been possible without the support and fruitful discussions of many experts in various areas and stakeholders. We cordially thank the following people for their encouraging help:

Dipl.-Vw. Klaus Lütjens, Institute for Air Transportations Systems, Head of Department Air Transport Operations, German Aerospace Center

Dr.-Ing. Alexander Koch, formerly Institute for Air Transportation Systems, German Aerospace Center, Hamburg

Dr.-Ing. Karl Echtermeyer, Manager Aircraft Assessment and Airline Fleet Planning, Lufthansa

Prof. Dr.-Ing. Jan Delfs, Head of Department Acoustics, DLR Institute for Aero- and Fluid dynamics

Dipl.-Ing. Alexander Lau and Dipl.-Ing. Niclas Dzikus, Institute for Air Transportation Systems, German Aerospace Center

Contents

Chapter 1
The Air Transport System

Abstract This chapter provides a broad entrance to transportation and the high level aspects of air transport. Starting with a description of the air transport system and its surroundings, the passenger expectations concerning highly attractive air transportation are explained. Further, the development of mobility and the principle transport chain are presented. Based on the global economic development of populations, the evolution of air transport and the general impact on climate are given. An introduction to high level global challenges as given in ACARE Vision 2020 or NextGen follows. A systems-based view of air transport and definitions of the roles of the most relevant stakeholders provide the way of thinking presented in this book. The chapter ends with a description of how performances can be described and measured to improve the air transport system.

1.1 Introduction

Transport defines all activities, which allow movement of people or goods from one location to another. There are various modes of transport like road, rail, water and air. But also pipelines, cables and space transport can be considered for special purposes.

A transport system is built on infrastructure, vehicles and operational procedures. Transport and travel are elementary drivers to develop civilization bringing people together and exchanging goods. As the air transport system is one of the major pillars of modern transport Fig. 1.1 provides a first insight into this complex system.

Since air transport is intended to move *passengers* and *cargo*, these elements are placed into the centre of the system.

Aircraft like fixed wing transport aircraft, rotorcraft, unmanned systems, etc. developed and produced by the *manufacturers* are the vehicle platforms for air transport. Aircraft are operated by *airlines*, which provide air transport as a service product. In order to enable this service product safe and efficient *Air Traffic Management* (*ATM*) performed by Air Navigation Services (ANS) has to ensure safe and scheduled aircraft flow around the world. *Airports* are understood as the interface between land and air transport, which provide the infrastructure for this interface.

D. Schmitt and V. Gollnick, *Air Transport System*,
DOI 10.1007/978-3-7091-1880-1_1

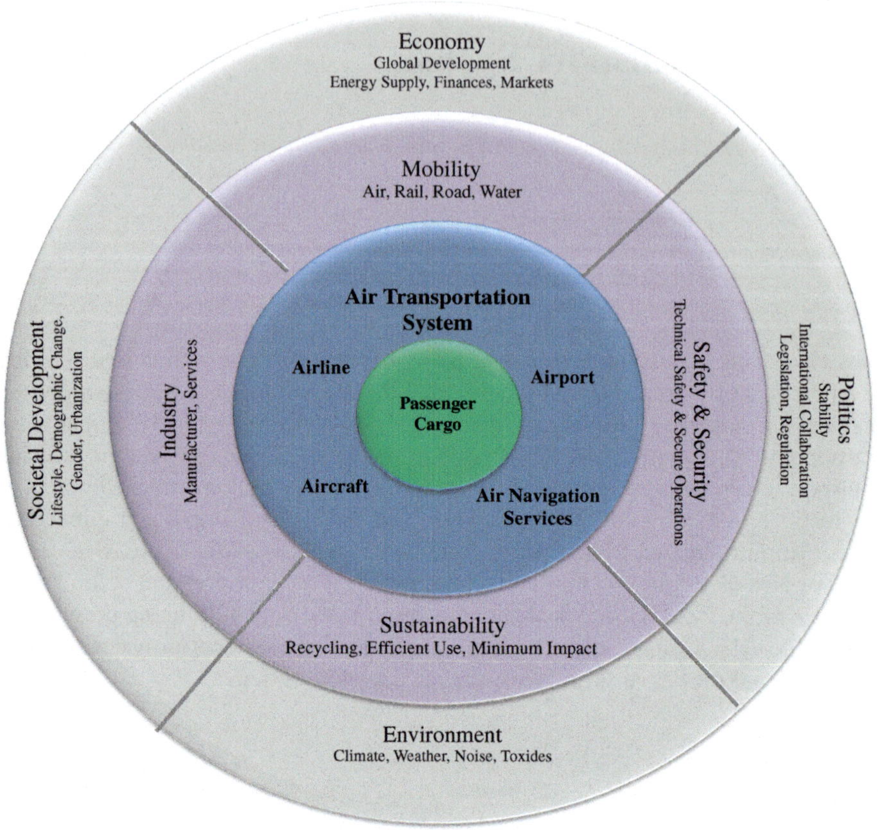

Fig. 1.1 The air transport system and its environment

Beside these main stakeholders in civil air transport, travel agencies, ground services or maintenance, as well as military and general aviation are further operators in the sky. In order to limit the focus of this book, these stakeholders are not explicitly considered.

All aircraft operations, civil as well as military and general aviation are mainly influenced by *society's* expectations and developments. *Politics* in general, represented by authorities develops and sets the legal and regulatory framework to enable air transport. *Economy*, as a key for people's prosperity and welfare influences air transport. *Other transport systems* like rail, ship or automotive are operated complementary in multi-modal operations with air transport, but they are also competing. At last, *environmental responsibility* mainly in terms of climate and noise impact has become a major influence on air transport today [1].

This brief overview gives a first impression of the main elements, which compose and affect the air transport system.

The purpose of this book is to introduce the different *stakeholders* and their ways of acting in the system. Further, it is intended to provide some awareness and understanding of the various *interactions and interdependencies* between the stakeholders. For these reasons, the major relevant *technical systems* and their principal characteristics are presented to provide the capability to assess new technologies and the impact to the overall system. Also the *main processes of air transport*, including the business models are described.

The readers, experienced professionals as well as students of mechanical and aerospace engineering, also logistics and civil engineering, shall be able to get a comprehensive technical and operational understanding and overview of the air transport system.

1.2 Passenger Expectations

Since the beginning of the twentieth century, aviation has tremendously affected mobility of people. During the last 100 years, the technical performance of the air transport system reached a very high level of maturity till date, Chap. 2.

Every day, when people make use of the air transport system, they discover some elements of discomfort and inefficiency. People complain, for example, about delays, uncomfortable seats in the cabin, toxic air in the cabin, environmental pollution, too long travelling times or too high ticket prices. It is not to be discussed here, whether these complaints are actually entitled or not; they only give an indication, that everybody has some aspects, which can be improved.

Engineers always tend to find solutions for problems or invent and develop new things. Aerospace engineers also try to improve the air transport system continuously.

Making it better will mean

- to advance *quality* and affordability
- to improve the technical *performance*
- to reduce cost and to increase *profit*
- to increase the *environmental* compatibility.

At the end, the air transport system shall be more attractive for people and be more accepted. This is the basic motivation for all stakeholders to improve, because the fulfilment of customer expectations provides market share, revenue and profit.

The **users** of the air transport system are able to use it in the most efficient way to achieve their individual goals, like travelling or sending goods between two points.

Air transport, by nature is an abstract service, performed by various contributing stakeholders, like the airline, the airport, the ANS and the aircraft and its manufacturer respectively.

The passenger or sender of freight as customer, cannot request for restitution or conversion in case of deficiencies. He pays for a service in advance, hoping for an orderly performance.

The creation and consumption of the product travel coincide. An offered seat on an aircraft, which is not used, is a loss for the airline. Therefore, an airline always targets at a high load factor (LF), which describes the relation between the offered and the occupied amount of seats.

Beside the pure travel service, supporting activities like check-in, security check, refreshments, lounges, etc. are also part of the experience of flight.

These global expectations and challenges are addressed by international targets as they are described for example in the ACARE Vision 2020 and further on Flightpath 2050, which will be introduced in Sect. 1.4, and regarding the future challenges in Chap. 11 [2, 3].

1.3 Transport and Mobility

From the beginning of human era, mobility was a fundamental prerequisite to survive and evolve. During the centuries, human mobility also became an essential pillar for prosperity and welfare (Fig. 1.2).

Mobility itself is the people's ability to move from one location to another. It can be performed by different transportation systems and measures. People can move individually or in groups either by walking or for example taking bikes, cars or aircraft.

With the development of technical features people were able to travel longer distances and reach locations much quicker. In the beginning, about 50 km could be overcome within 9 h travelling time per day, today during the same time most of the continents can be reached.

It is of paramount importance to distinguish between people′s mobility and movements of transport vehicles.

Transport Vehicle Speed (km/h)	*Distance (km)* 10 100 1000 10000	*Infrastructure*	*Destinations from Hamburg*
Walking Ca. 5 km/h	54 km (1h Rest)	Road	Lüneburg
Diligence ca. 12 km/h	70 km (1h Rest)	Driveway	Lübeck
Bicycle ca. 20 km/h	151 km (1h Rest)	Driveway	Hannover
Train ca. 70 km/h	285 km	Rail	Berlin
Ship ca. 40 km/h	770 km	Water	Dover
Automotives ca. 100 km/h	900 km	Highway	Paris
High Speed Train ca. 250 km/h	2500 km	High Speed Track	Madrid
Aircraft ca. 850 km/h	7500 km	Airport	New York

Fig. 1.2 Development of human mobility

A fixed amount of people as well as cargo can be transported, either by a large number of transport vehicles with limited capacity of payload or using fewer vehicles providing large storage capacity.

At this point, it is essential to understand, the capacity of the area where it is addressed is an essential design parameter, to set up an efficient transport system.

If capacity is associated with the transport vehicle, the required energy effort as well as fuel consumption and emissions could be shared by more people and cargo. On the other hand, when capacity is an issue of the transport flow, the frequency of vehicle movements and the capacity of the rail, road, air networks as well as air spaces, airports and railway stations become the essential design parameters.

In order to reach another location, people today often use different transport systems during a journey. This principle is called Multi Modal Transport (MMT). Each transport mission from door to door can be described by five phases [4, 5]. Further, if in a trip different transport systems might be used, this is known as Inter Modal Transport (IMT). It is possible to compare different multi-modal transport chains, using a Five-Phase-Model (FPM) with different main track transport vehicles in a transparent way as shown in Fig. 1.3.

The first phase, always beginning at home or in case of cargo transport at the production plant or logistics centre, covers the distance from this point to the border of the city. It is characterised by low speed and short distance up to 20 km approximately. Various transport choices are available like walking, taking public transport or automotive, which is a typical example of multi-modality. In case of rail or air transport on the main track, the second phase addresses the transition from those initial transport choices to trains or aircraft. Compared to all kind of automotive transport there is no transport performance for rail and air transport in the railway station or airport because no real distance is travelled! But in both cases, significant time is consumed to change from one system to another. Taking automotive transport as a reference, these systems overcome a distance of up to 100 km

	Phase 1	**Phase 2**	**Phase 3**	**Phase 4**	**Phase 5**
Mode	Departure	Transition Mode Change	Main Track	Transition Mode Change	Arrival
Range [km]	0 -20	<100	>>100	<100	0-20
Speed in [km/h]	0 - 50	0 - (100)	100 - 900	0 - (100)	0 - 50
Road	City	National Road	Motorway	National Road	City
Rail	Car, Taxi, PT	Railway Station	Highspeed Train	Railway Station	Car, Taxi, PT
Air	Car, Taxi, PT	Airport	Air Transport	Airport	Car, Taxi, PT
Water	Car, Taxi, PT	Harbor	Ship Transport	Harbor	Car, Taxi, PT

PT = Public Transport (Metro, Tram, Bus, ..)

Fig. 1.3 Five phases of multi-modal transport [4]

approximately to reach the highway for the main cruise track. The cruising speed is around 70–100 km/h until the highway is reached. Further, the third phase covers the main track, which is intended to overcome the longest distance as quick as possible. Here, all transport systems use their maximum speed. Phases four and five are the reverse phase of phase two and one.

Aviation in this context provides the unique capabilities to be the fastest and offers the largest range performance compared to the other transport systems. Further, it is not limited to any continental border. Therefore, aviation can connect cities on most continents directly without being hindered by oceans or mountains. However, air transport requires normally a mode change before and after the air phase (phase 3), which might last between 30 min and 2 h typically. This “loss” of time is the reason why air transport is only efficient at distances longer than 500 km. Here the geographic situation, i.e. the density of transport networks influences the attraction of a transport system significantly. At last, the main transport systems rail, aviation, automotive and ship are facing an increasing competitive situation, which will be discussed in Chap. 11. For the design of future air transport concepts, it will become more and more relevant to identify the individual advantages and disadvantages of all elements in order to integrate them in the most efficient way.

1.4 The Air Transport System Today

Mobility as a whole and air transport in particular have grown dramatically during the last decades. This development is driven by man’s wish to move quicker and further away. Mobility around the world is state-of-the-art today.

Air transport as a whole has a significant economic relevance. Almost 15 million jobs globally are associated directly or indirectly with the aviation industry [6]. 7,80,000 people are directly working in the aerospace industry, while 2 million are associated with airlines around the world. At the airports, about 2.7 million employees are engaged, which in summary lead to 5.5 million jobs, which are directly created by the aviation industry. These figures indicate strongly the welfare impact of aviation. More the countries are developing, the more people's mobility increases and the economic power grows.

As shown in Fig. 1.4 from a certain level of Gross Domestic Product (GDP) of about 25,000 no further increase of mobility is observed. Consequently in these regions only marginal increase in passenger movements and aircraft movements are to be expected.

This growth will be heavily driven by the growing economies in Asia, especially India and China, while the highly developed countries like USA and Europe will face certain saturation in air traffic mobility, Chap. 3.

For those markets, the competitive situation for air transport is becoming stronger, especially since high speed trains with cruise speeds up to 400 km/h strengthened their advantage to link cities at their heart. Compared to this situation airports are mostly located in the surrounding of cities, which requires more travelling time.

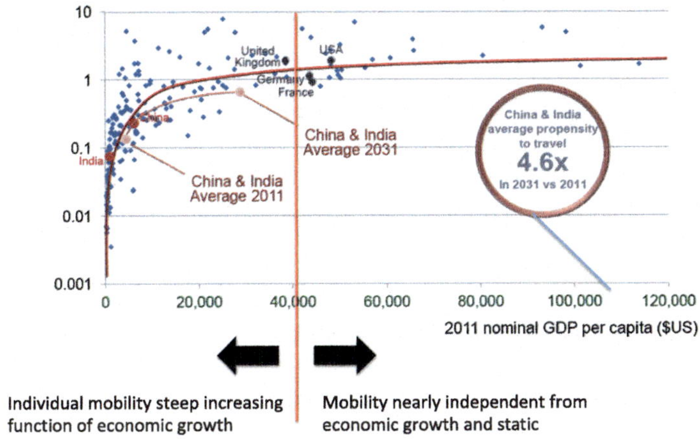

Fig. 1.4 Global mobility development depending on GDP [7]

For growing and developing countries, where passenger mobility is lower by a factor of roughly ten, there is a strong demand for more aircraft payload capacity as well as for aircraft movements. Here also airspace and airport capacity becomes essential.

From the 70s of the last century until 2000, air transport grew up to 3 billion passenger kilometre (Pkm), Fig. 1.5. In the decade 2000–2010, passenger air transport increased again from 3.5 Billion passenger kilometre to approximately 5.7 Billion passenger kilometre globally. Also for the next decade a global increase from 5.7 to 9 Billion passenger kilometre is expected.

Detailed analysis has shown that most of the world's aircraft and engine manufacturers came to the same perspectives [9]. However, this development is heavily depending on future global economic and political development. As shown in the Fig. 1.5 global events like the gulf crisis in 1990 or the 9/11 tragedy did not significantly affect the global trend. However, they shifted the progressive increase

Fig. 1.5 Expected global passenger air mobility trend 1970–2020 [8]

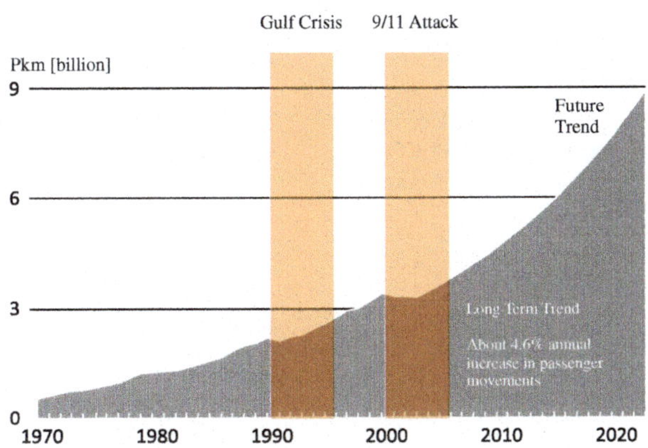

to later maxima. Nevertheless, this development leads to an average global passenger air transport growth of about 5.2 %.

A further aspect to be analysed refers to the development of the aircraft LF. As mentioned before, the LF of a transport vehicle describes the share how much available seat capacity is used on a trip.

In the last step, considering the aircraft movements under Instrumental Flight Rules (IFR), which are typical for civil passenger aviation, an increase from 8,500,000 to 9,500,000 movements per year, at least in Europe is observed between 2000 and 2010. Under the impression of the financial crisis in 2008, a lower increase up to 1,000,000–1,200,000 movements per year is prospected [10].

Comparing both trends, passenger movements are growing faster than aircraft movements, which is in line with the observation that the LF of the world fleet as well as the seat capacity are increasing. Therefore, with the same amount of aircraft more transport performance is provided to serve people mobility.

When this trend of growth of air transport will go on, the demand for fossil kerosene and the emission of CO_2 will increase proportionally.

The first is conflicting with the limits of crude oil causing high prices, the latter is threatening our environment and health leading to climate changes, see Chap. 10.

Normally one would assume that this development would be visible in the global air transport energy effort. Looking at Fig. 1.6 there is a significant increase of energy effort due to an effect, which is called "Rebound Effect" [11], meaning that on global level all individual effects like reduced engine fuel consumption are overcompensated by an increase in aircraft movements. On individual aircraft level, such improvements like structural weight saving are overcompensated by additional equipment for comfort, e.g. cabin entertainment systems.

This brief look at the development and status of the air transport system today has shown its social relevance to provide mobility and economic growth. On the other hand, due to its high level of maturity the ATS is facing technical limits and new breakthroughs are needed to evolve into the future. Going a step further, the established way of quantitative growth with more and more aircraft might shift to a new paradigm requesting for qualitative growth in air transport as raised by the Club of Rome in 1972 [13]. This way ahead will be discussed in Chap. 11, which is about the future challenges.

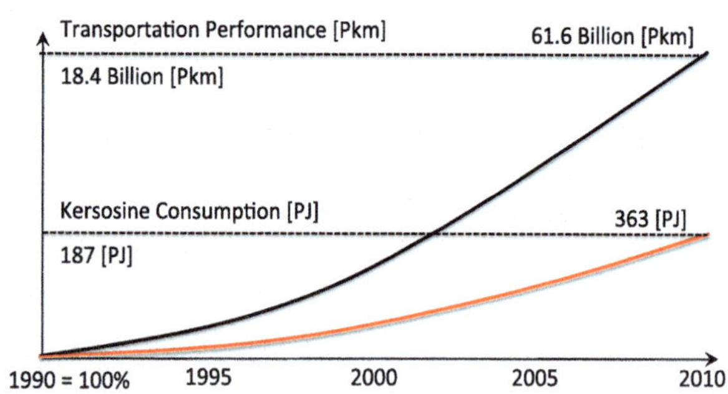

Fig. 1.6 Trend of decoupling air traffic growth and CO_2 emissions due to technologies [12]

1.5 Current Challenges of the Air Transport System

Summarising the global developments, previously described air transport grew tremendously in terms of passenger and aircraft movements. The latter is based on a significant increase in the amount of aircraft. The amount of aircraft causes limitations in airport and airspace capacities, especially in Europe and Northern America [7–9]. In the growing regions, those capacity limits are not yet reached, but need to be considered for future developments.

Responding to these challenges in 2001, the Advisory Council of Aeronautical Research in Europe (ACARE) has defined high level targets for future improvements, to make the global air transport system competitive and attractive for the twenty-first century. These high level targets are listed in Chap. 11 Table 11.1, known as the ACARE Vision 2020 [2]. Also in the United States targets for the future air transport have been formulated. Here on the operational field the *NextGen* programme especially defines objectives for more efficiency in air transport flow. The American *N+3* project driven by NASA additionally sets requirements on improved aircraft performance. Comparing both approaches the European Vision 2020 can be understood as more holistic, while the American *NextGen* ATS addresses more technologies to increase the throughput of aircraft in the airspace and at the airport. These goals are set to be achieved until 2020 and refer to the ATS performance of 2000 as the reference. It is essential to notice, that all these targets are related to a single aircraft performance of newly developed aircraft. Since there are thousands of older aircraft also in service in 2020, the entire world fleet will not be capable to come close to these targets.

A mid-term resume, however, indicated in 2011 that not all of these goals could be achieved until 2020 [14]. While the environmental goals concerning CO_2 and NO_x emissions are achievable by more than 50 %, an extension of the airport and airspace capacity as well as the improvement of punctuality are hard to reach until 2020. Further, actual research on climate impact of aviation has raised the question whether the percentage requirements on reduction of emissions are the right one, because the impact on global warming in terms of contribution to ΔT seems to be more appropriate. This metric covers interdepending effects in a better way and will be discussed in Chap. 10.

Therefore, only an integrated approach merging incremental contributions allows achieving the global goals for new air transport systems.

Following the ACARE vision, a new European revision on the future goals has been developed in *Flightpath2050* [3].

The potential reductions, which various technologies are considered to contribute, are understood as single disciplinary contributions [15–17]. It is therefore mandatory to understand the air transport system and its complexity as a whole and to

- analyse and identify *weaknesses in the entire system* as well as on substructure and subsystem level
- develop *future integrated concepts* as proposals for new solutions rather than single technology solutions

- improve *air transport processes* on global chain level and also on subsystem level.

For this purpose, the next section provides a system engineering approach for a holistic air transport system description.

1.6 A Systematic Description of Air Transport

There are different approaches to define and structure the air transport system. One proposed by *Wensveen* is driven by a management view [18]. *Wensveen* uses an economical view to address the organisational elements of air transport like regulators and associations. But he also addresses the different markets and economical influences. Further on, he describes the air transport system from airline perspective and its different business models.

Mensen provides a more organisational vision on the air transport system, focussing very much on the ATM/control and the regulatory organisations [19]. From his point of view, all institutions and procedures, which contribute to run the ATS define it. Also *Mason* and the *MIT* built the ATS description on organisational aspects.

To approach such a complex system, Systems Engineering (SE) is an appropriate method to define and structure the various elements.

A system generally consists of elements, which are related to each other, Fig. 1.7. Major characteristics of a system are its boundaries, which separate a system from its environment or other systems.

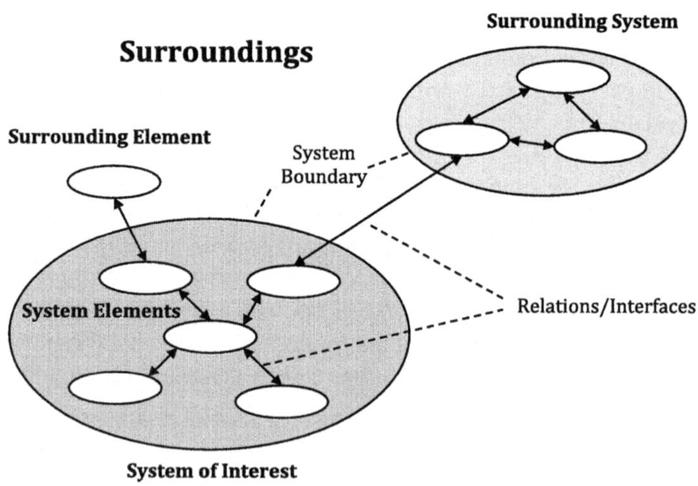

Fig. 1.7 Principle of system definition

The definition of these boundaries allows a separated analysis of a system and only the direct cross references to the outside world need to be considered. Such a system, considering also the outside impacts through interfaces or boundary conditions is understood as an open system, which is the Air Transport System. The global impacts as addressed in Fig. 1.1 need to be taken into account in further discussions. In order to provide an understanding of this approach, the societal environmental awareness should be considered in terms of CO_2 emissions. The reduction of these is a requirement for the overall aircraft and also for the engine. In this way, the aircraft and also the engine have to be considered as open systems. However, if the power supply of the electronic engine control system (EECS) is in the focus of research and development, this element is neither directly nor indirectly related to CO_2 emissions from operational perspective. Therefore the EECS can be considered as a closed system without these outer influences.

This approach simplifies the analysis and design. A system itself can also contain various substructures, which commonly affect the higher system level.

From this perspective, the air transport system is understood as a system of systems, which covers for example the aircraft, the airport and ATM as substructures.

Following the system engineering philosophy, the air transport system is hierarchically structured into the system, substructures, subsystems and components:

- the overall air transport system as the *system* is composed of
- aircraft, airlines, air traffic infrastructures, airports as *substructures*, Chaps. 5, 7–9 while
- e.g. wing, avionics, etc. of an aircraft, or e.g. surveillance radar, air space structures of the air traffic infrastructure, or the terminal, the APRON of the airport are *subsystem*s of one substructure and
- e.g. flaps and slats are components of the flight control subsystem of the aircraft, while antennas and receivers are parts of the radar subsystem of ATM, check-in areas, gates are *component*s of the airport terminal subsystem, etc.

Such an approach is suitable to develop balanced optimisations among the main substructures of the air transport system, in order to achieve multidisciplinary or global goals like those of ACARE.

Generally every stakeholder in the ATS provides some infrastructures and holds some processes to make the system run.

While this view is mainly technically driven, the stakeholder's perspective on the air transport is a different one:

- here the *aircraft manufacturer* is in charge of developing the aircraft based on various system and stakeholder requirements;
- the *airline* provides the core product air travel by operating the aircraft;
- *military and general aviation* which are also parts of the ATS occupy resources of air traffic control, airspace and airport capacities.

In addition,

- *general public* which is on the one hand the customer of the ATS and on the other hand requesting for social compliance;
- governmental and non-governmental *organisations*;
- *customer* as a passenger or one who is shipping goods

are stakeholders of the ATS representing needs, expectations and requirements, which should be fulfilled as described in Fig. 1.1.

Further on there are surrounding influences, which interfere with the air transport system. These are physical environments like natural laws, geographic conditions as well as meteorological and climatic conditions. Also, social implications like public employment and purchase power, travel demand, medial opinions or fear about terrorist attacks affect the air transport system.

At last, economical influences, e.g. world economic growth, raw material and oil market development or regional transport, economical and business situation drive the ATS.

According to other authors, there might be further stakeholders, e.g. like ground service provider, meteorological services, travelling agencies, research organisations [18–20].

Most of the stakeholders, except the general public, provide some sort of product or service like aircraft, regulations, ensuring safety, navigation performances, etc. to make ATS operational. The general public, as customer and affected community in contrast is using the ATS and raising expectations. The customer's view on air transport is quite individual. He wants to move between two points at the moment, which is very specific. He wants to move quick and comfortable, because he wants to be active at the final destination to spend his holidays, to do his business, or just to enjoy his leisure. Therefore, the passenger as a customer is always looking at air transport as a process. Typically he is not looking at a certain technology itself, but at seamless integrated performance of elements along his travel. The same is also true for air cargo transport. Also in this case the dispatcher and receiver of goods expect a seamless service and do not care about deficiencies in any technology being used, where the customer does not care about nor has any preferences.

There are two conclusions to be drawn from this observation. First, the customer does not care about who is responsible for a deficiency during the travel chain. The second issue is related to the technologies being used.
Here a technology is defined either as

- a *physical principle* being used in a sensor or machine, etc. like a laminar flow on an aircraft or a radar-based scan at a security check-in a terminal or
- a rule-based standardised *procedure*, which describes a certain sequence, like an approach and landing manoeuvre of an aircraft, or
- a *process*, which describes the chain of activities, like a production sequence during aircraft assembly, or a cargo moving process, comprising customs activities, transport activities, security checks, etc.

Especially since a lot of physical principles are known well and have reached a high level of maturity improvements in efficiency are expected to be made by investigating and developing new procedures and processes, where given physical principles are put together in a new and better way.

The development of new solutions for the ATS follows the roadmap of a V-model like it is well known from software systems development, e.g. Mil-Std 2197, IEEE 1220 (Fig. 1.8).

The system is decomposed to the relevant level of detail. On the lower level, (substructure, subsystem, component) the decomposition stops if all relevant interdependencies between the other system elements are addressed. On this level, a new solution is to be developed [21].

This leads to the aspect of integration, which is a key characteristic of a system. *Integration* of technologies in the aforementioned way can be done in different ways to create systems:

- *intellectual or descriptive integration*, merging physical principles and/or procedures to processes in a theoretical, functional way;
- *IT-based integration*, where different models for calculation and simulation are put together in order to set up a virtual system, which allows calculation, layout and simulation;
- *physical integration*, where the real hardware, operational software and procedures are put together to setup the real system;

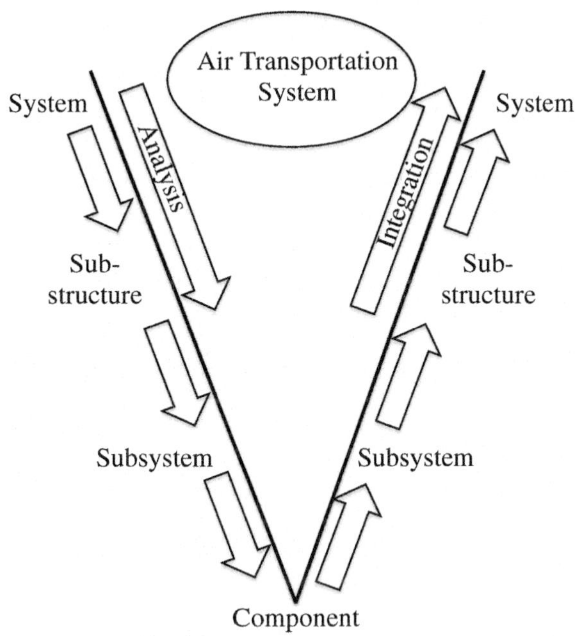

Fig. 1.8 V-Model for analysis and integration of the air transport system

All three stages of integration appear during the development and analysis of the air transport system. While the first provides a first insight to interdependencies of newly defined system architecture, the second brings out interactions between the systems elements, which have not been considered before, for example due to the huge amount of potential solutions. The physical integration at the end provides the ultimate way to merge different physical principles like hardware and software, mechanical and electrical solutions.

However it must be emphasised that currently all stakeholders follow individual interest and strategies to maximise their business instead of collaboratively contributing to an overall seamless air transport system.

1.7 Air Transport System Performances

Any kind of modification of the various air transport systems is intended to improve the entire system leading to more efficiency.

On a Meta-level, efficiency itself describes the relation between a requested benefit or target, like the movement of a certain amount of passengers and the effort and potential disadvantages which are associated with this target. Such an effort can be described as the amount of energy or fuel, which is needed to perform the transport task between A and B. Associated cost, for e.g. staff, fees or supporting services are understood as effort to be spent. Related emissions and noise, also required land use can be described as potential disadvantages, because these effects are not wanted.

In this context, it is necessary to discuss efficiency and effectiveness [22]. A popular distinction between these two performances, describes efficiency as doing things right, while effectiveness is understood as doing the right things [23]. In the context of air transport this definition means, that for example the manual assembly of an aircraft is less effective than the assembly using automation, which allows much quicker and higher quality assembly.

On the other hand, efficient air transport can be seen as the movement of passengers with as less fuel and time as possible.

As a basis for these considerations air transport work (ATW) is defined as the amount of passenger or goods being carried over a given distance, i.e.:

$$\text{ATW} = \text{pax or goods} * \text{distance } [\text{Pkm}] \text{ or } [\text{tkm}] \tag{1.1}$$

Referring to the goals of the *Vision 2020* efficiency determines the resulting transport performance in passenger kilometre or tonnes kilometre related to the effort to be spent in terms of overall travelling time, energy effort, cost and environmental impact.

The requested air transport work is related to the time and effort necessary to be spent, i.e. energy, cost and associated environmental impact.

Transport efficiency therefore is characterised by balancing the requested transport work and the required efforts in terms of cost, energy, emissions, noise, and land use.

Although these global parameters are applicable to all stakeholders in air transport, the detailed impact and characteristics differ. Eurocontrol, in 2006 first published an approach to describe efficiency and effectiveness in air transport [24]. Here, Key Performance Areas (KPA) and Key Performance Indicator (KPI) have been defined to describe and quantify the performance of air traffic, especially. Key Performance Areas in this context have been defined, like

- Capacity and delays
- Cost effectiveness
- Environment
- Airports
- …

These KPA are extended to those agreed by the 11th ICAO conference adding:

- Access and Equity
- Global interoperability
- Predictability
- Security

To determine these KPAs, it is not sufficient to use one parameter each only. This is the reason why different KPIs have been defined to characterise the KPA. Moreover, each KPI needs to be defined in particular for its individual environment of application.

Taking the KPI for capacity as an example, those characteristics have been chosen which influence the capacity of the air space in terms of IFR flights handled by the European ANS.

Increasing amount of take-off and landings depending on the available runway capacities are characterising airport performance, as another example.

If one tries to apply this philosophy of performance areas and indicators to other air transport stakeholders like the aircraft, the following indicator can be used. Aircraft capacity is described by seat capacity on an aircraft. Distinguishing between long and short range aircraft, more seats at the same aircraft size can be used as a KPI. Cost effectiveness as a further performance area might be described as the amount of cockpit and cabin crew cost as well as maintenance and fuel cost. The latter should be related either to a single flight and to the entire life cycle. Aircraft efficiency can be defined in two ways. First, the design efficiency in terms of the maximum payload capacity related to the operating empty mass can be used to characterise the efficiency of the design. Second, the fuel burn is a further economic characteristic of the aircraft. At last, environmental performance of aircraft is characterised by the amount of emissions and the noise carpet developing during take-off, cruise and landing. At this point, one may wonder about the missing physical aircraft performance in terms of range and speed. These parameters seem to be not really useful for performance indication, since their value is depending on the individual real mission. Cruise speed and range itself provide the capabilities of an aircraft for flexible operations on various missions.

For airlines, those performance areas may address the fleet's wide amount of emissions as an emission indicator as well as the relation of the amount of aircraft to the annual flown kilometres, which indicates the efficiency of the operated fleet. Also, aircraft availability is a useful indicator for airline effectiveness and flexibility. The amount of accidents and incidents related to an airline fleet and flown kilometres will indicate the level of airline safety.

At last it has to be noted, that airport specific performance indicators are still addressed within the ATM performance areas.

Reflecting this discussion about performance areas and indicators, there are various measures to characterise the performance of the different main stakeholders in air transport. It has been shown, that the definition of these indicators is depending on the individual stakeholder's interest and perspective. In order to make such an assessment comparable, at least the performance areas should be defined in the same way, while the indicators should be set up in a similar physical description.

References

1. Janic, M.: The Sustainability of Air Transport, 1st edn. Ashgate publishing company, Farnham (2007)
2. ACARE: European aeronautics: vision for 2020. www.acare4europe.org/docs/Vision%202020.pdf (2001). Accessed 27 Feb 2011
3. European commission: flightpath 2050—Europe's vision for aviation. http://ec.europa.eu/transport/modes/air/doc/flightpath2050.pdf. Accessed 02 July 2013
4. Gollnick, V.: Comparative assessment of different transport systems, Ph.-D. Thesis, Institute for Aviation Technologies, TU Munich (April 2004)
5. Gollnick, V.: Potential for transport efficiency improvements of aviation transport systems. In: Paper 99, 25th ICAS World Congress, Hamburg, 3–8 Sept 2006
6. ATAG: the economic and social benefits of air transport 2008, Air transport action group, 22 Route de l`Aèroport, P.O. Box 49, 1215 Geneva 15, Switzerland
7. Airbus: global market forecast. www-airbus.com/en/corporate/gmf2009. Accessed 28 Feb 2013
8. Boeing: current market outlook. http://www.boeing.com/boeing/commercial/cmo/. Accessed 25 Nov 2013
9. Nolte, P., Gollnick, V.: ACARE2020—A half time resumee, 2nd symposium about future air transport, Institute of Air Transport Systems, German Aerospace Center at the Technical University of Hamburg, Hamburg (Sept 2011)
10. Eurocontrol: Eurocontrol—seven-year forecast, Eurocontrol. http://www.eurocontrol.int/documents/eurocontrol-long-term-forecast-flight-movements-2010–2030 (Sept 2012)
11. Madlener, R.: Saving energy through improvements in efficiency is an illusion in a growing system. Energiewirtschaftliche Tagesfragen, **62**(8), (August 2012) (in German)
12. Pfeiffer, U.: Report2012—energy efficiency and climate protection. Bundesverband der Deutschen Luftverkehrswirtschaft, Berlin (2012) (in German)
13. Meadows, D.H., Randers D.L., et al.: The Limits to Growth. Universe Books, New York. ISBN:0-87663-165-0 (1972)

14. NASA: NASA & The Next Generation Air Transport System (NextGen). http://lp.ncdownloader.com/eb2/?q=nextgen%20whitepaper%2006%2026%2007%20pdf. Accessed 26 July 2006
15. Gollnick, V., Szodruch, J., Stumpf, E.: ATS beyond 2020, EREAnet forum the green air transport system, Bonn, Germany, 31 Oct 2007
16. Gollnick, V.: Environmental aspects of air transport future technologies & prospects, Presentation at the Kreditanstalt für Wiederaufbau, Frankfurt (Sept 2007)
17. Gollnick, V.: Air transport systems, Lecture Series, Technical university Hamburg-Harburg (2011)
18. Schilling, T.: A systems engineering approach to define the air transport system, Institute of Air Transport Systems, Technical University Hamburg-Harburg, IB-328-2009-09, Hamburg (2009) (in German)
19. Mensen, H.: Handbuch der Luftfahrt (Aviation Manual), 1st edn. Springer Publishing, Berlin (2003). (in German)
20. Plath, F.: Analysis and synthesis of civil aviation market forecasts in a database, Institute of Air Transport Systems, Technical University Hamburg-Harburg, IB-328-2008-06, Hamburg (2008) (in German)
21. Gollnick, V., Langhans, S., Stumpf, E.: A holistic approach to evaluate the air transport system. In: 26th ICAS World Congress, Anchorage (Sept 2008)
22. N.N.: Definition of efficiency and effectiveness. http://en.wikipedia.org/wiki/Efficiency. Accessed 13 Dec 2012
23. Wensveen, J.G.: Air transport—a management perspective, 6th edn, Ashgate publishing company, Farnham (2007)
24. Eurocontrol: single European sky (SES) regulations—regulatory report for performance review, 2.0 edn. Eurocontrol, p. 17 (August 2006)
25. Langhans, S.: A systems-engineering based methodology for economic ATS concepts assessment, Ph.-D. Thesis, Institute of Air Transport Systems, Technical University Hamburg-Harburg, DLR Research Report DLR-FB-2013-04, Hamburg, ISSN:1434-8454 (2013)

Chapter 2
Historical Development of Air Transport

Abstract The historical development of air transport starts with a short review of myths and legends, the Dream of flying, which is as old as mankind. The next part covers the physically based approach of flying, starting from Da Vinci and his drawings of flying vehicles, via the Montgolfier's hot air balloon, Sir George Cayley and his principles of flying. The part about the technically based approach covers briefly the different attempts from Clement Ader, Otto Lilienthal up to the Wright brothers, who finally in 1903 managed to fly with a vehicle heavier than air. It follows the beginning of commercial air transport in Europe and US between the two World Wars. In the 1950s, the jet age in civil air transport started with a disaster of Comet, but all lessons learned from these air accidents helped other companies to start successfully these new jet engine types of civil transport aircraft, which are still flying today. The aircraft design parameters of speed, range, size and fuel efficiency and their development of the last century are shortly addressed to extract the standards and the maturity of today's air transport system. A brief review of the airline development follows with the example of KLM. It follows a short airport review, where the airport development of Atlanta—the biggest airport today—is taken as example.

2.1 The Dream of Flying

The dream of flying is as old as mankind. In all civilizations (old and new like Greek, Chinese, Roman, Inca, Celt et alii.) Gods have certain capabilities to fly and pass easily between earth and heaven. Some courageous people tried to copy this capability by intensively watching the flight of birds and adapting certain mechanisms from them. The Greek mythology tells about the genius Daedalus, who was at his time an excellent artist and innovator. As the king of Crete named Minos wanted to keep his capabilities as architect just for his personal and own profit, Daedalus decided to escape by constructing and building a flying vehicle, which

D. Schmitt and V. Gollnick, *Air Transport System*,
DOI 10.1007/978-3-7091-1880-1_2

consisted of feathers, "fixed by thread and wax, thus constructing the wings with a certain camber just like the birds." [1].

In China, Kites were constructed and also played some mystic role as element between heaven and earth. But details about their technical efforts and achievements are not so well documented.

The Christian religion knows also some persons with flying capabilities, angels and devils, who can—with the help of wings—travel between heaven and earth and underworld/hell. An excellent description of these old myths and first attempts of flying is given in [2], where certain myths about flying attempts in nearly all culture have been found. These ideas and legends of flying are part of cultural or religious habits.

Behind the imagination of flying, which can be found in all old cultures and civilizations, there are also the basic emotional elements of mankind about freedom and mobility. Being capable to fly like a bird means to escape from your local area and discover new islands, a better world, finally the paradise.

But the reality of successful flying attempts has not been reported until the beginning of the "Renaissance" and immediately the name of the famous artist Leonardo da Vinci appears also on the engineering/technical scene.

2.2 Physics Based Approach

Leonardo has postulated "that human beings would be capable to depart into the air with the help of machines with large wings, which had to be designed to overcome the air resistance". A lot of drawings are showing different principles of his flying vehicles: some show a human being, lying horizontally in his apparatus and hands and feet are fixed or controlling some cables or bars; others are showing a person controlling a flapping mechanisms to move the wings up and down; others show a sort of screw, which can be rotated by a filament movement and which will be lifting off vertically when sufficiently accelerated (Fig. 2.1). Also a parachute system can be found in his archive of drawings. So a lot of different flying principles were shown in his drawings and it seems that he had constructed also a lot of models to test his principles. More details can be found in [3, 4].

The next step can be seen with the Montgolfier brothers, who by some chance and luck developed the hot air balloon. They had constructed a balloon and discovered the principle of hot air balloons. The flight of their hot air balloon in front of the King in Versailles in 1783 is reported as a sensation and huge spectacle, having seen the first three passengers being lifted up, a coq, a sheep and a dog (Fig. 2.2).

The principle of hot air balloons was immediately seen as a very good chance to be used for military services. But the disadvantage became also very soon apparent: the balloon was not controllable. He just followed the wind without the possibility to give him a specific direction of flight. So the interest for balloons disappeared quickly.

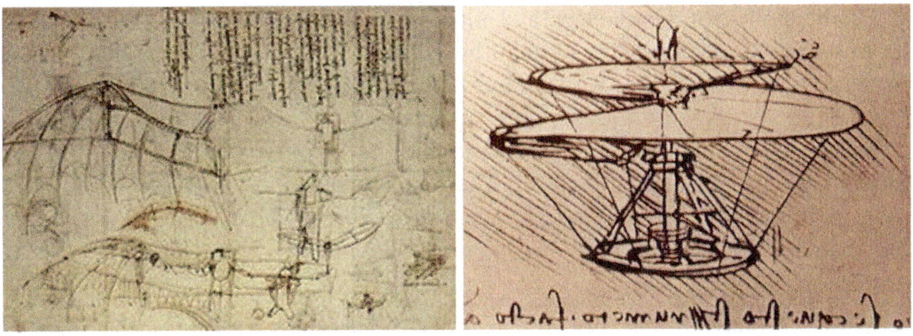

Fig. 2.1 Drawings about flying vehicles from Leonardo da Vinci

Fig. 2.2 The Montgolfiere hot air balloon

It took some further years, before Sir George Cayley (1773–1857) defined and developed some elementary principles fundamentally important for future success of flight vehicles [5].
He postulated the principles of flight in his paper "The art of flying, or Aerial Navigation".

- Separation of forces acting on the wing in lift and drag (vertically lifting and horizontally drag forces)
- Stability and controllability as basic principles for a flying vehicle
- Lift to compensate the mass; leading to light weight structures
- Independent thrust to compensate the aerodynamic drag.

He constructed a lot of models, which were quite successfully demonstrating these postulated principles. Some historians are seeing in Cayley the father of modern aircraft. But it has to be stated, he was just constructing models and he had not yet the final idea about the right propulsive force.

The beginning of the nineteenth century saw a lot of efforts to try to develop the steam engine as a propulsive system, but all efforts, to use steam engines for the flying vehicles failed. This was a false direction with no successful design layouts [6].

The physical scientific community, looking at and commenting the efforts for human flying were also not helpful. Their clear statement was, it would be physically impossible to have flying organisms/vehicles, which are bigger than eagles and vulture. The German physicist Hermann Helmholtz stated 1873 [7]:

> … in developing the large vultures, nature has found the limits, where with muscles operating organisms and by best conditions of alimentation have achieved the maximum size, which by its own wings and for longer time can stay in the air and keep flying. Under these circumstances it is rarely probable, that a man - even with the most sophisticated wing mechanisms - can be in a position to lift up his own body mass and keep it there by just using the force of his muscles.

So no hope and encouragement could be expected from the scientific community.

Nevertheless there were still continuous efforts and a lot of passion to develop a real flying vehicle, which was controllable. The demand from the Emperors, kings and rulers of the world for such sort of vehicle for military and surveillance purposes were still obvious, providing—as we would say today—"the market demand".

2.3 The Technically Based Approach

At the second half of the nineteenth century, a lot of efforts were still underway to overcome all the pessimistic view from the scientists about the "dream of flying". There can be seen two different and competing philosophies in the nineteenth century: Flying following the principle "Lighter than air" and flying following the principle "Heavier than air".

The flying concept "Lighter than air" ended in the development of airships, which had a propulsive unit and could be controlled. This principle, first being successfully tested by the Montgolfier brothers, culminated later on in the development of big airships by Graf Zeppelin. His Zeppelins finally managed to cross the Atlantic between 1931 and 1937 with quite an impressive passenger load of ~50 persons. However, with the disaster of the Zeppelin ZL 129 on 7 May 1937 in Lakehurst, the commercial transport with airships ended immediately.

The principle of flying "Heavier than air" was seen as more problematic. The scientific community classified this principle as impossible for mankind and was providing no support and help. All persons, who still were convinced that flying with machines "heavier than air" was possible, were seen as "fools" and hopeless utopists. The enthusiasts working on the concept of "heavier than air" were following two different principles:

- a sort of flapping wing like the flight of birds or
- a fixed wing but with a strong propulsive unit to accelerate the vehicle.

Some encouragement was seen, when the big steam motors appeared, developed for the railway and the big steamships. But the steam engines were too heavy to be used in the flying vehicle.

In France, two engineers have to be mentioned, who contributed significantly to the development of flying machines with fixed wings, Alphonse Penaud and Clement Ader. In 1876, Penaud patented a design for a large amphibious aircraft with such innovative features as retractable wheels, a glass-enclosed cockpit, a single-lever control for both the rudders and the elevators, and twin propellers driven by an engine concealed in the fuselage. The design was amazingly ahead of its time, but no engine existed that was light enough and could make such an aircraft fly.

Clément Ader (1841–1926) focused on the problem of heavier-than-air flying machines and in 1890 built a steam-powered, bat-winged monoplane, which he named the ***Eole***. It is reported that he flew it a distance of 50 m. The steam engine was unsuitable for sustained and controlled flight, which required the gasoline engine; nevertheless, Between 1894 and 1897 Clément Ader built a larger but still 'Eole-like' twin screw machine which he named the ***Avion***. Interrupted after an accident in 1897, the work was not continued due to a lack of financial resources.

During this time period between 1850 and 1900, a lot of important developments have been made, not only in France but also all over the world, in Brazil, Australia, UK and USA. However, it is not the place here to be exhaustive about the historical details, but [4–6, 10, 11] are giving further details.

A major breakthrough started with Otto Lilienthal. He and his brother Gustav were fascinated from storks. They discovered that young storks—when trying to take off—were always starting against the wind, a very important lesson learnt which we are still using today in our daily air operations.

Otto Lilienthal discovered the importance of forward speed, being similar necessary for lift like the flapping mechanism. He developed a circular rotating device named "Rundlauf", where he tested the wing shapes, first flat plates, than by copying the wing profiles of storks, wing profiles with camber and with incidence and finally complete wings.

All his systematic approach and research about wing profiles was finally published by him in a book in 1989 with the title "Der Vogelflug als Grundlage der Fliegekunst" [8]. It is the first time, that an inventor published his own knowledge openly, which was financed privately and therefore, helped other inventors and competitors in the race for the first successful flight.

In 1890, Otto Lilienthal started to develop his first "gliding vehicle", with cambered wings. The practical gliding tests started 1891 from a hill close to Berlin (see Fig. 2.3). Fortunately, Otto Lilienthal was a successful engineer and entrepreneur, who earned his living with his own company, producing boilers and heating machines, and could therefore finance all his private flights, his technical research and necessary tests by himself! There was at this time no military or research program available, to ask for a research budget!

Fig. 2.3 Lilienthal's "Sturmhügel" Flying base 1894 and a gliding flight

In total Lilienthal developed 18 different gliding vehicles, did close to 300 gliding flights, the longest flight was more than 250 m. He also tried to integrate a light engine, but the right engine did not exist for him. His sudden death after a flight accident stopped his approach. But all his knowledge and discussion with important persons like Langley, Joukowsky and others inspired other inventors like Ader and the Wright Brothers to continue and use the experience, developed by Otto Lilienthal. Important to mention is also the fact, that Otto Lilienthal was the first real pilot of his gliding vehicles. He took the risk to enter as a pilot and get the feeling for the lift and wind forces and also experienced the basic principles of flight control including stability.

With his openness of publishing and communicating his experience, with the role as pilot of a gliding vehicle and with his enthusiasm, to finance all his research and test efforts, Otto Lilienthal can be seen as one of the central engineers, who had prepared the flight of man.

Successful were then the Wright brothers in Virginia, who managed to develop a flying vehicle, capable to lift off and land with a pilot onboard.

The **Wright brothers, Orville** (1871–1948) and **Wilbur** (1867–1912), were two Americans who were inventing and building the world's first successful airplane and making the first controlled, powered and sustained heavier-than-air human flight, on December 17th, 1903. In the two following years, the brothers developed their flying machine into the first practical fixed wing aircraft (Fig. 2.4).

The brothers' fundamental breakthrough was their further development of three-axis control, which enabled the pilot to steer the aircraft effectively and to maintain its equilibrium.

Their first U.S. patent, 821,393, did not claim invention of a flying machine, but rather, the invention of a system of aerodynamic control that manipulated a flying machine's surfaces [9, 12].

With the news, that the Wright brothers had demonstrated the first autonomous flight with a machine heavier than air, a new impulse was given to all enthusiasts in all countries.

Fig. 2.4 The Wright brothers—Orville and Wilbur and their Kitty Hawk

In 1909, Louis Bleriot, a French aviator, made the first airplane crossing of the English Channel.

Within only 10 years, a lot of new flying machines were developed, very different concepts, different tail configurations, multiple wings, different propulsive engines and engine integrations.

Also the national bodies/governments started to get interest in these flying vehicles. National research started and national military sponsors appeared on the scene.

Figure 2.5 is showing the timeline with the major milestones of bringing the flying vehicles to real flight.

During World War I (1914–1918), it is reported that over 80 000 flying vehicles have been constructed and have been used [6, 9]. However it is also agreed by all specialists, that the flying vehicles have not been a decisive element during this war despite this enormous investment in air vehicles and despite the big progress within 15 years from the first flight in 1903 to the end of the 1st World war. Some examples are shown in Fig. 2.6. In 1918, the biggest bomber aircraft of WW1 (Gotha bomber and Handley Page bomber) had a takeoff mass of more than 5 tons [13, 14].

2.4 The Beginning of Civil Air Transportation

The civil air transport started after WW I parallel in different areas.

The Junkers F 13 was the world's first all-metal transport aircraft, developed in Germany by Hugo Junkers at the end of World War I. It was an advanced cantilever-wing monoplane, which could accommodate four passengers as shown in Fig. 2.7.

The Junkers F 13 is one attempt to use the experience of all the military vehicles and develop out of this knowledge a commercial transport. Hugo Junkers, the creator of F 13, had the vision that there is a big chance to use the aircraft as

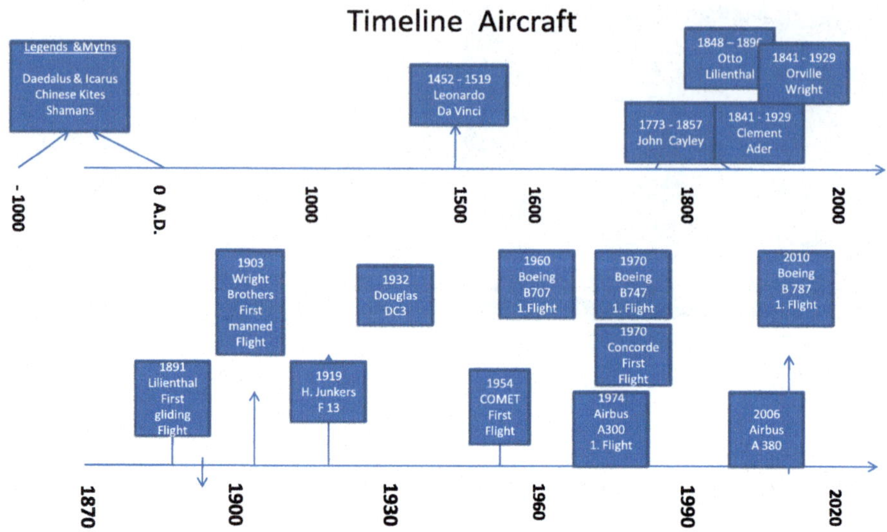

Fig. 2.5 Time horizon with milestones of major flying achievements

Fig. 2.6 Examples of World War 1 aircraft, a Fokker triplane and a Shorts waterplane

transportation means. Surprisingly, the F 13 has all the typical characteristics of today's aircraft. It has already a single cantilever wing, a classical tail, two engines with propellers, and a reasonable fuselage cabin. So only 16 years after the first flight by the Wright brothers, a nearly perfect configuration for air transport has already been developed with all the typical characteristics of a transport aircraft, as we know them today:

- an unobstructed cabin,
- a front cockpit,
- a fuselage to accommodate the payload (not yet pressurized!)

Fig. 2.7 Junkers F13

- a classical tailplane for control and stability,
- one engine mounted in front of the fuselage (certification rules were not yet invented!)

The F13 has been only slightly successful, as the market was not yet ready and the acceptance and infrastructure for air transport had still to be developed. Nevertheless 360 units from the F13 were built. Other aircraft constructors like Anthony Fokker [15] also started to develop commercial aircraft (Fokker F.VII trimotor), but were also not very successful.

The real push for a commercial air transport did not yet start. Reliability and safety have been still a very difficult subject and not yet satisfactorily solved. The infrastructure with airfields well positioned over the continents was not available. Passengers did not really believe on the reliability of the air vehicles and the demand from the public for commercial air transport was not strong enough.

Statistics show, that pilots in general in this time had only a lifetime in average ~ of less than 10 years.

In the world and specifically in the US, the aircraft was primarily used for mail transport.

A big push for air transport started in 1925 in the US where the government withdraw the air mail from the official "post office" and outsourced it to private competitors in order to reduce cost. This was a first push to reduce mail travel time. A next step followed in 1926, with the US "Air Commerce Act", which put air navigation, licensing of pilots and air vehicles as well as the investigation of air accidents under governmental control. This was a first step in pushing a "safety system" in place.

Aircraft had not yet enough range to travel between Europe and US. Charles A. Lindbergh opened the new transatlantic area with his direct solo flight in 1927 from New York to Paris. This spectacular flight got a lot of public interest and also helped a lot to show the new capabilities of modern aircraft and make flying more popular for ordinary people.

The Australian Charles Kingsford Smith was the first to fly across the larger Pacific Ocean in the Southern Cross. His crew left Oakland, California to make the first trans-Pacific flight to Australia in three stages to Brisbane in 20 h, where they landed on 9 June 1928 after approximately 7,400 miles total flight.

Direct mail routes from Europe to Africa and South America were opened in 1930. In 1930 appeared the Boeing "Monomail" model, which had already a retractable undercarriage and was aerodynamically a very proper design, reducing fuel consumption considerably.

The next step was expected from the "high altitude aircraft", which should fly above the normal clouds (thus improving travel comfort) and also increasing air speed without major fuel burn increase. New engine concepts (air charger for piston engines) and better and more reliable instrumentation to fly through clouds were developed and helped this purpose.

Around 1935 the first long range aircraft appeared on the market.

A statistic from all German airports in 1938 shows that 315 000 passengers, 9725 mail and 7165 t of freight have been transported, giving a percentage of air transport of 72,1 % for passenger, 16,1 % for mail and 11,8 % for freight transport [16].

In 1939, World War II started in Europe and all engineering efforts were related to military air vehicles. Speed and range increase and better maneuverability were the dominating factors for aircraft development. The first jet engines appeared in Germany with the ME 262,

The first swept wing concepts for high speed flights were developed in 1937 by DVFLR (A. Busemann), allowing higher speeds up to Mach Numbers of 1, the speed of sound! [17, 18].

The military aircraft became the dominant factor in the superiority of World War II, with speed and maneuverability as dominating performance characteristics.

After WW II, the military efforts first seemed to be reduced, but with the road blockage for West Berlin in June 1948 by Russia, the Cold War started between West and East and military aircraft development was still very high. However, the blockage of road transport from West Germany to Berlin gave a push to civil air transport and the Western Allies managed successfully an 11 months support for the city of Berlin only by an outstanding continuous air transport between Western Germany and the isolated City of Berlin ("Luftbrücke Berlin"!).

On the civil side, the recovery from the difficult and poor years of war took some time, as first a new economic push was needed to develop long-term stability in the economic area and secondly political stability was mandatory before confidence in a longer period of peace between the major countries could be established. This started in the Western world in the beginning of the fifties. The economic growth asked for more travel in the western world and air travel used its chance of drastic time savings between US and Europe, the new Western block. The new demand asked for new air transport vehicles. All the technologies, developed before and during the World War II were now also available for the civil air transport and a lot of new aircraft concepts appeared on the growing market. The biggest push came from the engine side, where the jet engines allowed flying faster and also higher. Compared to the propeller driven aircraft, the jet aircraft increased the speed by nearly a factor of 2, leading to considerable time reductions in the intercontinental routes.

The jet engines consume more fuel per thrust and hour, as shown later in Fig. 2.11 and explained in Chap. 6. But this disadvantage of higher fuel consumption was compensated by the higher speed (Ma = 0,8, compared to Ma = 0.5) and the higher altitude capability, allowing flights over the clouds and avoiding thus critical weather conditions.

The big change in civil air transport—jet age—started with the COMET, developed from the British company de Havilland. This new transport aircraft allowed a better way of flying, especially the time reduction for long range routes, were very quickly accepted from the passenger side. However, some completely unexpected aircraft accidents during the cruise phase of Comet I led to a very critical situation for air transport. British authorities did the utmost to clarify the root cause of these accidents and were building a big hangar which could simulate the external cruise flight conditions! Finally, it was discovered that fatigue characteristics of the fuselage material were a main reason for these air disasters. The lessons learnt revealed that the windows in the fuselage, designed as rectangular elements, were one major cause, where after several air cycles some cracks started to develop, leading to fuselage disintegration and a total aircraft loss during cruise phase.

A new discipline was born in aircraft design: material fatigue as a major design element for the fuselages of aircraft. As all tests and examination were discussed very openly, the other manufacturers profited from these lessons learnt and especially the American companies Boeing and McDonald-Douglas were profiting from these lessons learnt. The company de Havilland constructing Comet aircraft and having the merit to have built the first civil jet transport aircraft, has even in the updated design as Comet IV not taken any profit for their courage and innovative design. The future business of developing good civil jet aircraft was taken by the American companies Boeing and Douglas.

But it also has to be mentioned, that the fast development of always new and improved aircraft concepts was only possible, as there was a large community of international scientists and engineers who were intensively involved in important research programs, developing new aerodynamic profiles and wing design concepts, new materials and structural design methodologies and also developing the control systems in a way to drastically reduce the pilots work load. Some important names should be mentioned here like Ludwig Prandtl, Theodore von Karman, Dietrich Küchemann, William Boeing and Wolfgang Wagner amongst others.

2.5 The Jet Age

The civil air transport with jet engines started with a big failure, the Comet disaster!

The courage of European/British excellent engineering talents was not rewarded by a successful market acceptance.

After the failure from Comet, the American manufacturers Boeing and McDonald-Douglas developed their jet engine powered aircraft, the B-707 and the DC-8 about in parallel and both became fairly successful on the market. Both were designed for about 175 passengers, thus increasing the payload by roughly 75 % compared to the older long range aircraft like DC-7 and Lockheed L-1049, better known as "Super constellation". Jet aircraft were however more noisy during takeoff and landing. But this was not seen as a major drawback as this was also representing the new dynamic optimism and new positive economic push after WW II.

With the bigger cabin, direct operating cost went down by about 15 % in comparison to the older aircraft like DC-7 and DC-6.

The air transport across the Atlantic Ocean became faster and within 6-8 h east coast of US and west coast of Europe (London, Paris) could be reached which meant a travel between North America and Europe could be done within one day!

Already in 1956 the American airlines transported more passengers than the railway. But it has also to be reminded, that the railway in US was not so well established compared to Europe. But air transport offered more flexibility and it was easier to install some new airports instead of investing in heavy infrastructure for railway tracks. In 1957 more passengers travelled across the North Atlantic by air than by ship. The large fleet of cruise ships suffered considerably.

The next development steps came quite naturally. The bigger cabin pushed for bigger aircraft and the larger aircraft needed bigger wings, which allowed having larger fuel volumes, leading to more range for aircraft and thus offering more direct routes over water like South America routes to Africa and Europe similar like the Trans-pacific routes (see Fig. 2.9). The further development of the jet engines to provide more thrust and parallel having a higher bypass ratio with less fuel consumption allowed to further reduce fuel consumption and offered the possibility for even larger aircraft. The B-747—originally a military design for a large military transport aircraft—gave a big boost for air transport capacity and improved travel cost. Most airports were not really prepared to accommodate these new big "Jumbo-Jets". New air terminals had to be provided at the airports. New procedures for air traffic control, aircraft separation procedures, etc. had to be developed to ensure a safe and regular air transport system. Air transport became a more international business and ICAO, the International Civil Aviation Organization was becoming more powerful to establish international rules for the ever increasing air transport worldwide and defining international standards for all participants and shareholders of air transport.

Figures 2.8, 2.9 and 2.10 are showing the history of the main aircraft design parameters like speed, range and size from the 30s till the year 2020.

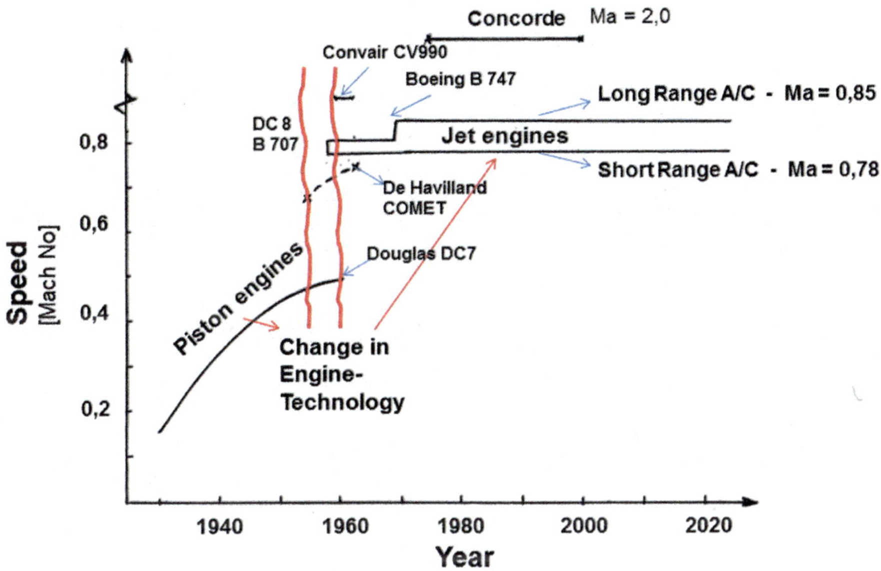

Fig. 2.8 Development of the aircraft design driver "Speed"

It becomes obvious from Fig. 2.8 that **speed** seems to have reached a certain stable standard (which is Ma = 0.74 – 0.78 for Short range aircraft and Ma = 0.82 – 0.86 for Long-range aircraft. (More reasons and details will be given in Chap. 4).

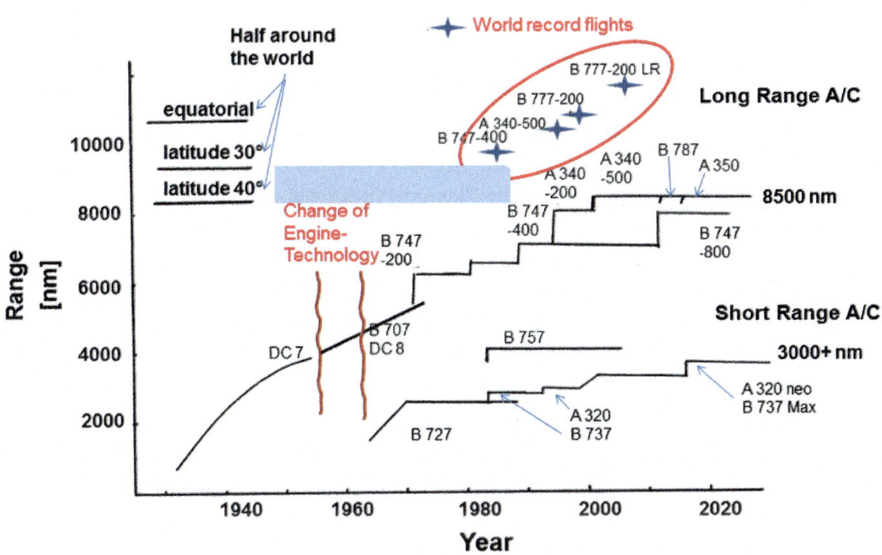

Fig. 2.9 Development of aircraft design driver "Range"

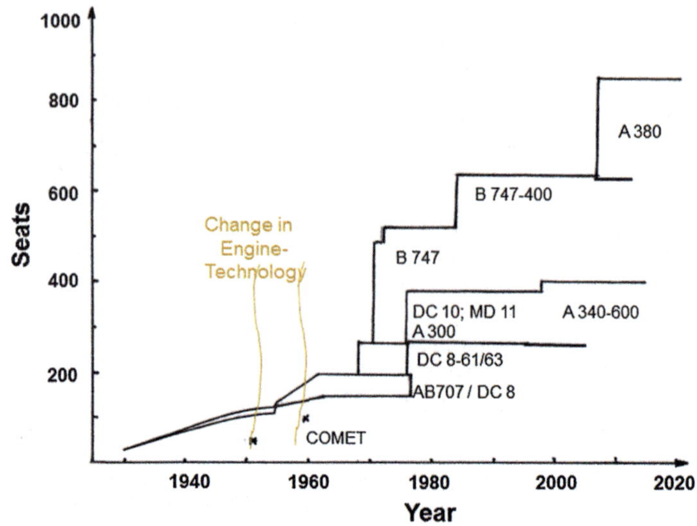

Fig. 2.10 Development of design driver "Seats versus Time"

As can also be seen from Fig. 2.9 **range** has been consistently increased, starting with 4000 nm with the introduction of the jet aircraft to provide today ranges of 8500 nm. Some nice competitive battles between Boeing and Airbus started at the beginning of the twenty-first century, establishing always new world records for the ultimate long range travel. (A340-500 and B777-300 ER have claimed several world records, but all this is and was of no real market interest, more an interesting marketing gag!). There are very few destinations, which are really located opposite on our earth (Singapore to New York, London to Sydney) and where it would be reasonable to install a direct flight route. But these routes are exceptions and it does not make sense to design an aircraft just for this very long range routes. In [21] the longest flown routes are given and in 2013 Singapore airlines is operating a flight from Singapore to New York, taking around 19 h. Are the passengers really using or demanding for such long direct flights? Independent from the passengers demand, it can be stated that range as design parameter and design driver has also come to a natural limit.

Figure 2.10 shows a design parameter, which has not yet reached its technical limits, the **aircraft size**. The A380 with a certified passenger capacity of 852 passengers is the biggest civil aircraft today. A fuselage stretch of nearly 6 m increase in fuselage length is still possible, leading to a capacity of approximately 1000 passengers. There is today no technical limit to design even bigger aircraft. The more important question is, whether there is still a market interest for such big machines and whether the passengers and the operators are interested to use such big planes? In this point opinions are quite different and dependent on the stake holders' interest. (Some comments are given in the Chaps. 3, 8 and 11).

Figure 2.11 is showing the most important parameter for aircraft design, the relative seat mile cost and how this parameter has been constantly improved over time.

The introduction of jet engine have increased the seat mile cost by about 22 % compared to the former piston engine aircraft, however, the advantage of increased speed, altitude and allowing bigger aircraft size were the overwhelming arguments. Today seat mile cost (smc) is the only real driver for all new aircraft designs. Partly, a decrease in SMC can be obtained by designing bigger aircraft, allowing a cost decrease due to size effects. The other part is coming from new technology elements, as they are just been introduced in the new aircraft designs from Boeing and Airbus (B787 and A350), which then have to lead to a real benefit in seat mile-cost (smc). The smc improvement should be at least in the order of 10 %; so that the airline has a clear advantage in the operation and can cover all cost which are dependent on the introduction of a new airplane in the existing own fleet (see Chap. 7).

Steiner [19] is describing the important step of Boeing in the sixties and seventieth, which led to the domination of Boeing as civil aircraft manufacturers. In [10] is the European answer described with all the existing engineering capabilities but the lack of cooperation and the willingness to overcome national egoisms, which have led to the establishing of Airbus as a competent aircraft manufacturer in competition to Boeing.

Schmitt [20] is defining the new challenges of future transport aircraft, which are no longer size, range and speed, but will be cost, low emission and green features to keep the positive mood and acceptance of the travelling public.

The duopoly of today between Boeing and Airbus seems to be a well-established market situation, where it will be difficult for new entrants, to challenge these 2 big aircraft manufacturers. There are, however, several new possible entrants (Embraer in Brazil, Bombardier in Canada, AVIC in China with the COMAC 91, Mitsubishi

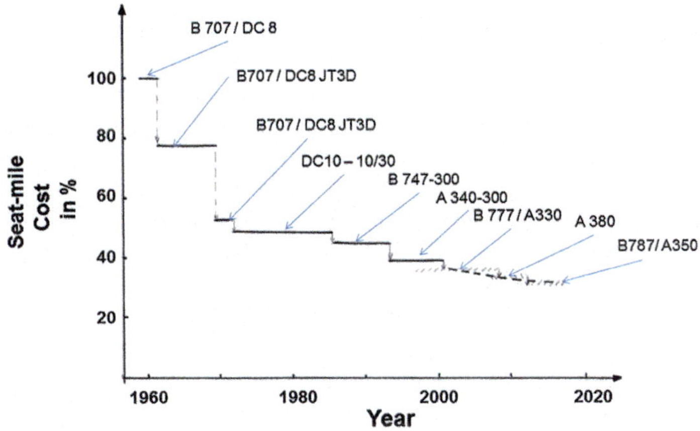

Fig. 2.11 Development of Seat Mile Cost "smc" over time

with the M21 and Russia with the "Superjet" from Sukhoi) which are preparing new designs for civil transport aircraft in the Regional class (90–140 passengers). All these new players are expecting to participate in this multi-billion dollar market and will become major challengers for the two big established players Boeing and Airbus. It will be very interesting to see, how these new aircraft manufacturers will manage their market entrance and market acceptance. But the airlines will for sure support the new aircraft manufacturers, as they will bring new ideas to the market. As we can see from today's situation where Airbus had major problems with the industrialization of their latest A380 aircraft, where the airlines had to accept major delivery delays of more than 2 years. A similar situation was happening with the latest Boeing design, B787 aircraft, which was also about 3 years late and had to be grounded for 2 months in 2013. In this respect, the airlines will highly welcome some new aircraft manufacturers in the market to increase competition [24, 25].

2.6 Development of Civil Transport Operation (Airlines and Airports)

2.6.1 Airlines

At the beginning of air transport, the airship was used for civil transport operation. The first company, who started with regular air transport was DELAG (*Deutsche Luftschiffahrts-Aktiengesellschaft*). It was founded in 1909 with government assistance, and operated airships, manufactured by the Zeppelin Corporation. Its headquarters were in Frankfurt. The idea was to establish regular air transport between major cities in Germany. In 1914—before the beginning of the 1st World War—DELAG operated seven airships on roughly ~1500 routes with a total range of 175.000 km and transported 18.500 passengers without major fatalities [10, 16].

Transportation of Mail stands at the beginning of the fixed wing commercial aircraft operation. In the US the Post-office started the first regular post transport between Philadelphia and New York. Also in Europe transport of mail started the commercial operation after WW 1. In 1920, the first transcontinental airmail service began and the first night flights started a year later. However, accident rates were still high and normal passengers did not yet rely on and believe in air transport.

The four oldest airlines that still exist but using fixed wing aircraft are Netherlands' KLM, Colombia's Avianca, Australia's Qantas, and the Czech Republic's Czech Airlines. KLM first flew in May 1920, while Qantas (which stands for *Queensland and Northern Territory Aerial Services Limited*) was founded in Queensland, Australia, in late 1920 [22, 23].

The real intercontinental and international air transport started at the end of the 1930ies. New aircraft designs like the DC4, B307, He 111, FW 200 and Ju 90 had increased considerably their speed and range capability, making air transport more attractive for the passenger and the airlines. World War II stopped a lot of these

KLM - development of a typical European airline

- **Oct 7, 1919** Dutch Royal Airlines for the Netherlands and its Colonies (KLM) was founded.
- **Oct 21, 1919** The first KLM office opened on Heerengracht in The Hague.
- **Apr 4, 1921** KLM resumed service with its own pilots and aircraft: the Fokker F-II and F-III
- **Oct 1, 1924** KLM initiated its first intercontinental flight, from Amsterdam to Batavia (Colonial Jakarta) in a Fokker F-VII.
- **Dec 1933** KLM flew Christmas and New Year's cards from Amsterdam to Batavia in a record time of just over four days in a Fokker F-XVIII Pelikaan.
- **Dec 1934** KLM made its first transatlantic flight, from Amsterdam to Curacao in a Fokker F-XVIII Snip.
- **Sep 1945** KLM resumed service following the Second World War, starting with domestic flights.
- **May 21, 1946** KLM initiated scheduled service between Amsterdam and New York using the Douglas DC-4 Rotterdam.
- **Nov 1, 1958** KLM opened its Amsterdam-Tokyo service, flying over the North Pole using the Douglas DC-7 "Caraïbische Zee".
- **Mar 1960** The Jet Age began with the introduction of the Douglas DC-8.

Fig. 2.12 Development of a typical national airline (KLM)

civil transport developments as all engineering skills went into military aircraft design.

After WW II all the aeronautical engineering Knowhow was transferred back to the civil air transport. The jet engine was introduced for civil air transport. At the end of the sixties, the aircraft Boeing B-707, Douglas DC-8, Sud Aviation—Caravelle, Tupolev Tu-104, appeared on the market and established the dominance of jet aircraft in short and long range flights and the newly established national airlines were interested to buy and operate them and develop their international network (see Figs. 2.8, 2.9, 2.10, 2.11). But international agreements had to be developed to build confidence for the travelling persons (see Chap. 4).

A typical development of a classical "flag carrier" or national airline can be seen in Fig. 2.12 with the development of KLM, starting in 1919. Common elements are, to use national aircraft design (Fokker), national pilots and start to connect with the own empire (colonies, when still existing). Here the air transport gave a new dimension to better connect these colonies with the homeland.

2.6.2 Development of Airports

The development of airports followed the need, that some operators wanted to offer transport services between two points and therefore needed the necessary infrastructure. This started with a green plane field, some hangars or light buildings to prepare the formalities for the flight. Most of these fields had not a dedicated runway, but provided a large round circle field, where aircraft could start and land in whatever was the preferred direction related to the wind conditions at the airfield.

Paved areas were created first at those positions, where the passengers were embarking and disembarking. Later on paved runways were installed to allow landings and takeoffs in nearly all weather conditions and during day and night.

The following short history of airport development is based on data from [16, 26–28].

The title of "world's oldest airport" is disputed, but College Park Airport in Maryland, US, established in 1909 by Wilbur Wright, is generally agreed to be the world's oldest continually operating airfield, although it serves today only general aviation traffic. Pearson Field Airport in Vancouver, Washington had a dirigible land in 1905 and planes in 1911 and is still in use. Bremen Airport opened in 1913 and remains in operation till today. Amsterdam Airport Schiphol opened on September 16, 1916 as a military airfield, but only accepted civil aircraft from December 17, 1920, allowing Sydney Airport in Australia—which started operations in January 1920—to claim to be one of the world's oldest continually operating commercial airports. Rome Ciampino Airport, opened 1916, is also a contender. Increased aircraft traffic during World War I led to the construction of several new landing fields. Aircraft had to approach these from certain directions and this led to the development of aids for directing the approach and landing slope. Following the war, some of these military airfields added civil facilities for handling passenger traffic. One of the earliest such fields was Paris—Le Bourget Airport in France. The first airport to operate scheduled international commercial services was Hounslow Heath Aerodrome in August 1919, but it was closed and supplanted by Croydon Airport (UK) in March 1920. In 1922, the first permanent airport and commercial terminal solely for commercial aviation was opened at Flughafen Devau near what was then Königsberg, East Prussia, Germany. The airports of this era used a paved "apron", which permitted night flying as well as landing heavier aircraft.

The first lighting used on an airport started during the latter part of the 1920s; in the 1930s approach lighting came into use. These indicated the proper direction and angle of descent. The colors and flash intervals of these lights became standardized under the International Civil Aviation Organization (ICAO, see Chap. 4). In the 1940s, the slope-line approach system was introduced. This consisted of two rows of lights that formed a funnel indicating an aircraft's position on the glideslope. Additional lights indicated incorrect altitude and direction.

Following World War II, airport design became more sophisticated. Passenger buildings were being grouped together in a central unit, with runways arranged in groups around the terminal and taxiways to connect the runway and the terminal area. This arrangement permitted expansion of the facilities. But it also meant that passengers had to move further to reach their plane (see also Chap. 9).

Airport construction boomed during the 1960s with the introduction of jet aircraft traffic. Runways had to be extended out to 3000 m (9800 ft). The fields were constructed out of reinforced concrete using a slip-form machine that produces a continual slab with no disruptions along the length. The early 1960s also saw the introduction of jet bridge systems to modern airport terminals, an innovation which eliminated outdoor passenger boarding.

Brief History of Atlanta Airport (US)

- April 16, 1925 Mayor Walter A. Sims signs a five-year lease on an abandoned auto racetrack and commits the City to developing it into an airfield.
- April 1929 The City pays $94,400 for the land and changes the name to Atlanta Municipal Airport.
- December 1930 Eastern Air Transport inaugurates passenger service from Atlanta to New York.
- March 1939 The Airport opens its first control tower.
- 1957 Atlanta is the busiest airport in the country with more than 2 million passengers.
- May 1961 Atlanta Municipal Airport is entering into the "Jet Age", parallel with the opening of the largest single terminal.
- June 1978 Sabena - Belgian Airlines - becomes Atlanta's first foreign international carrier.
- September 1980 Atlanta International Airport opens the world's largest air passenger terminal complex, accommodating up to 55 million passengers /year.
- December 1984 A fourth parallel runway was completed. An expansion of an 12,000-foot runway started, capable of handling the largest commercial airplane in development.
- 1988 MARTA's Airport station linked the Airport to Atlanta's rapid transit system.
- June 1996 The new Master Plan -- Hartsfield - 2000 + Beyond was proposed.
- March 2000 Hartsfield is the World's Busiest Airport, accommodating more than 78 million passengers and more than 900,000 landings and takeoffs for 1999.
- July 2005 Hartsfield-Jackson celebrates its 80th birthday

Fig. 2.13 Development of Atlanta airport in USA

Figure 2.13 shows a short summary of a big airport (Atlanta US), which is given to illustrate the constant development and increase in runways, terminal buildings, access to city and all the new technological improvements, necessary to follow the constant increase in passenger demand and societal expectations.

References

1. Naso, O.: Metamorphosen. Zürich (1958)
2. Behringer, W., Ott-Koptschalijski, C.: Der Traum vom Fliegen, German edn. S. Fischer Verlag, Berlin (1991). ISBN 3-10-007106-9
3. Galluzzi, P.: Leonardo da Vinci, Engineer and Architect, Montreal Museum of Fine Arts (1987). ISBN 2891920848
4. Ludwig, H., Dibner, B., Reti, L: Leonardo the Inventor. McGraw-Hill, New York (1980). ISBN 0070286108
5. Cayley, G.: On aerial navigation. Nicholsons J. Philos. **XXIV** and **XXV**, 1809/1810
6. Gibbs-Smith, G.H.: The Invention of the Aeroplane. Taplinger Publishing Comp, New York (1965)
7. Cahan, D. (ed.): Hermann von Helmholtz and the Foundations of Nineteenth-Century Science. University of California, Berkeley (1994). ISBN 978-0-520-08334-9
8. Lilienthal, O.: Der Vogelflug als Grundlage der Fliegekunst (1889)
9. A century of flight in: http://www.century-of-flight.net/Aviation. Accessed 1 Dec 2014
10. Roeder, J.P.: Evolution of the Art of Flying since Lilienthal, DGLR congress 1991 in Berlin, 100 Jahre Menschenflug—Otto Lilienthal, invited lecture, Berlin (1991)
11. McMasters, J.H., Cummings, R.M.: Airplane Design and the Biomechanics of Flight, AIAA 2004-0532. Reno, Nevada (2004)
12. Wright brothers in: http://en.wikipedia.org/wiki/Wright_brothers. Accessed 1 Dec 2014
13. Aircraft of World War 1: http://www.theaerodrome.com/aircraft/by_nation.php. Accessed 1 Dec 2014

14. Importance of aircraft in WW1: http://www.historylearningsite.co.uk/aircraft_world_war_one.ht. Accessed 1 Dec 2014
15. Dierikx, M.: Fokker: A Transatlantic Biography. Smithsonian Institution Press, Washington, DC (1997). ISBN 1-56098-735-9
16. Treibel, W.: Geschichte der deutschen Verkehrsflughäfen. Bernard & Graefe Verlag, Bonn (1992). (in German)
17. Meier, H-U.: Die Pfeilflügelentwicklung in Deutschland bis 1945. Bernard & Graefe Verlag Bonn (2006). ISBN 3-7637-6130-6 (in German, English version in preparation)
18. Hirschel, E.H., Prem, H., Madelung, G.: Die deutsche Luftfahrt—Luftfahrtforschung in Deutschland. Bernard & Graefe Verlag, Bonn (2001) (in German, English version in preparation)
19. Steiner, J.E.: How Decision are Made—Major Considerations for Aircraft Programs, AIAA, 1982; ICAS 1984
20. Schmitt, D.: Bigger, faster, further, greener?? ICAS Congress 2004, invited lecture. Yokohama (2004)
21. Longest flights: http://cruisinaltitude.com/2009/10/14/top-10-worlds-longest-flights-by-distance-flown/. Accessed 1 Dec 2014
22. Airline history: http://en.wikipedia.org/wiki/Airline. Accessed 1 Dec 2014
23. Aviation history Australia: https://sites.google.com/site/aviationhistoryaustralia/Home/airlines-operations. Accessed 1 Dec 2014
24. Boeing history: http://www.boeing.com/history/chronology/chron04.html. Accessed 1 Dec2014
25. Airlines Financial situation: http://www.airlinefinancials.com/AF_Stock/BTS_PDF/UA-BTS.pdf. Accessed 1 Dec 2014
26. Neufville, R., Odoni, A.: Airport Systems. McGraw-Hill, New York (2013). ISBN 978-0-07-177058-3
27. Airport history: http://en.wikipedia.org/wiki/Airport. Accessed 1 Dec 2014
28. History of Atlanta airport. http://www.atlanta-airport.com/Airport/ATL/Airport_History.aspx. Accessed 1 Dec 2014

Chapter 3
Market Aspects

Abstract This chapter describes the strategic importance of aerospace, the link between military and civil transport, the strong US dominance in the civil market in the 1960s and 1970s and the creation of Airbus in Europe, leading finally to a duopoly in the civil transport market for aircraft with more than 120 seats. Specific aspects of the aeronautical industry are the very long development cycle of an aircraft, where invested money will only be recovered after 12–20 years. National support is therefore needed, leading however to a constant fight between the US and Europe in front of the World Trade Organization WTO. New entrants are on the horizon to challenge the duopoly from the lower market area. Market forecast methods are described in detail and the outlook from industry for the next 20 years is presented, identifying a huge growth market with a doubling of the number of aircraft for this period. The air freight market with its specific elements is outlined, and the refurbishing of elder passenger aircraft into freighter aircraft is a main driver in the passenger market. Reflections about the importance of cost and commonality aspects conclude the chapter.

3.1 The Strategic Importance of Aerospace

Following the historical development of aerospace activities, it is obvious that aviation and space have become after WWII a very dominant area for all big countries, especially for the four allied countries, the winner of this war. During the cold war (between 1948 and 1990) military and space developments were in the focus of the two dominating blocks, the Western NATO block with the United States of America and its European Allies on one side and the Eastern bloc with the Soviet Union and its allies on the other side. Big military budgets were available allowing the development of continuously novel aeronautical and space vehicles. Focus may have been strictly on military usage, but technologies to improve the thrust to weight ratio for military aircraft could also be directly applied to civil aircraft vehicles and the big aeronautical industry in the USA with Boeing, Lockheed, Douglas,

D. Schmitt and V. Gollnick, *Air Transport System*,
DOI 10.1007/978-3-7091-1880-1_3

Fig. 3.1 Boeing B747—100

Mc Donnell, and others who looked to how they could improve their civil air transport business by using and transferring all their military aerospace technology. As stated in [1] the big engine manufacturers in the US developed their first big engines for the B747 and other large aircraft like DC-10 and L-1011 on the basis of a military core engine. The CF6 engine from GE as well as the PW JT9D engine from Pratt & Whitney largely profited from military programmes in the US, especially the CX-HLS (Heavy Logistics System), where a specification for a heavy military transport aircraft was awarded to Boeing, Douglas and Lockheed in 1964. Lockheed finally won the production contract in Sept. 1965 and started the development of the big military transport C5. Boeing, which has lost the competition for the CX-HLS military transporter then started to use the experience gained and developed a civil big transport aircraft named B-747. In April 1966, the biggest airline in the world—Pan American Worldwide Airways (Pan Am)—ordered 25 B747. As the basic development work for these aircraft and engines was mainly financed by military contracts and a proof of concept was already done on the military version, the technical and financial risk was then much reduced to develop out of this military transport concept a commercial version, the famous B747, which became a very good commercial product, Fig. 3.1. The placement of the cockpit above the nose is still the element, where an element of the military design is visible. But also, a lot of new technical design features were introduced at this time such as fault tree analysis, structural redundancy design, dual control surfaces, etc.

3.1.1 From a US Monopoly Status to a Duopoly Situation

Starting from the 1960s till the 1980s (1955–1980) the American civil aircraft manufacturers were dominating the worldwide civil aircraft market. The Boeing Company had successfully designed the B707 aircraft and by using the same fuselage cross-section for the development of their short and medium range aircraft, the B727, B-737 and B757, Boeing developed a complete aircraft family with 2- 3- and 4 engines by always using the same fuselage cross-section. The B747 had a difficult commercial start. The very big size, offering more than double the capacity of the well-established B707 and DC-8 was offered to the market during the critical period, when the first oil crisis in 1973 was shaking the technological enthusiasm of the sixties. OPEC was established and the crude oil price per barrel was increasing by a factor of nearly 4, which had a major impact on all energy and transport sectors in the world. Pan AM as launch customer for the B747 suddenly had problems to fill the big aircraft and the euphoric move to bigger, faster and further range demand got a first shock. Boeing was suddenly confronted with a big crisis, as all their investment for producing the B747 was challenged and the airline was no longer able and willing to buy this proposed big "Jumbo-aircraft".

The European industry was confronted with a similar shock. The Concorde consortium with the French "Aerospatiale" company and the British "British Aerospace" company had developed their civil supersonic transport concept called "Concorde", a 100-seater aircraft, which could fly supersonically ($Ma = 2.0$) between Paris and New York (see Fig. 3.2). The Concorde consortium had already received orders for about 100 aircraft, when the oil crisis started in 1973. The Concorde partners were suddenly hearing from their airline customers that they could no longer see a commercially viable operation of Concorde, regarding the drastic increase in fuel prices. Even when we look (with today's knowledge of fuel

Fig. 3.2 Concorde, the supersonic commercial aircraft (1978–2003)

prices) at the very low fuel prices in this period, the fact of the fuel price increase by more than a factor of three led to the complete cancellation of all Concorde orders. However, the aircraft had been built and was certified, so the French and British governments pushed at least their national air carriers to take the already produced aircraft (in total 16) and operate them. It is said that Air France and British Airways got their Concorde aircraft at a symbolic price of 1 $ each. The aircraft then were very successful in their operation; BA aircraft had up to 24,000 flight hours and Air France aircraft around 15,000 flight hours, all before Concorde aircraft was withdrawn from service [2]. The tragedy happened on Tuesday, 25 July 2000 with Concorde SN 203, F-BTSC outbound from Paris to New York. It crashed 60 s after takeoff after suffering tyre blow-out that caused a fuel tank to rupture. This started a sequence of events that caused a fire which finally led to two engines failing and the aircraft crashing. All 109 people (100 passengers and 9 crew members) on board were killed. This was the end, even when several attempts were tried to introduce new modifications which were asked for by the certification authorities. The Concorde program officially ended in 2003.

This commercial disaster of European Concorde was in line with some other European subsonic aircraft programmes. The European commercial aircraft designs of Trident from de Havilland, BAC 1-11, Fokker F 50 and F100 as well as Caravelle from Aerospatiale were designed mainly following the requirements of a national airline, but could not be reasonably sold on a larger world market scale, especially against the competing aircraft of the US manufacturers. Other designs like Mercure from Dassault and the German attempt of a 40 seater named VW 614 also failed to fulfil the market needs.

So the three US manufacturers Boeing, McDonald-Douglas and Lockheed were successfully dominating the world market for big transport aircraft. The next big battle was the competition between DC-10 and Lockheed L-1011, both three-engine aircraft for long range, in size slightly below the B747, which were trying to capture the market of the 300 seater for long range. The fierce battle and competition ended with a disaster for both companies. Lockheed, despite a wonderful engineering design with good performance for their L-1011, withdrew from the civil aircraft market and concentrated on the military market. In parallel, Rolls-Royce as engine manufacturer, mainly concentrating on Lockheed L-1011, also got nearly bankrupt and had to go through a lengthy restructuring process for many years, before being back on the market. The paper from John Steiner [3] describes in a very brilliant way the factors for success, which helped Boeing in this period, to become the world leader in civil aircraft manufacturing, despite the critical phase of B-747 at the beginning.

The European efforts on civil aircraft were regrouped in 1968 in a new consortium, integrating the French industry Aerospatiale, the German industry under "Deutsche Airbus" and the British industry of Hawker Siddley. This consortium was named Airbus and developed a new design for a twin engine widebody configuration, named Airbus A-300. In 1974, the first aircraft was delivered to Air

France, but the market acceptance was at the beginning fairly poor, as it was not very visible whether the consortium would be willing to support the aircraft over the next 20 years of operation.

There was however a strong push from the French manufacturer and the French political side to support and push the Airbus consortium to become a major aircraft manufacturer [4]. The next programmes followed in a 5-year sequence, the A310 in 1982, the A320 in 1988 and the A330/340 in 1992. The A320, developed with a new "glass-cockpit" and a "fly-by-wire" system, had some fatal accidents in the first 3 years of operation. But the analysis of these accidents showed no relation to any of the new technology features [5]. A320 has become the best-selling European aircraft programme ever and still has a bright future before it. The market segment of the 150-seater aircraft with a range of roughly 3000 nm is also called the "Single aisle" and "short range" airliner market. The two aircraft in the market are Boeing B737 and Airbus A320, both having a 6-abreast seating with only one aisle and making best use of the available floor space (see Sect. 5.2). It is the market element with the best-selling product today with a production rate of more than 40 aircraft/month for both manufacturers.

This historical review is intended to give evidence to some important elements of market forecast and market aspects.

In the 1960s there was still a strong link between military aircraft development and civil commercial aircraft development. Most of the R&T funding and development funding could be shared between both sectors. With the end of the Cold War, military funding has been reduced at least in Europe and the civil aircraft development has to be financed mainly from the civil business. Also, the technologies are now becoming more differentiated between military and civil business. The tendency in military aircraft is oriented actually towards unmanned vehicles, with technologies concentrated on autonomous flight capabilities. The commercial aircraft design is oriented towards continuous improvement in fuel efficiency, so using different technologies and strategies (see Chap. 5, Fig. 5.1).

In the commercial aircraft market with aircraft sizes of more than 120 passengers, there are only two big players acting in a duopoly. This is also a specific market situation, which partly pushes both actors to keep the balance of 50:50 market shares. There is also a strong push from the customer side to keep at least a good balance between both manufacturers that will at least guarantee in the long term a minimum of market competition. Three actors or even more would be better, drive the market and product development strategy.

But new manufacturers are coming up! They will all start in the class of Regional aircraft, ranging from 70 to 130 passengers. They start at the regional market but challenge the two big players from the bottom of the interesting 150-seater market and it can be expected, driven by national interest that China, Russia and Brazil will enter later into the airliner market (market from 120 pax upwards). This will then increase competition and will be beneficial for the market in the long term.

In summary, aerospace is seen by many countries as a strategic industry, especially by the dual use between military and civil aerospace technologies. This is clearly visible in USA, France, China, Brazil and Russia! UK, Canada and Germany are less clearly positioned, as their military expenses are happening at a smaller level [4]!

3.2 Specific Aspects of Aeronautics

Civil aircraft design and production does not correspond to normal market cycles;

Civil aviation is a very specific market. It is characterized by a very long development cycle (see Sect. 6.2). The normal development cycle lasts 5–6 years (recent developments like A380 and B787 have even by far exceeded these development times!), the product is active for more than 50 years and the Return on Investment or better the Breakeven point—this is defined as the point where for the first time all development cost will be balanced by the income from sales (see Sect. 6.5)—is normally of the order of 10–15 years and sometimes even longer! More details are given in Fig. 6.11! So aviation is a very specific market where normal financial institutions are not willing to involve themselves. When normal market forces are not interested in this business due to its very long and dangerous or better not predictable market behaviour, this business cannot be supported by classical financial institutions. Therefore, there is a massive interest of national states to develop their own aeronautical industry and keep it in a proper financial status. This means, in terms of crisis in the world economy, which affects directly the aviation sector, there is a need for a national support structure!

Very often the word "market failure" is mentioned in the context of the aviation business. Normal business processes for consumer goods, where a certain market behaviour may be seen as normal and market forces may act to keep the best and most efficient structure in a favourable business situation, these cannot be applied to the aviation industry! The history is showing that the aviation industry in nearly all aspects, i.e. the manufacturing part, the operational part, the infrastructure and the legal part is always very closely related to national state interests. This is absolutely clear with regard to military aviation. But this is also mainly the case in commercial aviation. The airports are under national control and rarely run by a purely business consortium. The airport is important for a region or a country as a gateway to the world and there is a national interest to control the political dimension of the business. In a similar way, each country is trying to keep a national airline as flag-carrier, to avoid the dependency on other foreign airlines business arrangements. The air traffic control agency is nearly always nationally controlled and organized.

Market forces are mandatory to run these businesses efficiently. However, many countries have no interest to open the national aviation stakeholders (airlines, airports, ATM-services, etc.) for a global market competition.

So market rules are of interest, to allow competition on specific international routes, which will help the travelling public. Especially, the development of new business models for the so-called "Low Cost Carriers" has had a big impetus on the airline market and led to a considerable reduction in ticket prices (see Chap. 9).

3.2.1 WTO Role and Activities

As mentioned above, with the start of the Airbus consortium, it is difficult for a new civil aircraft manufacturer in the air transport market to gain confidence and establish himself as a strong player in a specific aircraft segment. On the other hand, the airlines like to have strong competition on the market in order to have sufficient room for negotiation of products and prices.

It took Airbus—the European manufacturer consortium—more than 20 years to gain a certain reputation on the market. With the introduction of Airbus' first long range aircraft (A330 and A340) in 1992—about 23 years after the establishment of Airbus as an industrial consortium with well-experienced partners like the French Aerospatiale, the British Hawker Siddley and the German "Deutsche Airbus"—the Airbus sales had just reached about 15 % of the market share, despite a whole fleet of developed and certified products like A300, A310, A320, A330 and A340.

It needs between 15 and 20 years of production of an aircraft programme in order to gain some money and to develop new aircraft with cash coming from the production of existing standard aircraft! (see also Chap. 6).

So the market entry barriers for new manufacturers are very high! At the start, a strategic national decision has to be taken in order to develop the national aircraft manufacturing industry and develop sufficient knowhow and engineering, commercial and industrial skills to stay successfully in the market.

The market forecast for the next 20 years in the aircraft market from 100-seater and bigger aircraft is expected to be of the order of 4.4 trillion $, a huge market where a lot of newcomers would like to participate (see Fig. 3.3).

Market Forecast for new passenger and freight aircraft for next 20 years (2013 – 2032)

Category	Seat capacity	Boeing CMO Current Market Outlook	Airbus GMF Global Market Forecast
Single-Aisle Aircraft	110 - 200	24.670	20.242
Twin-Aisle Aircraft	220 - 350	7.830	7.273
Very Large Aircraft	≥ 400	760	1.711
Total Number		33.260	29.226
Market value [B$]		4.840	4.400

Fig. 3.3 Civil aircraft market forecast and business volume [16, 17]

Several new approaches are visible in the market, where five new countries with their national industry would like to develop new aircraft types in the regional market (70–120 seater aircraft).

- Brazil and Embraer have developed the ER 195 family [6]
- Canada and Bombardier are developing the Canadair Jet C-1000 [7]
- Japan and Mitsubishi are developing the MRJ [8]
- Russia and Suchoi are developing the Superjet 100 [9]
- China and COMAC are developing the COMAC ARJ21 [10].

Governmental support will be essential for at least the next 20 years before the industry will be capable to continue on commercial terms, as the past example from US and Europe has shown.

Very often military programmes are used to give the national industry the necessary support, to develop military transport, trainings or fighter aircraft in order to develop the basic engineering skills for aeronautics. A good example is the recent decision from the American DOD to develop a tanker aircraft. The contract was given to the Boeing Company, which was clearly expected, even when EADS, the European consortium, proposed another independent offer.

This programme will be worth some ten billion $ in RTD, and another 100 billion $ in terms of business with also some profit (normally not less than 10 %) for the Boeing Company. It seems very realistic to assume that all the technologies, being developed during this program and all production skills, know-how and investment will also be beneficial for the commercial aircraft business of Boeing.

The worldwide agency called World Trade Organization [11] was established to ensure that international business is done in an open and fair manner. The World Trade Organization (WTO) deals with the global rules of trade between nations. Its main function is to ensure that trade flows as smoothly, predictably and freely as possible. Unfair subsidies for a specific industry or service sector have to be avoided and should not happen. So far the general terms of condition are from WTO. The aviation sector is very often sued by WTO, as there are often some partners accusing their competitors for unfair subsidies. There is a constant conflict between Airbus and Boeing about unfair subsidies. But the examples above show the difficulty. The American partners are taking profit from some military contracts (see above the military tanker decision!) which are not controlled by the WTO and are complaining about unfair subsidies for their competitors. In [1, 4, 11] are given details about some historical disputes between Boeing and Airbus or more generally USA and Europe.

With the new transport aircraft in the 100-seater market, the new entrance companies will need strong subsidies from their government to develop the product, establish the production and final assembly line, develop market and product support services, etc. The discussion of unfair subsidies will soon come up again.

3.3 The Instruments for Market Predictions (Market Forecast Methods)

Market predictions are mandatory in a long-term market like the civil aeronautical industry.

Market prediction methods are developed by the aircraft manufacturers. They have to identify the market drivers and long-term aspects, which may influence the air transport market, including other means of transport (High-speed trains, ships, etc.) and societal changes (Mobile phone, Smartphone, Virtual travelling, etc.). Engine manufacturers, suppliers and research institutes are also developing their own market forecast methodology in order to get a better understanding and feeling about the future. This is important to develop an own product strategy with all financial risks and chances.

All forecast methods differentiate between cycles and trends. General aspects of market forecast methods can be found in [12] (Fig. 3.4).

"Trend" is defining the long-term tendency, independent of short-term aspects, caused by political or economical perturbations. The long term trend—normally a period of 20 years—is important for:

- Investment-Analysis
- Evaluation of possibilities for new products
- Business models and market forecast
- Legislative aspects and environmental tendencies
- Industrial Organization.

Cycles are determined by "Short term"—influences, perturbations from political or economical side—normally between 1 and 2 years—but important for the following analysis:

- Adaptation of production rate of aircraft
- Financial planing

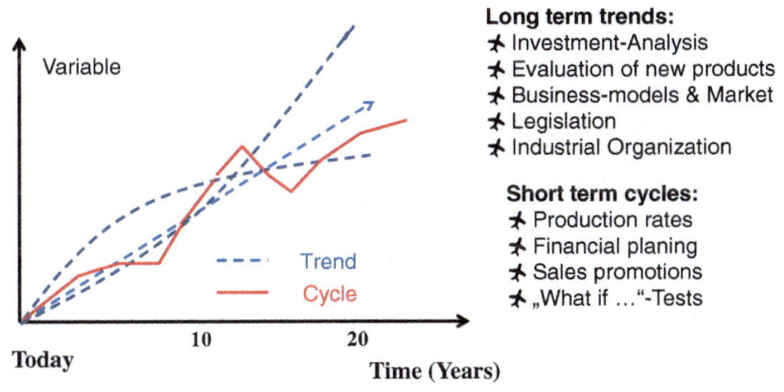

Fig. 3.4 Difference between trend and cycle

- Sales promotions
- "What if …"-Tests

The Global market forecast is defining the long-term trend.

There are also some independent institutions (Research Centers like NASA and DLR as well as commercial institutions) that are providing and offering their market forecast for air transport.

There is no unique and generally viable method.

Generally two different approaches for market forecast exist:

Top-down approach
Bottom-up approach

In addition to these two approaches, scenario techniques are used to identify specific risks and potential benefits for future market trends.

3.3.1 Top-Down Approach

The method of "Top-down-approach" starts from the market development of the last 10–20 years and assumes a continuation of the long-term trend with an adaptation of some main parameters such as:

- Global growth product
- Yield from the airlines (as defined from IATA et alii)
- World Economic growth:
- RPK-development: (worldwide or per region)
- Ticket price trend: in real terms
- Fuel price development:

Figure 3.5 is showing the development of the crude oil price and subsequently also the price for kerosene during the last 20 years. It is remarkable to notice that the steps in oil prices increase are getting larger and appearing in shorter sequences. Such a development is typical for a more and more destabilizing trend, especially if the amplitudes of increase are also getting larger.

For the top-down approach, the world will then be separated into several geographical sectors (North America, Europe, Asia-Pacific, Near East, etc.) and the main routes and traffic flows between these regions and within these regions will be identified and analysed.

The top-down approach consists of the following steps (see Fig. 3.6).

1. Select a region A and determine the available capacity expressed in ASK (available seat kilometres)
2. Assume an average traffic growth factor for the next 20-year period
3. Obtain as a result the expected traffic volume (RPK) for the year N+20
4. Convert the RPK into ASK by an assumption about the development of the load factor

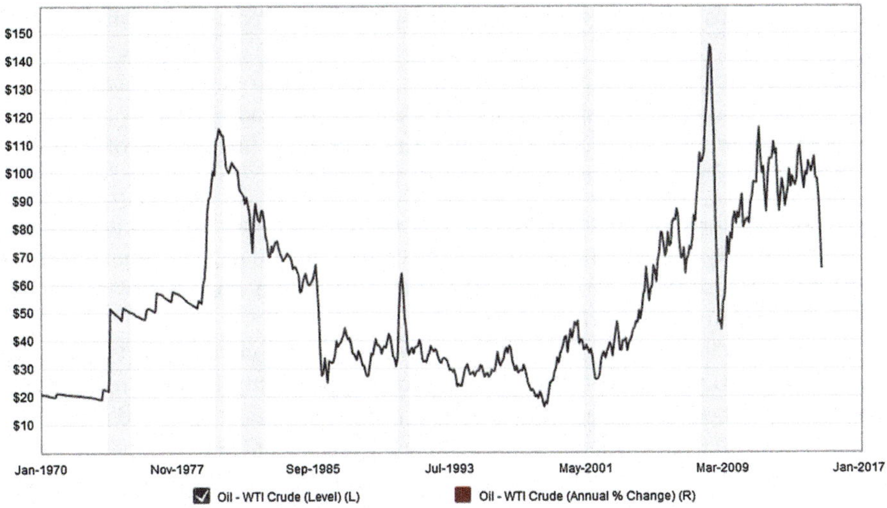

Fig. 3.5 Crude oil price development 1970–2014 [26]

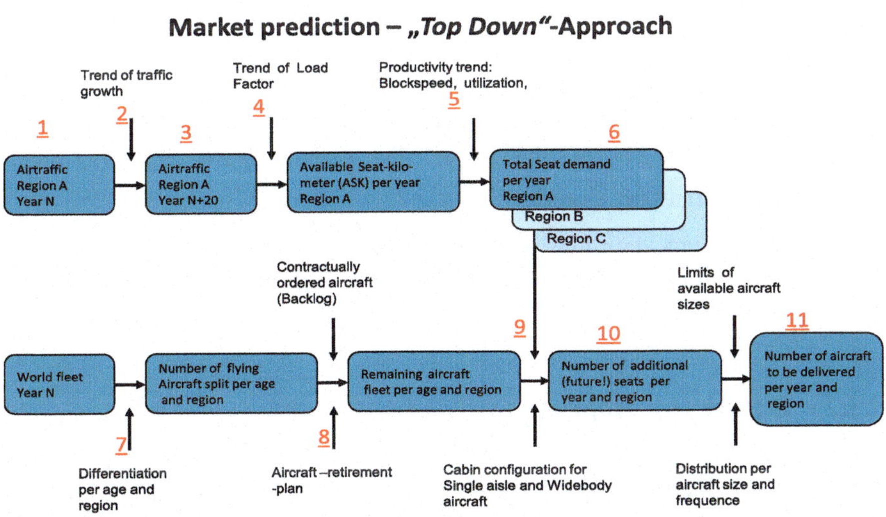

Fig. 3.6 Methodology for top-down market forecast

5. Make reasonable assumptions regarding the productivity trend (block speed, seat capacity, etc.)
6. Look at the world fleet of today (day of analysis) in terms of size and a/c category
7. Split the world's fleet into region and age of aircraft

8. Assumptions for retirement of ageing aircraft (retirement or conversion to freighter)
9. Subtract the existing aircraft still flying in year N+20 from the demand in step 5
10. Define the number of future seats required per year and region
11. With assumption on aircraft size and route frequency, the number of aircraft to be delivered per year and region can be obtained.

As in all forecast methodologies, the results that can be calculated very straightforwardly can only be as good as the chosen input data. The input data have to be checked carefully and have to be varied in certain boundaries in order to assess the sensitivity of each parameter.

It is obvious that this method is a fairly conservative approach, based on the data from the last 10 years and there is quite a lot of uncertainty in the forecast methodology, as nobody can imagine all possible events that could happen during today and the next 20 years. But at least this gives a first global estimate of the global aircraft demand in the foreseen time horizon. Normally, the time frame of 20 years is selected.

Some additional assumptions may also be integrated and quantified as:

- Growth of population:
- International trade development: (may stimulate long range and cargo traffic?)
- Political factors:
  - low liberalization of markets,
  - environmental concerns may reduce the demand for air transport,
  - fuel taxes may increase ticket prices and reduce transport demand.
- Competitive transport systems:
  - Telecommunication may reduce business trips and
  - high-speed trains (ICE, TGV, ..) may replace partly short range air routes.

3.3.2 Bottom-Up Approach

The bottom-up approach starts by analysing individual national or regional airline situations. For each airline the operational aircraft fleet of today is used as basis and the airline's specific development plan is taken as basis for the forecast of aircraft requirements for the next 20 years.

Figure 3.7 shows the methodology used for the bottom-up process:

Step 1 The bottom-up process starts from the actual airline situation, the actual fleet, the route network, the load factor, the aircraft retirement plan, the backlog and the financial situation of the airline.

Step 2 The actual flight plans of all airlines of a country/region will be taken and analysed.

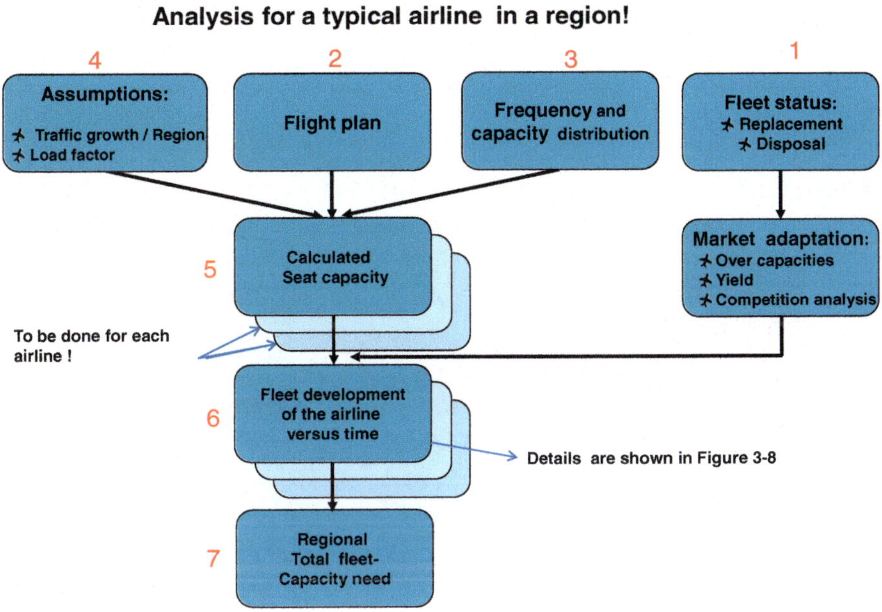

Fig. 3.7 Market forecast methodology "Bottom-Up—Approach"

Step 3 The flight capacity and frequency of all aircraft in operation in this region by each airline have to be identified.

Step 4 Assumptions about the traffic growth over the next 20 years, the economic development of the region, the development of load factor and the integration of new airlines in this market have to be assumed

Step 5 For this region and the airlines under investigation, all available seats per each route, per period (day, week, month, year), per airline can be listed. This defines the actual "status quo".

Step 6 From the situation in the region, each individual airline and its fleet development over the next 20 years can be developed.

Step 7 The addition of all airlines with their capacities and development plans have to be integrated to define the future capacity needs of this region.

In order to illustrate this process in more detail, an example of a typical European airline carrier (we call it "TEA") is chosen.

Figure 3.8 shows the fleet of this TEA airline, which consists in this example actually of 58 aircraft with four different aircraft types. The airline has already ordered and placed contracts for 18 additional aircraft which will be delivered during the next 4 years.

These additional 18 ordered aircraft are called "backlog", shown in Fig. 3.8 in green. These aircraft enter the service over the next 4 years.

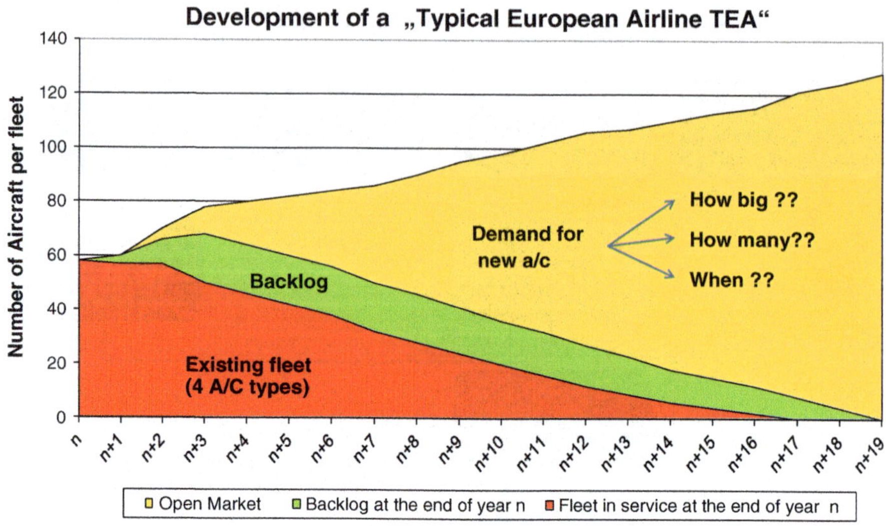

Fig. 3.8 A/C fleet development for airline TEA (typical European airline)

Based on this actual airline situation, the particular airline and their future strategy has to be analysed and assumption about the development of their fleet (retirement, replacement, disposal, backlog, new aircraft purchase/leasing) as a function of future growth and strategy with respect to capacity adaptation/increase on the existing routes, development of new routes, market competition analysis have to be established.
The following detailed questions have to be answered:

- What is the age of all registered aircraft?
- Define a reasonable plan for aircraft replacement and retirement (red line), average replacement of new aircraft is assumed to take place in ~12–15 years.
- Which new routes will be planned (SR or LR?)
- What is the financial situation of the airline?
- What is the competitive situation?
- Is an increase in airline fleet envisaged? (yellow line in Fig. 3.8)
- Are there national plans to support the national carriers? (e.g. traffic rights/ICAO)
- Is there a tendency towards more liberalization in this region/area?

In our example of Figs. 3.8 and 3.9 the retirement plan for this airline is shown by the red line. 12 aircraft will be taken out of service during the next 6 years. After 15 years no aircraft of the existing fleet will still be in service in this example.

But there is also an assumption for the future development of the fleet of this airline, which is assumed to more than double during the next 20 years.

In the year n+19, the total fleet will consist of 128 aircraft.

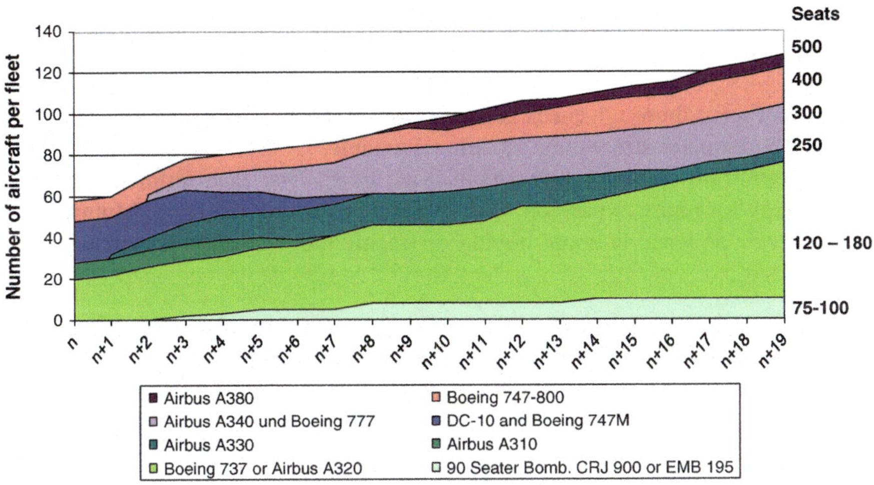

Fig. 3.9 Long-term fleet planning with allocation of aircraft size

Figure 3.8 shows only the three major elements of the future aircraft development plan

a. the existing aircraft fleet diminishes with time,
b. the backlog of undelivered/ordered aircraft will increase the fleet size during the next 4 years and will also be phased out before 20 years,
c. the new aircraft to be still purchased in order to follow the planned and expected fleet increase.

The airline forecast can also be performed in a way to indicate the increase in RPK, but then at the end a breakdown in number and size of aircraft types will follow.

Figure 3.9 shows such a possible aircraft size allocation. It can be seen that two aircraft types will be taken out of service (A310 and MD11) and new models will be introduced into the fleet like A330 and even some A380.

The mentioned aircraft are possible examples and have been the result of a student's project applying the a.m. methodology.

The bottom-up approach is principally more precise compared to the top-down approach, but this approach is also more complex and has more risk factors included. The risk in the methodology is that the expansion plans of the airline are often not very realistic. Often, they overestimate their individual growth potential, because even the competitors will increase their capacities. The competitive situation will anyway lead to some compromises in terms of capacity growth for each individual airline and all ambitious strategic visions will have to face a realistic trend.

With this information and the equivalent assumptions, the air traffic growth prediction for each airline can be generated and a plan for the fleet size development over the next 20 years (see Fig. 3.8) per individual airline is feasible.

The next step is the aggregation of the data at the level of the region, afterwards the same process has to be performed for the continent and finally for the global market.

At each level a thorough check is needed to verify that the overall growth rate for the country/region/continent, etc. is still reasonable.

The bottom-up approach is normally used to analyse each airline, understand their strategy for future expansion, their specific market conditions and future needs and also their philosophy about buying or leasing the necessary aircraft fleet.

3.3.3 Scenario Techniques for Risk Assessment

Scenario techniques are a way to analyse complex systems and their future development.

Scenario techniques are widely used in several domains where long-term conditions have to be investigated, drivers for change to be analysed and possible future strategies to be defined. Several industry sectors like oil, energy, transportation and mining—with specific difficult long-term strategies and where large investments are needed to secure their long-term future, are using scenarios to better identify the underlying risks and chances.

The great benefit of scenario techniques is the involvement and participation of different specialists from all disciplines, their input and the common discussion about the major influencing parameters, the common understanding about major drivers and the well-structured way into future possible worlds, which will have an influence on the future long-term visions. There is a combination of technological evolutions, breakthroughs, societal demands and changes, commercial aspects, financial investments etc. which all are influencing the future and many parameters are not deterministic and are difficult to be quantified.

For the aeronautical scenarios, the main parameters are kerosene prices, alternative energies, environmental challenges, societal acceptance and hindrances for air transport, alternative transport modes, commodities of air transport, etc.

The scenario technique has been described in detail in the following books and papers [13–15].

The scenario technique is a methodology, which is using normally five steps (see Fig. 3.10a–e):

1st step Focus and definition of the problem; definition of all parameters of interest and influence; definition of overall scope. Figure 3.10a illustrates this "view from above" and the definition of problem boundaries and involved main parameters.

Fig. 3.10a

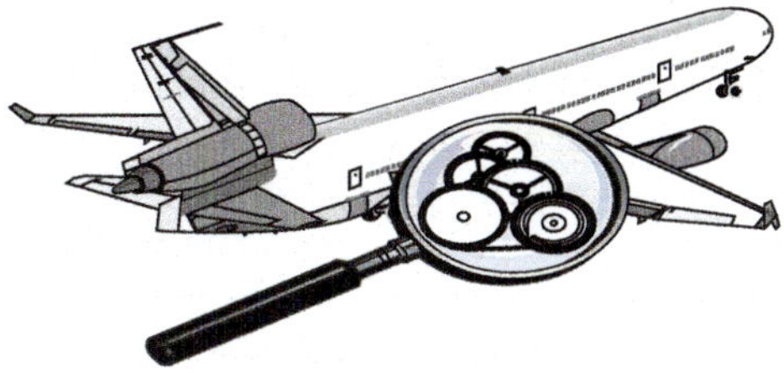

2nd step In step 2 all parameters of importance for the problem have to be defined (normally 30–50 parameters, like fuel price, traffic growth, GDP growth, etc.). An impact matrix with all parameters has to be established and then possible interactions and interdependencies/reactions between all parameters have to be quantified in rough terms.

Fig. 3.10b

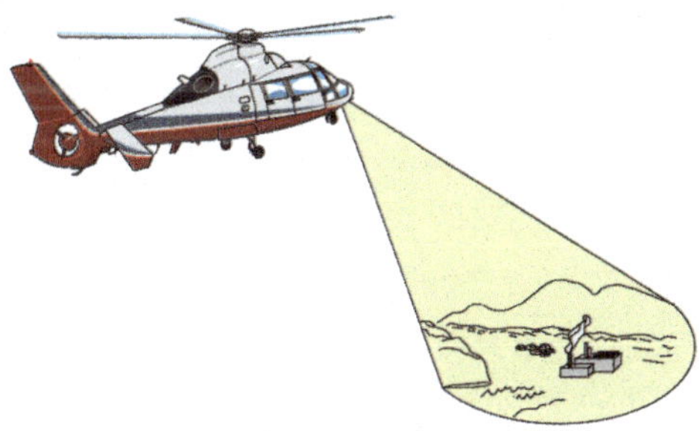

3rd step Step 3 includes the identification of alternatives and a critical review of all parameters and the establishing of reasonable extremes for all chosen parameters.

Fig. 3.10c

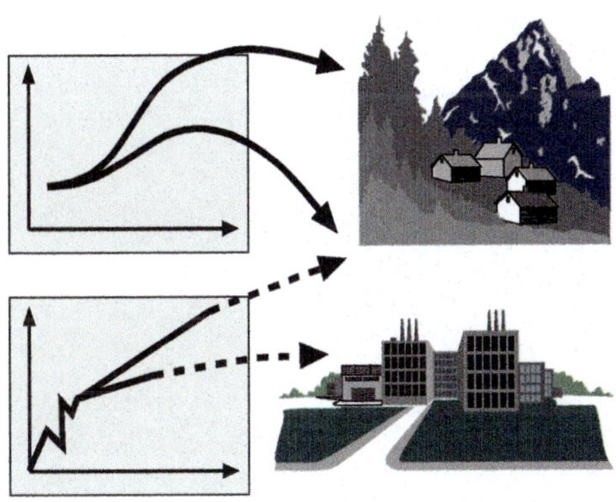

4th step Based on the impact matrix, a specific tool will help to define a large quantity of future scenarios. By clustering of several solutions and a critical review of all established scenarios, 3–5 main scenarios will finally be selected and specified in more detail, amongst them: one trend scenario and several adverse and alternative scenarios. For each of the finally selected scenarios, a clear set of assumptions, definition of main parameters and impact between them lead the way directly to this solution. The final scenarios have to be selected carefully and they have to be described in detail and depth. Especially the extreme scenarios are of interest to think and analyse what could happen and which combination of parameters and assumptions will lead to these scenarios! The picture below shows the typical funnel shape, indicating that the farer we look into the future, the possibilities will further increase and more scenarios will evolve.

Fig. 3.10d

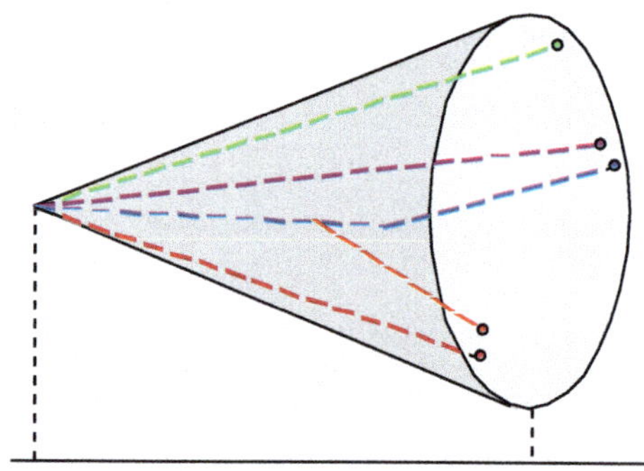

5th step The last step is then the analysis of the selected scenarios. Using the basic parameters, it is helpful to see which parameters are the drivers for the future and which are more dependent and "driven" parameters.

Fig. 3.10e

The analysis of the extreme scenarios is useful and mandatory to understand which parameters can change dramatically a scenario, which parameters can be used to influence the own business and what can be influenced, which actions can be undertaken with a certain hope for success. On the other hand, there are elements of importance in a scenario (like the fuel price development) where no own control, influence or change on these parameters is possible. The discussion among the specialists and the generalists from top management is one of the best results from each scenario process and will help to better understand the business environment and future risks and opportunities.

The importance for a successful scenario process is:

- a very competent group of several specialists from different disciplines (engineering, manufacturing, marketing, financial, human resources, communication etc.),
- the support from the top management and
- finally, a professional moderator to control and manage the scenario process.

In Europe, EADS, Shell and Daimler Benz are using and developing scenarios over more than 20 years for their long-term analysis of their business and strategy development.

The scenario technique is not another market forecast method. It is a complementary part of the classical market forecast tools as described before. But it has shown and proven to help considerably to understand much better the complexity of the interconnection of several contradicting parameters and get a much better insight into certain dynamics of the complex air transport system. Technische Universität München and DLR (Deutsches Zentrum Für Luft- und Raumfahrt) are using scenario techniques in their education and training systems. Students get a clearer understanding about the connectivity and interrelationships between all the players and stakeholders in the air transport system when they are participating in such a scenario Workshop [13–15].

As a conclusion for the Market Forecast methodologies it can be stated: No single method should be used, but a variety of instruments to identify risks, societal changes, environmental constraints and safety and security features for the future and thus reduce risk and avoid major market misinterpretations!

3.4 Passenger Aircraft Market

The two market forecast approaches "Top Down" and "Bottom-Up" are complementary and a combination of these two approaches has proven to be successful: The "Top-Down approach" will generate the global picture, the "Bottom-Up approach" will help to understand the strategy of each individual airline, their development potential and the regional situation. Verify that the sum of individual airline's forecasts will not exceed the global need (e.g. determined by the Top-Down method!)

The market forecast programmes used today by aircraft manufacturers are no longer split into top-down and bottom-up approach. With the capacity of large computers and the big know-how of the aircraft industry, an integrated approach is feasible today.

The following two scores (Figs. 3.3 and 3.11) are taken from the Airbus Global Market Forecast (GMF) [16], and show the input parameters and the final result for the year 2030.

The input parameters chosen cover an even wider range of social, political and economical development factors and will be adapted or increased, if a new situation (financial or political crisis) appears on the horizon.

The new elements in this GMF are the introduction of demographics, highlighting the population growth and rapid urbanization and also the new business models in the airline operation.

A typical result of such a GMF is shown in Fig. 3.4. The final overall result shows that around 29,000 new aircraft will be needed within the next 20 years and the total business in this period is of the order of 3 trillion $, an unbelievably huge market number!

Figure 3.3 provides a forecast of which sort of aircraft type the market will demand. Overwhelmingly, small short range aircraft will be demanded from the market counting for 80 %. However, looking at the value of the market instead of pure aircraft numbers, it becomes obvious that also the few numbers of big aircraft

Fig. 3.11 Main elements used within Airbus GMF [16]

represent a very big and important value of the market share, which explains the strong fight between the two rivals Boeing and Airbus in this market segment!

More details can be found in the forecasts published by Airbus [16] and Boeing [17].

3.5 Air Cargo Market

The air freight or air cargo is an own market segment, which is normally treated separately from the passenger market.

Air cargo is only one part of the global goods distribution network. Shippers demand that shipments arrive at their destination on time, undamaged, and at a reasonable price, regardless of the selected transport mode. Different transport modes like road, rail, maritime and air can often move the same goods. For intercontinental transport only sea and air transport are in direct competition. The primary benefit of sea transport is the fairly low cost; however, the transportation task takes a lot of time. Air transport offers the advantage of speed and reliability, but at a different and much higher cost level.

Speed is still a very important issue for all such goods which have a low weight but high value for transport goods, which are time critical like food, flowers, animals, etc. In addition to these time critical goods, a lot of maintenance parts, electronic components and complex mechanical parts are mainly carried and transported by air.

In reality, there are no dedicated civil freighter aircraft on the market. All flying freighter aircraft have been deduced from passenger aircraft (B747F, MD11F, A330F, B777F, etc.). But there have been some ideas about dedicated freighter aircraft and their characteristics [18, 19].

3.5.1 Cargo Operators

There are in general three different types of air freight operators:

- Integrators that offer a "door to door- service" like FedEx, UPS, DHL, etc.
- All cargo airlines are purely freight carriers and operate cargo aircraft exclusively
- Classical line carriers, which transport both passenger and freight.

The flowchart of cargo operation is shown in Fig. 3.12. There is a shipper who wants to send a cargo (this could be a mail, a small package, a box, a complete container) to a consignee. If the piece of cargo is small (a mail or a parcel) it is mostly given to an integrator. Bigger cargo pieces are normally handled by freight agents or freight forwarders, who—when receiving the cargo—are looking for an airline (Combi carrier or all cargo airline) and are negotiating the best conditions for

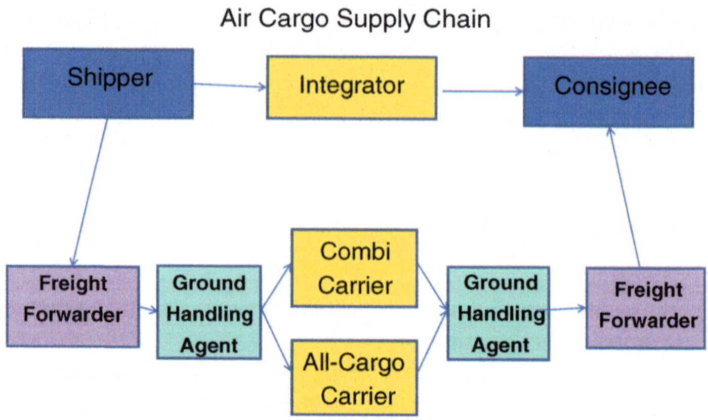

Fig. 3.12 Air cargo supply chain

the transportation with the airlines. To load and unload the cargo, there may be also some specialized ground handling agents involved, who are part of the overall transportation logistic.

Air cargo is normally multimodal. The shipper is normally not living close to an airport. Therefore, the cargo has to be transported by a truck from the shipper's area to an airport, where the cargo load will be taken, often repacked into a container or a palette and then transported by air to the next airport of the consignee, then the cargo will be reloaded, repacked to be taken by another truck to the final destination.

Integrators provide a specific service in air cargo. From 1977 through the early 1980s a dramatic new initiative in air transport was started with the "overnight-express business". The newly founded Federal Express company offered in the US an overnight service for all sorts of parcels, to be delivered till next morning at nearly each place in the US. This started an unbelievable success story and several other companies followed this business principle and offered this express service all over the world. The following statement is taken from the Boeing World Air Cargo Forecast 2010–2011:

Classical line carriers offer both services, passenger and cargo air transport. Cargo is normally transported in the lower cargo hold, using the space which is not used for baggage from the passengers. The continuing strong competition between airlines has made the classical line carriers (sometimes also called "Combination carriers") to focus on opportunities for additional lower-hold cargo revenues. On average, cargo revenue [20] represents about 15 % of total traffic revenue, with some airlines earning well over half of their revenues from this business. There is however a very strong competition today in the air cargo transport business.

All Cargo airlines are airlines that only operate pure freighter aircraft, where the cargo will be loaded on the main deck or/and in the lower cargo compartment.

Fig. 3.13 Operation of an all cargo aircraft

A classical all cargo operator is Cargolux, with its home base in Luxemburg, a member country of the European Union.

All cargo airlines like Cargolux, Atlas Air, ABX Air, Maersk, etc., are only operating freighter aircraft for international resp. Intercontinental and long haul air freight operation. They also often operate from an airport which is less heavily occupied by passenger transport and provide fewer airport fees, but which is well connected to the road/motorway system of the country or the region. Very often these airlines only operate with one type of aircraft, mainly B 747F (see Fig. 3.13), MD11F or converted passenger aircraft.

Freight forwarders play a major role in the air cargo business. Their essential function is to consolidate several small shipments into one larger consignment and tender it on behalf of the shippers, as one single unit to the carrier. Forwarders are thus able to offer cheaper freight rates than carriers are capable to the customer (shipper). Thereby they have to overcome the trade-off between holding a shipment for the best possible consolidation and moving a shipment without delays to the consignee in order not to satisfy the customer and lose future business. In addition, the forwarder offers additional services like collection, packaging of the shipment as well as documentation, customs clearance and final delivery. Forwarders normally manage a shippers supply chain, offering customized and industry-specific solutions.

3.5.2 Freight Market Forecast

Freight market forecast is provided similar to the passenger market forecast—by all major aircraft manufacturers and several independent institutions. The following summary is mainly taken from the "World air cargo forecast" WACF from Boeing, which seems a bit more detailed and very logically presented [17].

The summary says: "Over the next 20 years, world air cargo traffic will grow 5.2 % per year. Air freight, including express traffic, will average 5.3 % annual growth, measured in RTKs (Revenue Tonne Kilometres). Air mail traffic will grow much more slowly, averaging only 0.9 % annual growth through 2031. Overall, world air cargo traffic will increase from 202.4 billion RTKs in 2011 (down from its 2010 record of 204.2 billion RTKs) to more than 558.3 billion RTKs in 2031. Asia will continue to lead the world air cargo industry in average annual growth rates, with domestic China and intra-Asia markets expanding 8.0 and 6.9 % per year, respectively. Latin America markets with North America and with Europe will grow at approximately the world average growth rate, as will Middle East markets with Europe. The more mature North America and Europe markets reflect slower and thus lower-than-average traffic growth rates.

The number of airplanes in the worldwide freighter fleet will increase by more than 80 % during the next 20 years, as demand for air cargo services more than doubles. Since 2001, freighter airplanes have carried on average just over 60 % of the world's total air cargo traffic each year. The role of large freighters will increase as the large freighter share of the fleet rises to 36 % by 2031, compared to 31 % today and 22 % a decade ago. The significant efficiency and capability advantages of large freighters will enable carriers to manage projected traffic growth without increasing the number of airplanes proportionately. About two-thirds of fleet additions for airplane replacement and fleet growth will come from modified passenger and combi airplanes. Yet, production freighters will continue to play an important role because their superior reliability, operating cost, and capability can outweigh the significant on-ramp acquisition cost advantages enjoyed by conversions." [17].

The global economic downturn of 2008 and 2009—the worst economic contraction since the Great Depression—dragged down all modes of transport. Air cargo traffic fell 12.5 % between mid-2008 and year-end 2009, the worst decline since the beginning of the jet transport age. By mid-2009, however, worldwide industrial production began to peak up, pushing air cargo traffic toward recovery.

Rising fuel prices have been a factor in air cargo traffic slowdowns since late 2004, diverting air cargo to road transport and maritime modes, which are less sensitive to fuel costs. The price of jet fuel has tripled over the past 8 years, and prices are likely to remain volatile as the threat of supply disruptions persists.

On a positive note, however, oil and jet fuel prices are forecast to remain around mid-2012 levels or, in some scenarios, even decline over the next 3–5 years. Economic activity, as measured by world GDP, remains the primary driver of air cargo traffic growth. World economic growth averaging 3.2 % over the next 20 years, coupled with the forecasted stable fuel prices, will help air cargo traffic grow.

Freight yields have declined at an average rate of 4.2 % per year over the past 20 years.

Continuing profit challenges at passenger airlines have focused airline attention on opportunities to earn lower-hold cargo revenue. On average, cargo revenue represents approximately 15 % of total air transport revenue, with some airlines earning nearly 40 % of their revenue from cargo. Declines in yield for cargo and passenger services reflect productivity gains, technical improvements, and intense competition. While declining yield creates pricing pressure on all industry segments, it also helps stimulate growth for the industry by enabling lower shipping costs for the consumer.

Averaged over the past two decades, freight yield has declined by 4.2 % per year. The most recent decade saw a slight yield increase of 0.9 % per year, compared to the 9.0 % average annual decline recorded in the preceding decade.

3.5.3 Changes in the Aircraft Market

In the past aircraft manufacturers have been the main drivers in the market for innovative products. The airlines are constantly asking and pushing for new and better (more fuel efficient but reliable) products. The strong competition between more than 800 airlines and their small profit margins are leaving the innovation and the definition of new aircraft products to the manufacturers.

The recent strongest change came from "Low Cost carriers" (LCC). They have introduced new business models to reduce cost drastically and this has given a new impetus to the air transport business. New entrants like Ryan Air, Easy Jet, Air Asia, etc., have successfully entered the market. More details about the business models and those elements where the Low Cost Carriers are having their commercial advantages will be given in Chap. 7.

Another game changing element in air transport is introduced by the International Leasing Companies. The strong role of Aircraft Leasing Companies that are buying aircraft partly without having a clear customer to whom to lease the aircraft has changed the direct contact between manufacturer and airline. The leasing companies, which are purchasing sometimes more than 100 aircraft in one campaign, can therefore get specific prices and therefore provide good leasing deals to smaller airlines or LCC.

With the interim partner between manufacturer and operator—the Leasing Company—the manufacturer is no longer the only contract partner with the operator. He has to deliver a standard aircraft to the leasing company, without knowing the specific request from the later operator. Cabin changes should therefore be easy and simple in order to reduce the customization cost for these aircraft. Especially for large and long-range aircraft this interim partner can be difficult. Customization is a major issue and especially the A380 with all their cabin options is difficult to be reconfigured in the cabin without major adaptation cost.

This will be a difficult market segment for leasing companies and also for the aircraft manufacturer. More details will be given in Chap. 7.

Aircraft, crew, maintenance and insurance (ACMI) providers, sometimes called "wet lease providers" offer cargo operators the flexibility to obtain lift on a trial basis and provide service in markets that are highly seasonal, all this with no capital investment required. Large freighters in long range markets account for the most significant segment of the air cargo ACMI business. The ACMI business has become an established industry subsector since the early 1990s. Several providers, representing about 14 % of the large freighter ACMI traffic, exited the business in the recent years.

Another changing element in the market was introduced with the airline global alliances (see Chap. 8). The airlines of a global airline alliance are also pooling their demand for new aircraft purchases in order to receive better market prices due to a higher number of aircraft orders.

3.6 Cost and Commonality Aspects

Cost aspects are—as in all businesses—a major element for purchase decisions and are part of all business models.
In the civil aircraft business, there are—among several others—normally three main cost definitions which are also used in the technical environment:

- Direct operating cost
- Total operating cost
- Life cycle cost.

The operating cost models (DOC) will be discussed in detail in the airline part in Chap. 7. Here we briefly review the life cycle cost on an engineering level without details about economic factors like depreciation, interest rates, etc. [21, 22].

3.6.1 Life Cycle Cost

A simplified description of the life cycle cost is given in Fig. 3.14.
Beginning with the manufacturer on the left side, it can be found that he differentiates mainly between two big cost elements per aircraft:

- The development cost (one big block of cost per aircraft program, also called non recurring cost (NRC)
- The production cost per aircraft unit (named Recurring Cost—RC), including materials, production facilities (jigs and tools), aircraft assembly and flight testing.

The production costs are a function of the number of aircraft assembled. A certain production learning curve has to be assumed. The first aircraft, which is assembled will have higher cost than aircraft No 100 and aircraft No 500, which are

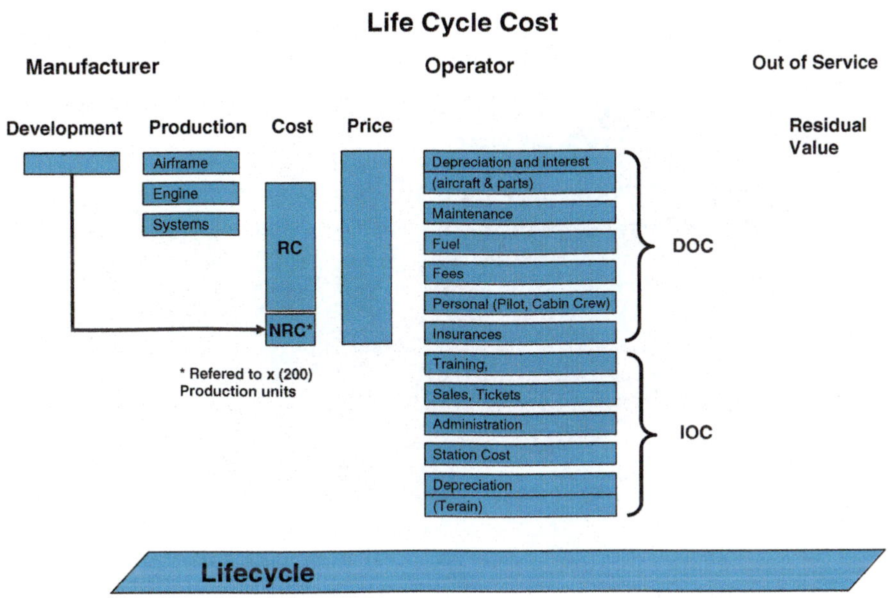

Fig. 3.14 Life cycle cost definition

down the "learning curve" and several production improvements have already been integrated in the production process (see [1, 20, 23]).

The manufacturer has to define a certain market price per aircraft, which includes the development cost, production cost, sales, administration and financial cost and also a certain profit margin. Figure 6.16 shows a graph, where the "rough order of magnitude" of aircraft prices are shown as a function of aircraft weight (MTOW). These prices are taken from the "Airbus aircraft list prices" [24]. These are the official list prices, but it is well understood that each airline may negotiate specific conditions depending on the market situation, the strategic importance vis-à-vis the competition and other factors.

The airline has a different view on their cost structure and in the view of the operator, the aircraft market sales price is only one cost element of the aircraft operating cost, however a major one!

- The purchase cost is normally depreciated over a certain amount of years. Historically, 15 years have been used as the time period for the depreciation of an aircraft. But today more sophisticated models and different considerations are introduced by leasing companies and also line airlines trying to buy and release the aircraft from some financial companies, which are then trying to refinance the aircraft via several private financial partners, so that suddenly 20 years and more appear for the aircraft life cycle and depreciation time.
- The classical direct operating cost (DOC) includes, besides the depreciation and interest for purchase of elements maintenance, fuel, fees (airport charges, ATM

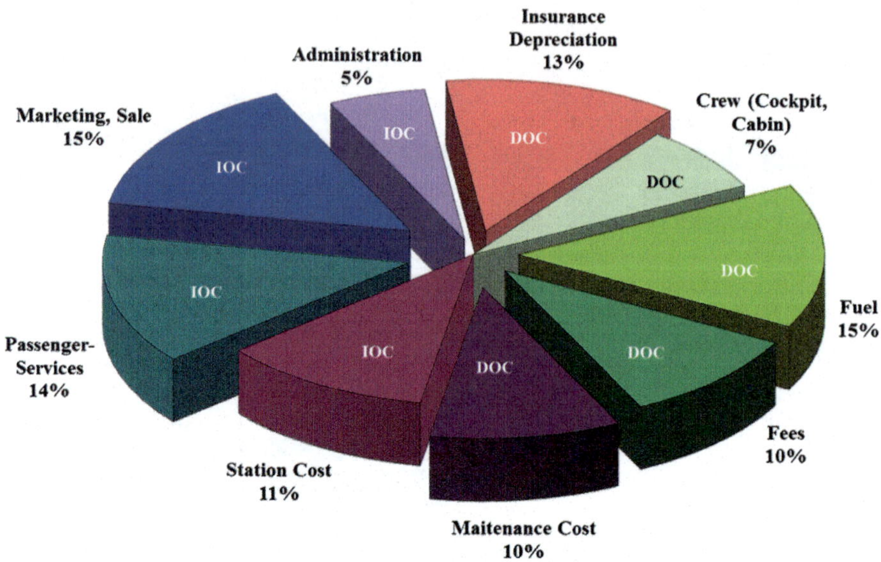

Fig. 3.15 Operating cost of an airline

fees and ground handling charges), personal cost for pilots and cabin crew and insurances. These are all cost elements directly related to a specific flight operation, therefore named DOC! (see also Chap. 7)

The Operator/airline has also indirect cost—not directly related to the operation of a specific flight—for their business, for example for training of crew and pilots, for sales and ticketing, station cost for the home base and abroad, rent or depreciation for buildings and services and general administration cost, all integrated as indirect operating cost (IOC). Figure 3.15 provides an example of DOC and IOC for an airline.

- At the end of the normal life of an aircraft, the aircraft still has a residual value, either as being sold to another operator (sold to operators in third level countries?) or being scrapped, but having still a certain value due to its components and material—the residual value.
- In the golden age of air transport, this residual value was always considered to be of the order of still 10 % of purchase price. But today's world with the domination of financial controllers in all sectors has overruled this simple and understandable approach and has put a lot of additional complexity into a fairly simple calculation methodology.

3.6.2 Family Concepts and Commonality Aspects

Besides DOC and IOC another element has shown to be of great importance for aircraft selection by an airline/operator: The aspect of *commonality* of the fleet. Commonality can include several aspects like:

- Commonality of several aircraft as part of a family concept (Fig. 3.18).
- Hardware commonality (same components like wing, tail, undercarriage, cockpit, equipment, etc.).
- Design commonality! (same cockpit interface and therefore common pilot interfaces with a lot of advantages for pilot training).

The family concept commonality can be seen in Fig. 3.16. as a common design goal, to develop a whole family of aircraft by keeping major elements like wing, tailplane and undercarriage the same and "just" changing parts of the fuselage (The reality is more complex, but there is an enormous cost saving aspect for both partners, the aircraft manufacturer as well as the operator!). As shown in this Fig. 3.16, four different aircraft types are developed out of a basic aircraft design, in our example, the Airbus A320 aircraft. (Boeing uses a similar family concept for their B737 family!) By just changing parts of the fuselage and inserting or reducing fuselage sections, four different aircraft types have been developed with different fuselage lengths and different number of seats. In reality, this is a bit more complex as also parts of the wing and tailplane have to be adapted and modified. The A321 has a different high-lift system compared to the A320 baseline configuration, which

Fig. 3.16 Family concept and commonality

Fig. 3.17 Modern aircraft cockpit with the sidestick controller (Airbus concept)

was needed to improve or adapt the takeoff- and landing performance for the modified aircraft version. It is also visible that the vertical tailplane (fin) of the A318 aircraft had to be increased to improve the lateral stability and controllability of this A318 version. Also, the engine has to be adapted to provide the necessary thrust for the different range and fulfil the Lowspeed (Takeoff and landing) requirements. But production cost, operating cost spare parts, maintenance cost can be reduced considerably by such family concept designs.

The hardware commonality defines all physical parts and components that are used in different aircraft types, for example, cockpit parts and instruments, engines, brakes and wheels, system components, equipment parts, cabin items etc. Commonality here means that common physical parts are used in an aircraft family like B737- 600, B737-700, B737-800 similar to the Airbus world for the A320 family members. This means that spare parts are common for the different family members and can be interchanged.

Design commonality is another aspect of aircraft commonality, which relates to a common design philosophy of an aircraft manufacturer. The most obvious example for design commonality is the cockpit design. Commonality here means that the cockpit layout is done in such a way that pilot training can be dramatically reduced while switching from one aircraft type (A320 family, Fig. 3.17) to another (A330/A340 family) and still have the same instrumentation environment and input controls, leading to considerable cost advantages in the training process (see Fig. 3.18).

Commonality is today a major design point in the definition and development of new aircraft. New technological developments are always introduced in each new aircraft concepts. But a lot of standardization still exists and has shown to be advantageous and good.

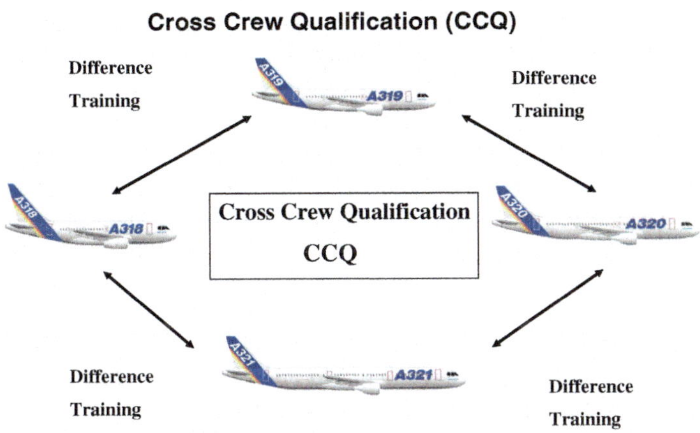

Fig. 3.18 Commonality in the cockpit and the benefit of cross crew qualification

The simplest way of getting the full potential of commonality is the philosophy to buy all aircraft from a single manufacturer and only one aircraft type! Then the design philosophy of all aircraft in the fleet is identical! (This is one aspect of some low cost carrier philosophies! See Chap. 7).

3.6.3 Cross Crew Qualification

The commonality in the cockpit design is a result of the digital glass cockpit, which has been developed for the A320 aircraft as part of the fly-by wire philosophy. The introduction of the sidestick controller in the A320 allowed several modifications for the role of the pilot such as:

- Change of the role of pilot from a continuous flying officer into a cockpit manager.
- The stabilization of the flight (like in cruise) is done by the aircraft flight control system, like a continuous autopilot function.
- The pilot inputs are commands to the flight control system to change the flight level or direction.
- The flight control system has some protections integrated, which will help the pilot in critical cases to keep the aircraft in safe flying conditions and protect the aircraft from entering into several critical low speed and critical high-speed conditions, but still informing the pilot about flight conditions of the aircraft.

The basic information—what the pilot really needs to see and be informed about—and in addition, what he would like to see from time to time to monitor his en-route flight data—has become more generalized and has now been well examined by the engineering community. This changes the role of the pilot from an active driver of the vehicle (similar to today's car driver, who continuously controls the car via wheel commands) to the role of a controller and system master, who monitors the well-functioning of the system and "just" provides commands to the system in order to change attitude, climb angle, speed, etc. But all his commands are then transformed into control signals for the aircraft.

Pilots now have a fairly standardized cockpit in front of them—similar to a private car driver when he is driving a car from different car manufacturers but will easily adapt with the steering wheel, the pedals and instruments. His driving interface—the car cockpit—is the same, the only difference may result from the motor and vehicle reaction which may be slightly different from his own car's reaction. But from the beginning, he feels himself comfortable and familiar with this cockpit (his car) and has no problem to adapt to the slightly different instruments and a different motorization of his new vehicle!

Figure 3.18 gives a short explanation of the training benefits for pilots in a common A320 family concept.

These benefits of commonality are larger, if an airline has a big fleet size.

Some papers and details about these aspects can be found in [21, 25].

But there is also a more fundamental aspect, especially in the cockpit design.

CCQ is a unique concept developed by Airbus, which gives pilots the possibility of transitioning from one Airbus FBW-equipped type to another via difference training instead of full type rating training. Even aircraft types like A320 (a short range aircraft) and A380 (a four engine long range aircraft) can benefit from the CCQ. The transition training from A320 Family aircraft to the A380 takes 13 working days, from A330/A340 Family aircraft it takes 12 working days, while a pilot with no Airbus FBW experience requires 24 working days to complete the A380 standard type rating course. These time savings lead to lower training costs for airlines and considerably increased crew productivity. The benefits of commonality extend from the flight deck into the passenger cabin as well, with a maximum use of similar systems, control panels and procedures within the various aircraft families. As a result, cabin personnel benefit from the familiarity aspects on various aircraft types from one manufacturer, while aircraft maintenance is eased with the high inter-changeability of systems and parts.

References

1. Lawrence, P.: Aerospace Strategic Trade. Ashgate, Aldershot. ISBN:0 7546 1696 7 (2001)
2. Concorde Aircraft. http://www.concordesst.com. Accessed 30 Nov 2014
3. Steiner, J.E.: How decision are made—major considerations for aircraft programs. In: ICAS (1984)
4. Lawrence, P., Braddon, D.L.: Strategic Issues in European Aerospace. Ashgate, Aldershot (1999)
5. A320 Aircraft Accidents. http://airsafe.com/events/models/a320.htm. Accessed 30 Nov 2014
6. Embraer Aircraft. http://www.embraercommercialaviation.com/Pages/default.aspx. Accessed 29 Nov 2014

7. Bombardier Aircraft. http://www.bombardier.com/en/aerospace/commercial-aircraft.html. Accessed 30 Nov 2014
8. Mitsubishi Aircraft. http://www.mrj-japan.com/. Accessed 29 Nov 2014
9. Sukhoi Aircraft. http://sukhoi.org/eng/planes/projects/ssj100/. Accessed 29 Nov 2014
10. Chinese COMAC Aircraft. http://english.comac.cc/products/ca/pi/index.shtml. Accessed 30 Nov 2014
11. World Trade Organization WTO. http://www.wto.org/. Accessed 30 Nov 2014
12. Elliott, G., Timmermann, A.: Handbook of Economic Forecasting, 1st edn. vol. 1, Elsevier, Amsterdam. ISBN:9780444513953 (2006)
13. Gausemeier, J.: Scenario Management. Hanser Verlag, München (1995)
14. Strohmayer, A.: Szenariomethoden im Vorentwurf ziviler Transportflugzeuge, PhD thesis, TU Munich, Dr. Hut Verlag (ed.) (2002)
15. Phleps, P., Kuhlmann, A., Eelman, S.: Environmental awareness and the future of flying. In: 14th ATRS Conference in Porto, Portugal (2010)
16. Global Market Forecast Airbus. http://www.airbus.com/company/market/forecast/. Accessed 30 Nov 2014
17. Global Market Forecast Boeing. http://www.boeing.com/boeing/commercial/cargo/. Accessed 30 Nov 2014
18. Schmitt, D., Roeder, J.: The Ecolifter—a new concept for a dedicated advanced cargo transport concept. In: ICAS Congress, Melbourne (1998)
19. Logan, M.: Future vision for global air cargo. In: AIAA 1998-0437, Reno conference (1998)
20. Anon.: Flight plan 2010 analysis of the US aerospace industry, US Dep. of Commerce (March 2010)
21. Bador, D., Seering, W., Rebentisch, E.: Measuring the efficiency of commonality implementation: application to commercial aircraft cockpits. In: ICED 07, Paris, 28–31 Aug 2007
22. Echtermeyer, K.: Designing the aircraft of tomorrow, Aeronautics days, Vienna, 19–21 June 2006
23. Schmitt, D.: Air transport system, lecture notes, LLT, TU Munich (2006)
24. Airbus Aircraft Price List. http://www.airbus.com/presscentre/pressreleases/press-release-detail/detail/new-airbus-aircraft-list-prices-for-2014/. Accessed 20 Nov 2014
25. Airbus, Importance of Commonality. http://www.airbus.com/aircraftfamilies/passengeraircraft/a320family/commonality/. Accessed 30 Nov 2014
26. Cruide Oil Price Development. http://www.macrotrends.net/1369/crude-oil-price-history-chart . Accessed 30 Nov 2014
27. Norton, B.: Lockheed Martin C-5 Galaxy. Specialty Press, North Branch, Minnesota (2003). ISBN 1-58007-061-2

Chapter 4
The Regulatory Framework of the Air Transportation System

Abstract This chapter gives an introduction to the global regulatory and organizational setup of air transport. Based on the freedoms of the air, which are given, the main international contracts are introduced as well as major organizations like ICAO. Here also the way how international regulations are transferred into national rules is described.

Flying through the air was always a potential risk and always caused some concerns in people's minds. However, fascination and excitation about the feeling to overcome physical boundaries and to explore new spaces have always dominated these concerns. Nevertheless, aviation safety is a fundamental prerequisite for people's acceptance of aviation as a major pillar of global mobility.

Additionally, aviation by nature is international and global and requests for coordinated global approaches to ensure safety.

These two elements, the relevance of safety in aviation and its global character led to the development of international and national standards and regulations for the development of aircraft and its operation.

Furthermore, security is a very important aspect of aviation. While aviation safety considers all measures to ensure technical and operational safe operation of aircraft, security addresses all aspects to protect aviation from human criminal and terroristic impacts.

The issues of safety and security are essential for successful air transportation.

In this chapter the major international organizations and standards are introduced to provide an understanding of the organizational framework of aviation. Further, the principal approach in using regulations for aircraft design and operations is described exemplarily.

4.1 The Freedom of the Air

Based on the "Atlantic Charta", where the nations announced in 1941 the "Freedom of the Seas", in 1944 the nations agreed upon the "Freedom of the Air", leading to the following rules:

D. Schmitt and V. Gollnick, *Air Transport System*,
DOI 10.1007/978-3-7091-1880-1_4

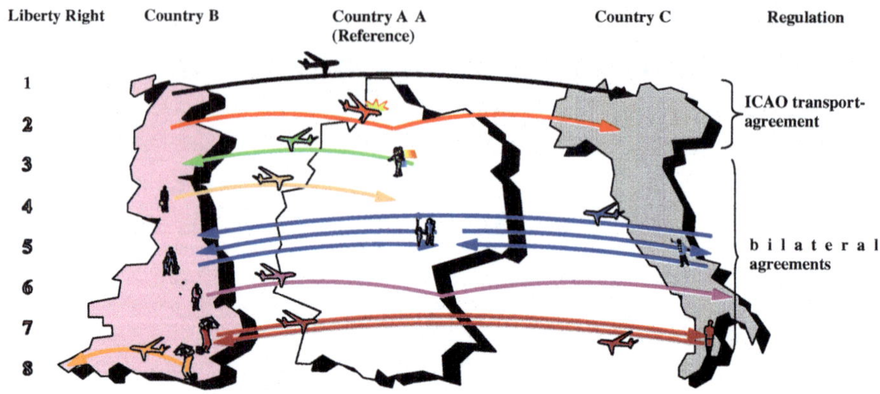

Fig. 4.1 The eight rights of freedom of the air

1. Right to cross a state airspace without landing (i.e. B to C over A in Fig. 4.1)
2. Right for intermediate stop for non-traffic purposes, e.g. aircraft problems, passenger health problems, refueling, maintenance, (i.e. B–C, but intermediate stop at A needed)
3. Right to carry people or goods from the home country of the airline, e.g. A to another state, e.g. C or B ➔ "coming from home"!
4. Right to carry people or goods from a foreign state, e.g. C to the home state of the airline, e.g. A ➔ "destined to home"
5. Right to carry people and goods between foreign countries, while the origin and destination of a flight is in the home state of the airline, e.g. flight from B to A and follow on flight from A to C or the other way round. ➔ "coming from and destined to = distribute"

In the agreement at Chicago the states committed to the first five freedom rights. However, the USA withdrew from this agreement in 1946, so that the binding character was lost, although the first two freedom rights are commonly accepted. Further freedom rights as shown in Fig. 4.1 are formulated but need to be supported by bilateral agreements.

4.2 Regulations for Transportation

Short after the First World War, when aviation became more and more relevant for commercial transportation, the "Convention of Warszawa" was signed in 1929 initially by 23 nations (in 2011 more than 130 nations) to harmonize and globalize the rules of liability. Two major issues were addressed, i.e.

1. a standardized and common look at the transportation documents and
2. a liable framework for air transport provider.

To declare air transport as "international" in terms of this convention, transport must be performed between at least two countries, which signed the convention. Transport itself under these conditions is defined by

- the origin or departure location of the transport,
- the final destination,
- potential intermediate stops,
- contracting parties, i.e. the countries, the airline, the passenger.

The characteristics named on the transportation documents are typically

- the flight ticket for passenger transport,
- the passenger baggage tag,
- the airway bill.

Referring to the "Convention of Warszawa" the air transport provider is liable for

- personnel damages by injury, damaged health or death of travellers,
- material damages by loss, destruction or damage of baggage or cargo,
- inconvenience and damages by delay by exceeding time limits for passenger and goods.

Based on these principles civil air transportation liability is established. However, damages on the ground have not yet been considered. This issue will become relevant for upcoming unmanned air vehicles, which are controlled from the ground and which have to be integrated into the air space.

Further, supplementary conventions were signed in The Hague, Montreal and Guadalajara.

4.3 International and National Organizations

There are numerous international organizations representing the interests of the different stakeholders globally and also regionally. For the purpose of this book to provide an understanding of interactions between the various stakeholders of air transportation, in the following the International Civil Aviation Organization (ICAO) and the International Air Transport Association (IATA) are introduced, which have a real constitutive role in air transportation.

Other organizations like the Airport Council International (ACI) or the International Federations of Air Traffic Controllers (IFATCA) or Airline Pilots (IFALPA) are more but not only acting as lobbyists.

4.3.1 The International Civil Aviation Organization—ICAO

Due to the global character of aviation, based on the 1944 contract of Chicago about the "international civil aviation", the International Civil Aviation Organization (ICAO) was founded and located in Montreal, [1].

Most of the states of the world have signed the contract and are committed to develop common recommendations and regulations to enable a harmonized and consistent air transport system.

The mission of ICAO—an intergovernmental organization—is the systematic, organized and safe development of international civil air transportation. Further, ICAO supports the development and operation of civil aircraft. Also, the evolution of air roads, airports and air navigation service systems is one part of the ICAO mission in order to improve continuously air transport safety and efficiency. As a governmental organization, ICAO looks at balanced cooperation among the member states.

The general assembly meeting of all member states every 3 years is the ultimate decisive institution of ICAO. Various technical committees are installed and are responsible for the elaboration of new regulations, technical requirements and procedures in the form of standards.

These recommendations normally called Standards and Recommended Practices (SARP) are becoming obliging when the individual state has transferred them to national law.

To provide an overview of the structure and contents of these SARP, the following table provides the list of the annexes to the ICAO convention:

Looking through Table 4.1 all relevant aspects of air transportation are addressed. Formal issues like licensing (annex 1) or information processes (annex 15) are also defined as standards as well as design standards for aircraft (annex 8) and airports (annex 14). These are only some examples of the work and role of ICAO.

Summarizing, ICAO has to be recognized as the most important global regulatory institution, which is accepted by most nations of the world.

4.3.2 National and European Regulatory Organizations

The international regulations need to be transferred to national law and subsequent orders to become operational since air law is under national authority. Figure 4.2 shows the flow down of global regulations agreed upon on ICAO level through national law to concrete standards, regulations and practices.

To ensure the maximum level of safety of aviation in Europe as well as in the United States, so-called "safety authorities" have been established in the past to ensure the sovereign responsibility of the various countries for safe aviation.

The Federal Aviation Administration (FAA) was founded in the United States in 1903, i.e. in the year of the worldwide first engine driven flight of the Wright

Table 4.1 List of SARP associated to the ICAO contract, [1]

Annex	Contents
Annex 1	Personnel licensing
Annex 2	Rules of the air
Annex 3	Meteorological service for international air navigation
Annex 4	Aeronautical charts
Annex 5	Units of measurement to be used in air and ground operations
Annex 6	Operation of aircraft—international commercial air transport—aeroplanes, general aviation—aeroplanes, helicopters (part I–III)
Annex 7	Aircraft nationality and registration marks
Annex 8	Airworthiness of aircraft
Annex 9	Facilitation
Annex 10	Aeronautical telecommunications—(surveillance radar and collision avoidance systems)—(volume I–V)
Annex 11	Air traffic services
Annex 12	Search and rescue
Annex 13	Aircraft accident and incident investigation
Annex 14	Aerodromes—aerodrome design and operations, heliports (volume I–II)
Annex 15	Aeronautical Information Services
Annex 16	Environmental protection—aircraft engine emissions and aircraft noise (volume I–II)
Annex 17	Security
Annex 18	The safe transport of dangerous goods by air

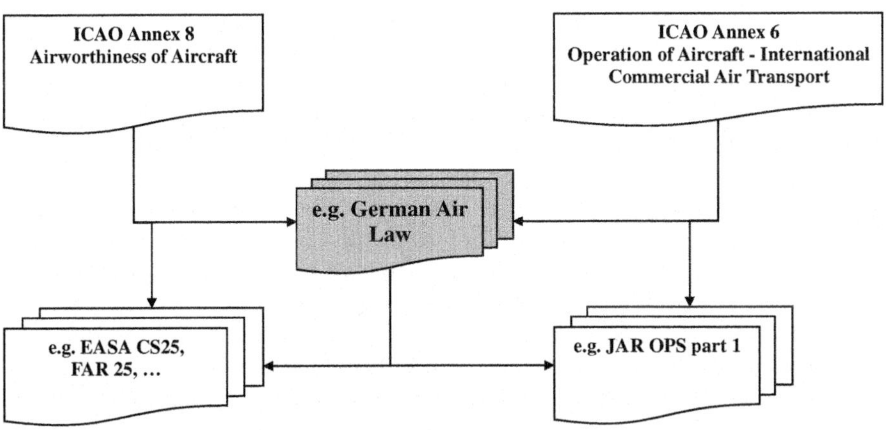

Fig. 4.2 Regulation applicability flow chain ICAO to national orders and standards

brothers, [2]. FAA is in charge of all regulations, processes and requirements to ensure safety of aircraft, airports and air traffic management.

Over the decades in Europe each nation had its own aviation safety agency, since air safety is a sovereign responsibility within the country borders.

For example, in Germany the Federal Aviation Office, called "Luftfahrt-Bundesamt", was in charge of all aspects of aviation safety.

Taking the German structure as a representative example the responsible ministry of transportation develops the national air law, called "Luftverkehrsgesetz", (LVG). This law defines the principles of air regulations in Germany. It is supported by national instructions and orders. The ministry also delegates the operational responsibility typically to lower governmental offices or organizations. In Germany the air safety authority ("Luftfahrtbundesamt", LBA) is in charge of aircraft certification and operation, among others, while the German Air Navigation Service Provider (GANSP), called "Deutsche Flugsicherung GmbH", (DFS) is responsible for safe and efficient air traffic operation.

While today the LBA has transferred most of its responsibility to the European Aviation Safety Agency, EASA, the DFS is still in charge of national air safety in operation, because this is still under national authority.

In 2002, the EU member states founded the European Aviation Safety Authority (EASA) merging and transferring their national responsibility on a European level [19]. The EASA is the more powerful successor of the European Joint Aviation Authority (JAA), which was established in 1970.

EASA is the European certification authority and shall promote the development of common standards in all relevant fields of civil aviation safety and also environment in Europe.

The national safety authorities like the German Aviation Authority LBA have changed their role in supervising and monitoring the compliance of the aviation stakeholders in the relevant countries with the European standards.

The complementary set-up and way of working of FAA and EASA can be easily shown by a comparison of the baseline safety standards given in Table 4.2.

Aviation safety philosophy is very much process oriented. Looking at the different standards, safety starts at company level, where Part 21 standards define a lot of requirements a company needs to fulfil until it is allowed to develop (EASA CS part 21J) and produce (EASA CS part 21G) civil aircraft. These requirements address organizational prerequisites as well as procedural ones in terms of set-up of the design organization, independency of the certification engineers, etc. For aircraft maintenance, Part 145 defines the quality assurance procedures and organizational requirements to be fulfilled by the relevant companies.

A second set of standards is dedicated to operational requirements concerning procedures (part AWO), minimum equipment required (Part OPS), noise (Part 36), etc. The last and largest group of standards addresses the technical airworthiness requirements for the different categories of aircraft and engines. These standards like Part 25 for large aircraft or Part 27 for small rotorcraft give detailed descriptions of the performances of the air vehicle itself and also its structures, systems and required documentation.

When in Europe the aeronautical industry (Airbus with the partner companies) wanted to certify A320, Airbus had to follow 22 different certification rules for each individual European country. Airbus Industrie was then strongly pushing the EU to harmonize these national rules and define only one European certification

Table 4.2 Comparison of European and American Safety Standards

CS-definitions	Definitions	JAR-1	Definitions and abbreviations	FAR-1	Definitions and abbreviations
EASA part 21	Certification procedures for aircraft, and related products and parts	JAR-21	Certification procedures for aircraft, and related products and parts	FAR-21	Certification procedures for products and parts
CS-22	Sailplanes and powered sailplanes	JAR-22	Sailplanes and powered sailplanes		
CS-23	Normal, utility, aerobatic and commuter category aeroplanes	JAR-23	Normal, utility, aerobatic and commuter category aeroplanes	FAR-23	Airworthiness standards: normal, utility, acrobatic, and commuter category airplanes
CS-25	Large aeroplanes	JAR-25	Large aeroplanes	FAR-25	Airworthiness standards: transport category airplanes
		JAR-26	Retroactive airworthiness requirements		
CS-36	*Aircraft noise*	JAR-36	Aircraft noise	FAR-36	Noise standards: aircraft type and airworthiness certification
		JAR-147	Maintenance crew training services	JAR-147	Aviation maintenance technician schools
CS-27	Small rotorcraft	JAR-27	Small rotorcraft	FAR-27	Airworthiness standards: normal category rotorcraft
CS-29	Large rotorcraft	JAR-29	Large rotorcraft	FAR-29	Airworthiness standards: transport category rotorcraft
CS-E	*Engines*	JAR-E	Engines	FAR-33	Airworthiness standards: aircraft engines
		JAR-P	Propellers	FAR-35	Airworthiness standards: propellers
CS-APU	*Auxiliary Power Units*	JAR-APU	Auxiliary power units		
		JAR-TSO	Joint technical standard orders		

(continued)

Table 4.2 (continued)

CS-definitions	Definitions	JAR-1	Definitions and abbreviations	FAR-1	Definitions and abbreviations
CS-AWO	All weather operations	JAR-AWO	All weather operations		
CS-VLA	Veiy light aeroplane	JAR-VLA	Very light aeroplane	FAR-103	Ultralight vehicles
EASA 145	*Approved maintenance organisations*	JAR-145	Approved maintenance organisations	FAR-145	Repair stations
		JAR-OPS Part 1	Commercial air transportation (aeroplanes)	FAR-121	Certification and operations: domestic, Flag, and supplemental air carriers and commercial operators of large aircraft
		JAR-FCL	Flight crew licening	FAR-61	Certification: Pilots and Flight Instructors
		JAR-OPS Part3	Commercial air transportation (helicopters)	FAR-12	Certification/operations of scheduled air carriers with helicopters

document, which became the JAR (Joint Airworthiness Regulations). The differences between the national authorities and their regulations were small, however, huge administrative effort was needed to demonstrate compliance of the aircraft with all specific national regulations and convince all member states to give up some of their national sovereignty. The establishment of a single airworthiness authority—the EASA—was the next consequent step in this harmonization process.

Worldwide, there are still some specific national regulations in place, but most countries have adopted the FAA and JAA regulations as national standard.

Although not mentioned here, the structure and philosophy of aviation safety, e.g. in Canada (Canadian Aviation Administration, CAA), Great Britain (Civil Aviation Authority, CAA) and France (Direction General l'Aviation Civil, DGAC) is very similar. Also, many other countries in the world follow the way of FAA and EASA to set-up aviation safety procedures.

In the next Sect. 4.4 the structure of the aircraft design standards is described and typical examples of compliance demonstration are given.

4.3.3 Air Navigation Services

Air Navigation Service Provider (ANSP) are in charge of performing the Air Traffic Management to ensure safe and efficient aircraft flights. From the organizational point of view

- Air Traffic Control
- Air Traffic Flow Management
- Air Space Management

are services performed by the Air Navigation Service Provider (ANSP) of the relevant nations [18].

In the United States this responsibility is associated with the Federal Aviation Authority (FAA), while in Europe the nations as well as Eurocontrol and European Aviation Safety Agency (EASA) share this responsibility.

For air traffic management (ATM) in 1960 Eurocontrol was established as a civil-military organization and European centre, both leading and supporting ATM improvements across Europe. Therefore Eurocontrol is mainly responsible for

1. The management of the European Air Traffic Management Programme (EATMP)
2. The operation of the Central Flow Management Unit (CFMU)
3. The operation of the European Upper Area Control Center Maastricht (UAC)
4. The performance of research and development for safe and efficient air traffic
5. Charging air navigation fees (route charges)
6. Developing the international Central European Air Traffic Service Programme (CEATS)
7. Establishment of Safety Regulatory Requirements (ESARRs) during its work in the Safety Regulation Commission (SRC).

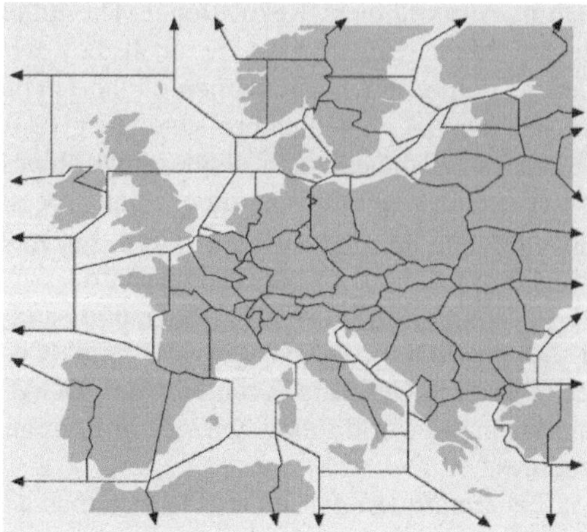

	Europe	USA
Air Space (million km²)	10,5	9,8
ATC Service Provider (civil & mil.)	47	1
Centers	58	21
Operating Systems	22	1
Programming Language	30	1
Flights (million)	9	18
ATC Cost per Flight (EUR)	742	386

Fig. 4.3 Airspace sectors in Europe and USA, [Eurocontrol]

In 2011, 39 member states belong to Eurocontrol. Focusing very much on ATM, the biggest challenge of Eurocontrol in the early years of the twenty-first century is the set-up of a Single European Sky (SES). Compared to the American airspace the European sky is very heterogeneous and consists of too many small sectors due to the history of national sovereignity. The current structure of the European airspace is given in Fig. 4.3.

The goal is to establish a more common airspace and standardized ATM procedures and equipment across Europe.

With the European regulation ((EC) 1108/2009) the tasks of the European Aviation Safety Agency were extended to the safety of aerodromes, air traffic management and air navigation services. Through this step the EASA responsibilities and organization, located in Cologne, Germany are now very similar to FAA. Like the FAA, the EASA is responsible for

1. Aircraft certification and safety
2. Airport compliance and safety
3. Air Traffic Management organization, safety and standards.

4.3.4 The International Air Transport Association

While the ICAO represents the world states community in aviation, the International Air Transport Association (IATA) is the federation of the aviation industry, especially the airlines of the world, [3]. Originally founded by the national flag carriers, which were mainly owned by the hosting countries, today about 240 airlines are members of IATA. These represent approximately 93 % of all worldwide international airlines. During the 1980s and 1990s of the twentieth century, so-called "low cost carriers" were developed in the liberalized aviation market, which are also members of the IATA today. Airlines, which only serve national markets, are invited to be associated members.

Originally founded in 1919, IATA was newly founded in 1945 (Havanna, Cuba), as a consequence of the Second World War.

IATA defines its mission by supporting safe, regular and economical civil air transportation worldwide, which sounds similar to the ICAO mission. Additionally, it is pushing for the collaboration of all companies involved in aviation, by coordinating the development of common technical and economical methods. For this goal IATA cooperates closely with ICAO and other aviation federations like ACI.

Comparing ICAO and IATA the latter has a clear economical focus to support the growth of the aviation industry.
Various committees at IATA perform the "technical" work of the organization, i.e.:

- The traffic committee
- The technical committee
- The financial committee
- The legal committee.

While ICAO has more a legislative character the power of IATA is based on the huge amount of industrial members. In practice, IATA can formulate recommendations to the states and governmental organizations.

As an actual example IATA—through its suborganization Air Transport Action Group ATAG—is strongly working and promoting the massive reduction of CO_2 emissions in order to further improve environmental compatibility of aviation, [4, 5] By organizing workshops and conferences, which are developing recommendations, CO_2 emissions could be reduced by regulatory, operational and economic measures.

4.4 Aviation Safety

Aviation safety is of paramount importance to ensure air transportation growth driven by customer's confidence.

Aviation safety addresses airworthiness of the air vehicles and related systems, as well as operations, while aviation security is dedicated to secure operation of air transportation. The latter is introduced in Sect. 4.5.

4.4.1 Aviation Safety Philosophy

In order to ensure aviation safety air vehicles must demonstrate airworthiness before being allowed to enter the airspace.

Airworthiness is defined as a safety standard of an air vehicle

- designed and built according to relevant requirements,
- operated in a defined environment within a quantified envelope,
- maintained according to the defined maintenance procedures.

Airworthiness is always a balance and compromise between mandatory requirements with respect to safety and economical acceptable effort.

As the basic principle to describe the risk of fatal accidents, the probability of death is used as described in Fig. 4.4. Over the lifetime of a human being the probability of death varies significantly due to medical and ageing reasons (babies, older people). Looking at the figure a probability of death of 10^{-7} or less is achieved between 2 years and 18. The probability of a fatal air vehicle accident causing death of passengers due to a single failure is therefore set to 10^{-9} fh the relevant design standards.

In order to provide an idea of what one fatal loss of 10^{-9} fh (i.e. 1.000.000.000 fh) caused by a technical failure means, the following example may be used:

Let us assume an aircraft type is operated by several airlines with 400 aircraft worldwide. Further, each aircraft will fly 1200 h per year. All these aircraft fly about 480.000 h/year. Consequently, one of these aircraft will have one fatal accident within 2083 years, which is defined as extremely improbable. For a world fleet of about 15,000 aircraft, flying 4000 fh/year one fatal accident will happen once in ~ 16 years. However, severe and fatal aircraft accidents occur more often in reality, which are mainly caused by maintenance, or piloting or operational failures—poor communication between pilots and air traffic controller, both often non-native English speakers—ending up in such a catastrophe.

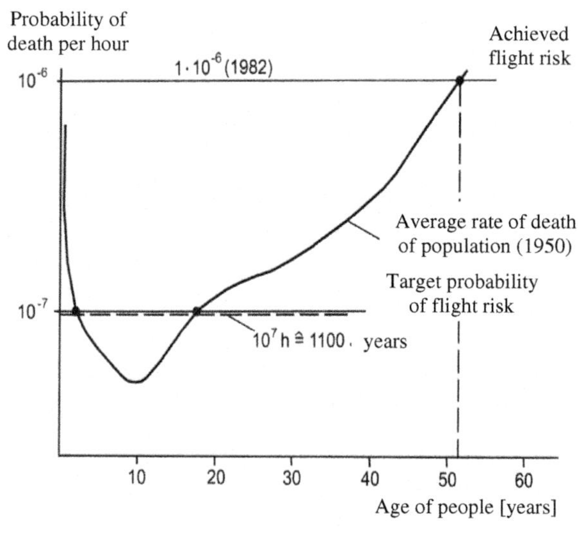

Fig. 4.4 Probability of death of humans beings

Table 4.3 Correlation of failure occurrence probability and consequences

Probability of failure occurence per flight					
	10^{-1} | 10^{-2}	10^{-3} | 10^{-4}	10^{-5}	10^{-6} | 10^{-7}	10^{-8} | 10^{-9} | 10^{-10}
Probability	Probable	Relatively improbable	Less probable	Improbable	Extremely improbable
Consequence	Minor		major	High risk	catastrophic
Description	Occurs potentially more often during aircraft lifetime	Occurs potentially less often during operation, but sometimes during aircraft lifetime	Occurs potentially not during operation, but may happen during lifetime of single aircraft	Must be considered as potential, but is considered as improbable during lifetime of a single aircraft	Extremely improbable, so that it is considered as not happening during aircraft lifetime

Table 4.3 shows how these probabilities of occurrence per flight—without considering external effects—are transferred into consequences of technical failures on an air vehicle.

The higher the probability of a failure the less its consequences are allowed to be. Fatal failures, which cause the loss of the aircraft and death of people on aircraft and/or on ground, must be extremely improbable. Therefore aircraft components, which are mandatory for flight, like the fuselage structure or the flight control system must provide a high level of reliability, while systems that are more used to provide comfort like inflight entertainment, are allowed to be less reliable, because in case of a failure safe flight is not affected but only reduced cabin comfort in this example (Table 4.3).

In order to define the level of importance for flight safety of a component or system, a fault tree analysis (FTA) is performed to assess the impact of a single system failure on the aircraft.

The applicant for a type certification, which is typically but not necessarily the aircraft manufacturing industry, has to demonstrate the acceptable probability by a functional hazard analysis (FHA) of the relevant system.

Due to these strong safety requirements, over the decades the amount of severe accidents decreased significantly as shown in Fig. 4.5.

It is visible that the worldwide accident rates (left ordinate) decreased from 50 accidents per million departures at the beginning of the 1960s down to less than approximately three accidents per million departures in 2010. This is more than ten times less at first glance, but it has to be recognized that in 2010, around 20 million departures happened which is much more than in 1959. The onboard fatalities, where passengers came to death onboard, depend heavily on the individual severity of an accident. Therefore, occasionally these rates increase, while it is lower in some years. Here also the growing aircraft capacity influences the fatalities.

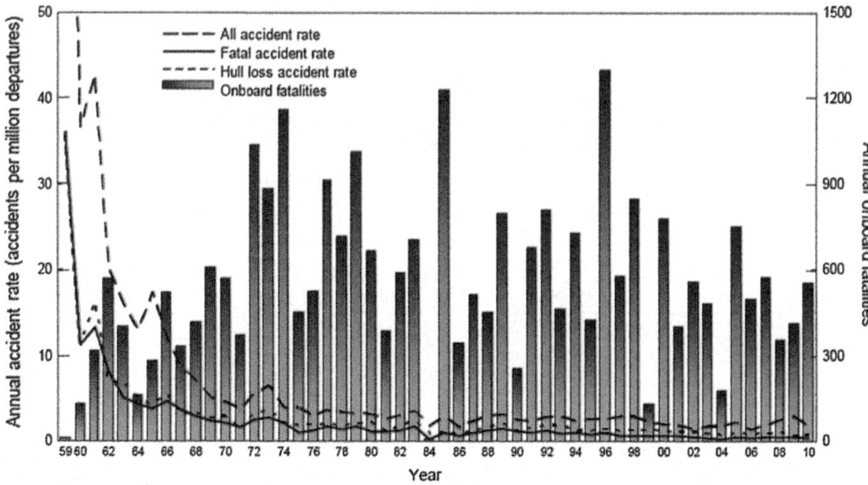

Fig. 4.5 Worldwide accident rate and onboard fatalities 1959–2010 [6]

Aviation safety is not only an issue of aircraft design and piloting but is also mainly influenced by the maintenance quality. The latter is the responsibility of the aircraft operator, which is the airline.

Taking the transportation performance, which is typically defined as passenger or cargo mass transported over a certain distance, and the amount of accidents and incidents as a measure for safety performance, Table 4.4 shows the top 15 airlines in 2011, being the safest.

The table presents the hull losses and rate of death of the top 15 airlines. For comparison, some airlines are listed which are far below the top 15. The main parameters which affect the safety performance are the number of accidents, the number of deaths and the time since the last accident, indicating how much an airline improved its safe operation. The lower the safety performance figures the safer an airline operates its aircraft. Looking at the lower rank some airlines show very few hull losses, however, associated with a high amount of people died (fatalities). This indicates that a single accident has much more effect than the number of accidents and the time factor. On the other side, Aeroflot airline shows more hull losses with less fatalities compared to other airlines. This relation is more representative for a lower level of aircraft safety.

Looking at the overall accident situation of aviation the worst annual accident rate was about 3300 *people* dying in 1972. Statistically, the worst year was not 2001 with the *9/11attack of the World Trade Center,* when 419 *people* died in aircraft, while most of the people (2752) were killed by the collapse of the World Trade Center.

In 2011 we have seen the lowest rate of 498 people died since the beginning of turbo jet commercial air transportation. As shown in Fig. 4.5 there is no clear tendency in the development of aviation accidents, but it has to be remarked that

Table 4.4 Top 15 of safe airlines in 2011 [16]

Rank	Airline	Begin of service	Hull losses	People died	Years since last accident	Safety performance*
1	All Nippon Airlines	1954	0	0	40	0,005
2	Finnair	1923	0	0	48	0,006
3	Cathay Pacific Airways	1946	0	0	39	0,006
4	Etihad airways	2003	0	0	8	0,006
5	Hainan Airlines	1993	0	0	18	0,006
6	Jet Blue Airways	2000	0	0	11	0,007
7	Emirates	1985	0	0	26	0,007
8	Virgin Blue	2000	0	0	11	0,007
9	Air Berlin	1979	0	0	32	0,007
10	Air New Zealand	1940	0	0	32	0,008
11	Qantas	1922	0	0	51	0,008
12	Lufthansa	1955	1	2	18	0,008
13	British Airways	1919	1	0	3	0,008
14	EVA Air	1991	0	0	20	0,01
15	Transaero Airlines	1991	0	0	20	0,01
…						
29	United Airlines	1931	4	247	10	0,059
35	Aeroflot—Russian Airlines	1992	10	182	3	0,107
39	Air France	1933	8	344	2	0,164
44	Japan Airlines	1951	3	544	26	0,233
50	Thai Airways International	1960	5	384	13	0,419

*Relation between flight performance and accident rate Status: 2011

since the end of the Second World War air transportation has grown from a negligible amount in 1950 to nearly 5.5 *billion* passenger kilometre in 2010. Due to the significant growth in air transport performance the relative accident rate has decreased dramatically, [6]. Further, the diffuse distribution of fatal accidents over time indicates that no real correlation between the accident rate and any global cause is visible. Therefore, further effort needs to be spent on all potential accident risks, which are technical reliability as well as procedural improvements to reduce human errors or misbehaviour.

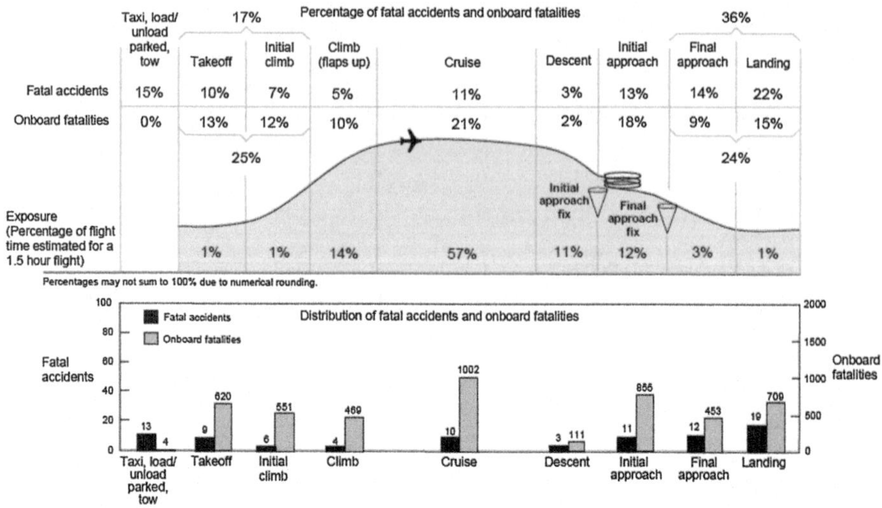

Fig. 4.6 Accident rates distributed over flight phases in 2010 [6]

Looking at the individual flight phases, it becomes obvious that during take-off and initial climb (overall 17 %) and final approach and landing (36 %) most of the accidents happened, Fig. 4.6.

Individual accidents in cruise cause significant onboard fatalities, because here the probability to survive is much lower compared to the take-off and landing phases.

Based on these observations in the following sections the principal set-up of responsible organizations and their way of working is described. Further, the principal process to realize aircraft and aviation safety is exemplarily described.

4.4.2 Establishing Aircraft Airworthiness

Airworthiness—as defined in the previous section—needs to be established in the industrial design process, in the production process of an air vehicle, in the operation of an air vehicle and also in its maintenance procedures. In this section the principal processes and regulations are introduced to demonstrate the philosophy of airworthiness.

Airworthiness is based on four pillars:

1. Type certification of an aircraft type
2. Certificate for operation of an individual, single aircraft
3. Operational regulations for aircraft types
4. Approvals for acting companies for design, production, maintenance, and operations.

Within this section, the first three pillars will be introduced, since they are directly associated with the air vehicle itself.

Taking the German aviation legislation as an example an aircraft is allowed for operation in the German airspace only if it has an individual certification and is listed in the "German aircraft list".

The individual certification of an aircraft for operation in the German airspace is provided, if

(a) a type certification of the relevant aircraft type is given,
(b) safety to traffic for the individual aircraft is formally confirmed,
(c) the operator of the aircraft holds an insurance for the aircraft,
(d) a noise certificate for the aircraft is provided.

The type certification is owned by the aircraft designer, which is typically a company and will be given by the airworthiness authority confirming that the aircraft design has been proven to be safe for operation.

For each individually produced aircraft the readiness for operation is declared by the airworthiness authority in front of the aircraft manufacturer, when the assembled aircraft is proven to fulfil the type certificate and the production process fulfils all quality assurance requirements.

The insurance for aircraft is requested because the keeper but not the owner of an aircraft is liable for all losses caused by the operation.

In the military world the situation is different, because here the national government holds the type certificate. For each military activity the government is in charge of global liability and in case of an accident it has to pay for any compensation.

Finally, the legislator requests for less noise affection of the environment as possible. In practice, each aircraft in operation has to have a noise certificate, which should be compliant with the limits given by the ICAO annex 16. If an aircraft fails to fulfil these limits, higher landing fees or operational restrictions may be set.

To achieve a type certification for an aircraft type processes are mutually defined on ICAO level (annex 8), where FAA, EASA and equivalent agencies develop design standards to be applied for the design of an air vehicle, e.g. EASA CS airplanes (23, 25), rotorcraft (27, 29), see Table 4.2.

All these standards show a very similar structure as given below for the EASA design standard for large aeroplanes, CS25 (Table 4.5) [20].

Looking at the principal structure of such a design standard the different subparts cover special requirements for the overall aircraft (subparts A and B), and also respective subsystems like the power plant, the structure and the systems (subparts C, D, F). Further, subpart G addresses overall aircraft issues again, which are especially dedicated to the required operation and documentation. The aircraft documentation contains the aircraft flight manual, including the flight procedures, preflight checks and the flight limitations. These refer to the definition of airworthiness, which is defined for a quantified envelope. Requirements for aircraft maintenance, which are also called continued airworthiness, are listed in subpart F.

Table 4.5 Structure of aircraft design standard for large aeroplanes, EASA CS25

Section	Contents
Section 1	Requirements:
Subpart A	General
Subpart B	Flight (general, performance, flight characteristics, ground and water handling characteristics), Miscellaneous flight requirements
Subpart C	Structure (general, flight loads, control surface and system loads, ground loads, water loads, main component requirements, emergency landing conditions, fatigue evaluation
Subpart D	Design and construction
Subpart E	Powerplant
Subpart F	Equipment (general, instruments installation, electrical systems and equipment, lights, safety equipment, Miscellaneous equipment
Subpart G	Operating limitations and information (general operating limitations, markings and placards, aircraft flight manual
Subpart J	Gas turbine auxiliary power unit installation
Section 2	Acceptable means of compliance & interpretations (ACJ)
Section 3	Advisory material (AMJ)

Sections 2 and 3 of the design standard support the airworthiness compliance demonstration in defining accepted Means of Compliance (MoC).

The different means of compliance (MoC) as shown in Table 4.6 cover all practical ways to prove that the item to be checked fulfils the relevant requirement. While MoC 0-7 are quite logical, MoC 8 and 9 are of special interest for new technologies. Typically for a new technology, where no practical operational

Table 4.6 List of Means of Compliance for airworthiness demonstration

MoC	Type
0	Inspection by authorities
1	Production data (drawings, circuit diagrams, part lists, etc.), design reviews
2	Specified major aircraft component, (material, testing, process specifications, Declaration of Design and Performance)
3	Ground tests (component tests, test programmes/reports, simulations, etc.)
4	Flight tests, (test programmes/reports)
5	Calculations, (Demonstration by calculation: resistance, design loads, performance, flight characteristics, mass, centre of gravity, etc.)
6	Technical notes (system analyses, energy balance, reliability/safety analyses, test procedures)
7	Flight manual, maintenance and inspection guidelines
8	Assumptions and definitions
9	Empirical evidence on the basis of other aircraft types
(-)	Not applicable

*e-mails and internal memos are not acceptable as compliance demonstration documents!!!

experience is available, theoretical calculations and labtests are used to demonstrate a minimum reliability. The authority within very restrictive limits allows operational usage in those cases as systems are used as supporting equipment in parallel to established state of the art equipment. Further, there is often stated the obligation to document faults and incidents of such technologies during operations. By learning the operational reliability in this way, the envelope of operational usage is extended over the lifetime, stepwise. Regarding MoC 8, which addresses assumptions and definitions, these are the defining operational conditions in the flight manual.

In order to provide some examples of how airworthiness design requirements are formulated, an example for an overall aircraft requirement regarding stall speed and an example for systems and equipment installation are briefly discussed.

Example 1:
Section 1—Subpart B—Flight
CS 25.103 Stall speed

The reference stall speed V_{SR} is a calibrated airspeed defined by the applicant. V_{SR} may not be less than a 1-g stall speed. V_{SR} is expressed as

$$V_{SR} \geq \frac{V_{CLmax}}{\sqrt{n_{zw}}}$$

where

V_{CLmax} = Calibrated airspeed obtained when the load factor-corrected lift coefficient

$\left(\frac{n_{zw} \cdot W}{q \cdot S}\right)$ is first a maximum during the manoeuvre prescribed in sub-paragraph c) of this paragraph. In addition, when the manoeuvre is limited by a device that abruptly pushes the nose down at a selected angle of attack (e.g. a stick pusher), V_{CLmax} may not be less than the speed existing at the instant the device operates;

n_{zw} Load factor normal to the flight path at V_{CLmax};
W Aeroplane gross weight;
S Aerodynamic reference wing area; and
q Dynamic pressure

Reading this extract of the requirement one will derive that the stall speed to be demonstrated is first defined by the aircraft designer. Second the fulfillment of the requirement given by the equation has to be demonstrated. For this purpose the definition of the stall speed is covered by MoC 8, while the compliance demonstration is realized by calculation (MoC 5) and flight test (MoC 4).

As a second example the requirement for the installation of equipment and systems is presented here. This requirement achieved paramount relevance for modern aircraft designs, because it directly addresses the probability of a system or equipment failure and how to handle it. The key message (boldfaced) is that all systems and equipment have to be classified according to their safety relevance, which is typically done by a fault tree and functional hazard analysis. Together with the certification authority the aircraft designer defines the criticality of an equipment or system concerning its impact on aircraft safety.

Example 2:
CS 25.1309 Equipment, Systems and Installations (See AMC 25.1309)

The requirements of this paragraph, except as identified below, are **applicable**, in addition to specific design requirements of CS-25, **to any equipment or system as installed in the aeroplane**.

Although this paragraph does not apply to the performance and flight characteristic requirements of Subpart B and the structural requirements of Subparts C and D, it does apply to any system on which compliance with any of those requirements is dependent. Certain single failures or jams covered by CS 25.671(c) (1) and CS 25.671 (c) (3) are excepted from the requirements of CS 25.1309(b)(1)(ii).

Certain single failures covered by CS 25.735(b) are accepted from the requirements of CS 25.1309(b). The failure effects covered by CS 25.810(a) (1)(v) and CS 25.812 are excepted from the requirements of CS 25.1309(b). The requirements of CS 25.1309(b) apply to powerplant installations as specified in CS 25.901(c).

(a) The aeroplane equipment and systems must be designed and installed so that:
   (1) Those required for type certification or by operating rules, or whose improper functioning would reduce safety, perform as intended under the aeroplane operating and environmental conditions.
   (2) Other equipment and systems are not a source of danger in themselves and do not adversely affect the proper functioning of those covered by sub-paragraph (a) (1) of this paragraph.

(b) The **aeroplane systems and associated components**, considered separately and in relation to other systems, **must be designed** so that –

   (1) Any **catastrophic failure condition**

      (i) **is extremely improbable**; and
      (ii) does not result from a single failure; and

(2) Any **hazardous failure condition is extremely remote**; and
(3) Any major failure condition is remote.

(c) **Information concerning unsafe system operating conditions must be provided to the crew** to enable them to take appropriate corrective action. A warning indication must be provided if immediate corrective action is required. Systems and controls, including indications and annunciations must be designed to minimize crew errors, which could create additional hazards.

For classifying the criticality of an equipment or system the Table 4.7 might be used.

Table 4.7 Classification of single failure criticality on aircraft safety

Effect on air vehicle	No effect on operational performance or safety	Minor reduction of functional performance or reduction in safety margins	Significant reduction in functional performance and reduced safety margins	Extensive reduction of functional capabilities of the aircraft or significant reduction of safety margins	Loss of aircraft
Effect on crew	None	Minor increase in crew workload	High increase in physical and workload	Physical damages/distress or extreme increase of crew workload, which reduces flight performance	Death or mission completion impossible
Effect on passenger	Inconvenience	Minor increase in physical load	Significant physical loads and potential injuries	Severe or deadly injuries of single passengers	Many passengers come to death
Classification of failure	No effect	minor effect	Significant effect	Dangerous effect	Catastrophic effect
Permitted qualitative probability	None	Probable	Minor	Remote	Extreme remote
Permitted quantitative probability per flight hour	None	$<10^{-3}$	$<10^{-5}$	$<10^{-7}$	$<10^{-9}$
Classification of required SW-class	E	D	C	B	A

The criticality of a system fault is associated with its impact on the aircraft, the flight crew and passengers. If one of these three is mainly affected by a single system failure this determines the criticality level. In the next step the permitted probability of occurrence is derived in either a qualitative or quantitative way.

Originally, this requirement of §1309 was developed for mechanical, electrical or electromechanical systems. Since the early 1980s software-based systems have been introduced into air vehicles. In order to handle software in terms of safety a different way has to be chosen, because software by nature is never completely free of any bug or failure. To handle this, a close meshed system of quality assurance was developed, which is laid down in the RTCA-DO178 "Software Considerations in Airborne Systems and Equipment Certification", [21]. The more safety relevant a software failure is, the more quality checks in terms of software specification, software test and documentation are required. Therefore software is sorted into five classes A-E, where E characterizes the lowest level of software impact, while class A is typical for flight critical software like the flight and engine control system software. Today the effort for software specification, test and documentation is about 35–40 % of the overall development effort and cost. Especially, the part of software test includes also the activities of software integration, which covers a significant portion of time and effort.

These requirements give advice on how airworthiness has to be established during aircraft design. When all obligations of the certification programme are fulfilled, the applicant receives the type certificate, which confirms the airworthiness of the design.

If the applicant as an aircraft manufacturer is also certified according to, e.g. EASA part 21G he is now allowed to produce this aircraft. In order to assure aviation safety also during the production phase, dedicated quality assurance processes are to be fulfilled, laid down in Part 21G approval of the company. At the end of the aircraft production each individual aircraft receives an individual certificate for operation.

4.4.3 Standards for Safe Aircraft Operations

On operational level, two standards are introduced as examples. The ICAO Procedures for Air Navigation Services—Rules of the Air and Air Traffic Services", also known as document 4444 are introduced at first, [8]. It is not the purpose of this book to introduce every single regulation in detail but to provide some insight into the structure and philosophy, so that the reader is able to pick up and read the relevant standards. Further, it is intended to set the various standards in the right context, so that the role is understood. In this manner, Table 4.8 gives an overview of the addressed areas of the ICAO 4444 document. The document structure reflects the relevant operational tasks as they are organized at the ANSPs, e.g. area control, approach control, aerodrome control. Further, surveillance and information services like radar, flight information or controller-pilot data link communications are

Table 4.8 Overview of the ICAO 4444—Procedures for Air Navigation Services

Part	Contents
Part I	Definitions
Part II	General provisions
Part III	Area control service
Part IV	Approach control service
Part V	Aerodrome control service
Part VI	Radar services
Part VII	Flight information service and alerting service
Part VIII	Co-ordination
Part IX	Air traffic services messages
Part X	Phraseologies
Part XI	Controller-pilot data link communications (CPDLC)
Appendix 1	Instructions for air-reporting by voice communications
Appendix 2	Flight plan
Appendix 3	Air traffic services messages
Appendix 4	Air traffic incident report
Appendix 5	Controller-pilot data link communications (CPDLC) message set

addressed. The appendices provide some form sheets and formalism on how the various services are applied in a worldwide standard way.

As an example for an operational safety rule in the following the "Vertical Separation Minimum" is introduced, [8]:

Part III. Area Control Service
3. Vertical separation minimum
3.1 The vertical separation minimum (VSM) shall be:

(a) within designated airspace, subject to regional air navigation agreement: a nominal 300 m (1000 ft) below FL410 or a higher level where so prescribed for use under specified conditions, and a nominal 600 m (2000 ft) at or above this level; and
(b) within other airspace: a nominal 300 m (1000 ft) below FL 290 and a nominal 600 m (2000 ft) at or above this level.

Note—Guidance material relating to vertical separation is contained in the Manual on Implementation of a 300 m (1000 ft) Vertical Separation Minimum Between FL 290 and FL 410 Inclusive (Doc 9574).

The rule defines for flight levels above 41,000 ft (FL410) a vertical separation of 600 m for designated airspaces as well as for flight level above 29,000 ft (FL290). For flight levels below FL410 and FL290 for the respective airspaces a vertical separation of 300 m is worldwide defined. Although this rule seems to be simple and clear it is relevant, that this rule is worldwide applied. A different type of a safety related requirement addressing equipment performance is given in the following [8].

Part II. General Provisions
14.4 Contents of Automatic Dependent Surveillance (ADS) reports
14.4.1 ADS reports shall be composed of data blocks selected from the following:

(a) Basic ADS (Latitude, Longitude, Altitude, Time, Figure of merit)
(b) Ground vector (Track, Ground speed, Rate of climb or descent)
(c) Air vector (Heading, Mach or IAS, Rate of climb or descent)
(d) Projected profile (next way-point, Estimated altitude at next way-point, Estimated time at next way-point, (next+1) way-point, Estimated altitude at (next+1) way-point, Estimated time at (next+1) way-point
(e) Meteorological information (Wind speed, Wind direction, Temperature, Turbulence (if available), Humidity (if available)
(f) Short-term intent (Latitude at projected intent point, Longitude at projected intent point, Altitude at projected intent point, Time of projection). If an altitude, track or speed change is predicted to occur between the aircraft's current position and the projected intent point, additional information would be provided in an intermediate intent block as follows:
    - Distance from current point to change point
    - Track from current point to change point
    - Altitude at change point
    - Predicted time to change point
(g) Extended projected profile (in response to an interrogation from the ground system) (next way-point, Estimated altitude at next way-point, Estimated time at next way-point, (next+1) way-point, Estimated altitude at (next+1) way-point, Estimated time at (next+1) way-point, (next+2) way-point, Estimated altitude at (next+2) way-point, Estimated time at (next+2) way-point, [repeated for up to (next+128) way-points]

In this rule is the data structure and content of an Automatic Dependent Surveillance (ADS) system also known as a secondary surveillance system described for all aircraft equipped with this equipment. Going through the requirement the actual position (a) and the actual track (b) as well as the flight state (c) of the aircraft have to be communicated to the air navigation services. Further for coordinated flight planning the projected flight profile (d), the atmospheric conditions (e) being used for global weather forecasting and the short term flight profile need to be transmitted.

For handling different tasks in the air transport system either research, or design and development or operation those regulations from ICAO and EASA/FAA or equivalent authorities need to be taken into account. The obligation to apply these regulations is formulated in the national air laws.

4.4.4 Operational Safety Aspects

The safety standards discussed before were established to ensure a maximum level of technical and operational safety by regulatory means. Nevertheless in practice, accidents also happen when people's attention is decreased or technical equipment fails.

In the following some aspects are briefly introduced, which affect aviation in daily operation.

4.4.4.1 Master Minimum Equipment List

While the activities briefly described before direct mainly to procedural measures to ensure aviation safety, the Master Minimum Equipment List (MMEL) provides the minimum technical aircraft status, which allows for safe aircraft operation on a daily basis.

During operation single components of an aircraft may fail due to several reasons. However, these failures may not necessarily prevent the aircraft from operation. Functional hazard and fault tree analysis are performed during the development and certification process of an aircraft type to determine the criticality of an equipment failure according to Table 4.7.

The MMEL lists all equipment, which may be temporarily inoperative or subject to certain conditions, while maintaining an acceptable level of safety as intended in the applicable certification standard or equivalent requirement. Figure 4.7 shows as an example that one EFIS control panel ND can fail without grounding the aircraft, as long as the other is still operative. On the other hand, if both EFIS control panel TAXI may fail the aircraft can be operated as well.

Each MMEL is specific to an aircraft type and its specific design architecture and equipment, [9]. All items related to the airworthiness of the aircraft and which are not included in the list are automatically required to be fully operative. Non-safety-related equipment such as galley equipment and passenger convenience items need not to be listed in the MMEL.

U.S. DEPARTMENT OF TRANSPORTATION FEDERAL AVIATION ADMINISTRATION				MASTER MINIMUM EQUIPMENT LIST	
AIRCRAFT: A-380				REVISION NO : ORIGINAL DATE: 08/31/2009	PAGE: 22-7
SYSTEM & SEQUENCE NUMBER	ITEM	1.	2. NUMBER INSTALLED	3. NUMBER REQUIRED FOR DISPATCH	4. REMARKS OR EXCEPTIONS
22	AUTO FLIGHT				
81-10	EFIS Control Panel ND Mode Selector	C	2	1	(O)One may be inoperative.
		B	2	0	(O)Both may be inoperative provided one MFD FCU backups is operative.
81-11	EFIS Control Panel TAXI Pb	C	2	0	(O)May be inoperative.
81-12	EFIS Control Panel Pb Light	C	-	0	

Fig. 4.7 Example of A380 MMEL Auto Flight System Content [9]

4.4.4.2 Extended Twin Engine Operations (ETOPS)

A reliable aircraft engine is essential for safe aircraft operation. Further, it is a very complex and sophisticated machine due to its high temperatures and high internal pressure levels.

For a long time the engines have been the most critical component of the aircraft. This is still reflected in the certification requirements, which request for at least three engines installed on the aircraft on long-ange flights.

This requirement was valid till the beginning of the 1990s. With the considerable improvement in reliability of aero engines during the 1980s and 1990s, there was a strong and common push from the aircraft manufacturer's side and the airlines to open the old rule that twin engine aircraft were only allowed flying over water for max. 60 min.

Extended Twin Engine Operations (ETOPS) was established as a safety procedure to enable the operation of such aircraft over longer distances [22].

Flights from the United States to Bahamas is a typical mission where there was a clear benefit to enlarge the 60 min rule, in order to operate these routes by two-engine aircraft. In a first step the restriction was extended from 60 to 90 min. But once the first step was achieved, there was a clear intention to go further and allow twin engine aircrafts to operate for 180 min over water, which opens the transatlantic flights for two-engine aircraft.

Figure 4.8 shows two different routings on the North Atlantic between New York and London. The route with the 60 min circles is established in such a way

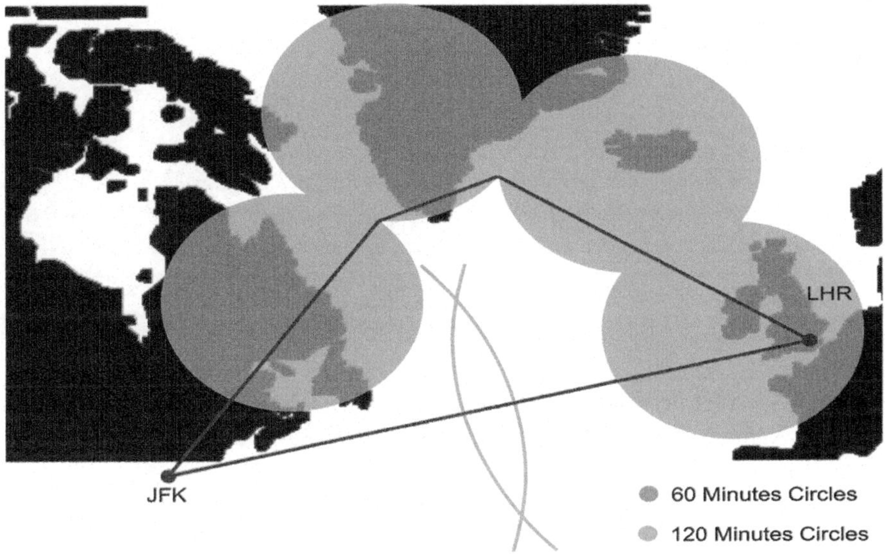

Fig. 4.8 Improvement of 120 min rules for 2 engine aircraft on the North Atlantic route from New York (JFK) to London (LHR)

that in case of an engine failure at each point of the route, the aircraft is able to reach an alternative airport in less than 60 min. As can be seen from Fig. 4.8 the route with the 60 min. circles is about 40 % longer than the direct route, which requires a 120 min permission to fly with one engine, in case of an engine failure.
This change in the regulation was only feasible with a lot of additional obligations for the aircraft, the airline and the engine reliability. These three aspects are briefly listed (for more details see [10, 11]):

- The aircraft must have a specific "ETOPS"-certification, which affects the system definition, requirements for onboard energy supply, a different MMEL (minimum equipment list) before taking off, a specific fuel reserve policy, etc.
- The engine must have an "ETOPS" certification, it means the engine has to have a certain statistical proven reliability. The "in-flight-shut-down" rate of the engine has to be demonstrated on statistical data from the flying fleet. The operational reliability over a certain period (~2 years) in real operation has to provide this evidence.
- The airlines have to show their capabilities, to operate an aircraft under ETOPS rules. This requires special maintenance procedures, new fuel reserve policies, specific pilot and crew trainings, route planning processes, weather infos, passenger recovery plan, etc. The airline has to prove to the authorities their concept and needs approval for it.

The ETOPS rules started to be mainly applicable for two-engine aircraft, but today, the ETOPS rules concern all types of aircraft. FAA has issued a new

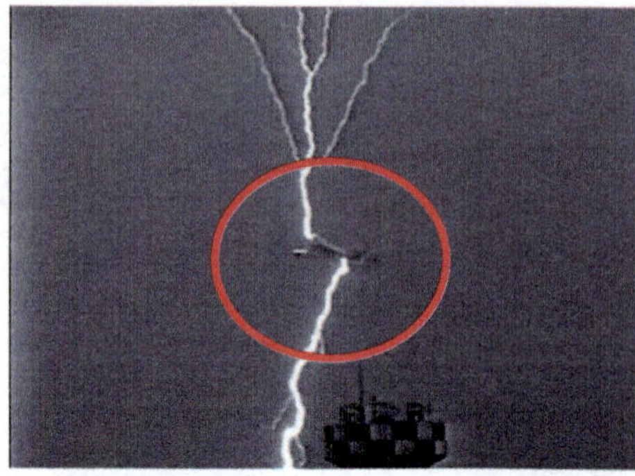

Fig. 4.9 Aircraft lightning strike on a flight between Manchester and Edinbourgh [14]

regulation in 2007. ICAO is defining the ETOPS regulation in their Volume 2: Air Operator Administration in Chap. 3. Further details are given in [12].

4.4.4.3 Weather Conditions Affecting Air Transportation Safety

Besides the technical operational issues various weather conditions like thunderstorms, hail, icing, snow, turbulences and gust heavily impact air transportation safety. As an example lightning strike as shown in Fig. 4.9 might severely damage an aircraft as well as heavy turbulences do.

In the recent years, volcanic ash became a serious impact on air transportation, which causes the downgrade of air transportation due to its impact on aircraft surfaces and potential engine blockages and damages. Research is required to assess the impact on the aircraft, but also on the operational impact to avoid these volcanic areas in the atmosphere.

These are only two examples how weather and atmospheric conditions may influence aviation safety.

4.4.4.4 Bird Strike

Every year several hundred thousand euros cost are caused by bird strike damages, see Fig. 4.10.

Here, especially single and twin engine aircraft are affected, because they are significantly reduced in their thrust performance if an engine fails.

The emergency landing on the Hudson River in 2009 is an example, where bird strike caused this accident with a successful outcome.

Looking at the statistical distribution of bird strike events in Fig. 4.11 around 90 % of all accidents occur at heights below 500 ft, which is mainly considered to be in the vicinity of airports. About 70 % of all bird strikes happen during take-off and final landing, indicating that prevention has to be performed on the airport.

Fig. 4.10 Aircraft collision with bird crowd

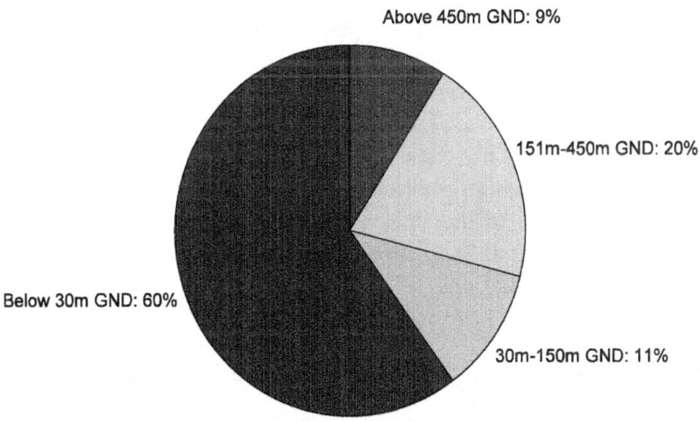

Fig. 4.11 Statistical distribution of bird strike events [7]

Consequently, bird strike prevention at airports is an operational task, which increases operational safety by hiring hunters, which scare off the birds. Curiously airfields due to their wide extension invite birds to rest in these fields.

In Germany about 150 bird strike events per year are noted. However, there is no global notification system and obligation for bird strike events, which would provide a comprehensive and global view on this safety aspect. The ICAO IBIS (International Bird Strike Information System), International Bird Strike Committee (IBSC) is an organization, which works on the establishment of such a notification system.

Fig. 4.12 Events causing fatal accidents [15]

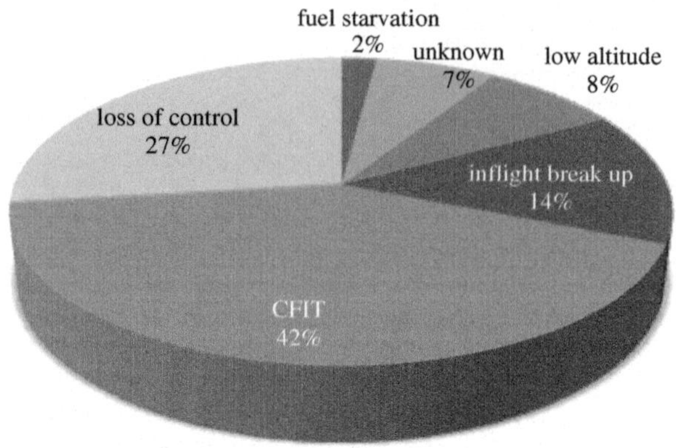

4.4.4.5 Controlled Flight into Terrain

Controlled Flight Into Terrain (CFIT) especially occurs in degraded meteorological conditions. Until the crash the aircraft is technically ok, but from the navigation point of view the crew is disoriented.

Looking at Fig. 4.12 CFIT covers approximately 40 % of all fatal accidents, but the cockpit crew does not only cause these accidents. Very often the initiation of a CFIT is originated by misinformation from the air traffic controller or weather forecast. For about 91 % of the fatal accidents, ATC was involved in the initial misbehaviour and only in 1 % of all investigated CFIT accidents the pilot acted opposite to the ATC instructions [17].

4.4.4.6 Human Misbehaviour During Ground Operations

Ground safety is the most important operational safety issue. Since a lot of actors are involved in the aircraft operations on the airfield, the risk of potential inattentiveness is high and collisions between different vehicles may occur (Fig. 4.13).

Fig. 4.13 Aircraft collision with cars (*left*)—aircraft–aircraft collision (*right*)

Fig. 4.14 Aircraft ramp crash due to weight and balance and loading mismatch

The consequences of these accidents are not only related to potential injuries of people involved or damages of vehicles, but also the operation of the entire airfield may be affected leading to delays and capacity reductions. The major reasons for such events are high traffic loads and insufficient communication between the acting people leading to workload exceed. Total Airport Management (TAM), which provides as an integrated communication and management system better awareness to all stakeholders is a technology that may reduce the occurrence of accidents.

Another example of a potential ground operation safety issue is given in Fig. 4.14 showing a tilted aircraft due to a mismatch between the real aircraft loading and the calculated weight and balance configuration. Consequently, the centre of gravity of the aircraft moved too far behind the main landing gear.

In this situation the passenger seating did not comply with the baggage loading of the aircraft, which let the aircraft dump to the aircraft tail causing damages.

Summarizing the previously mentioned safety relevant events in air transportation it has to be noted that continuous reporting of incidents is a valuable procedure to reduce future accidents and incidents. Incident reporting requires a neutral, confidential and comprehensive information reporting and data acquisition to get a substantiated view on the real causes and mechanism leading to such an accident. In order to establish confidence and trust in such a system pilots and crew behaviour must be handled in an anonymous way to prevent them from any disadvantages.

Human errors and misbehaviour as well as environmental effects like bird strike and weather conditions are major classes of safety aspects, which have to be considered in the safety regulations concerning design and operation of aircraft.

4.5 Security Aspects of Air Transportation

While aviation safety addresses the technical and operational reliability of aircraft and its operation aviation security is dedicated to the protection of air transportation against any kind of criminal and terroristic impact.

In the early days of aviation until the early 1970s, security of aviation was not considered as a significant issue. With the upcoming of worldwide terrorism and the increasing relevance of aviation in the 1970s, aviation security became more and more important. The world's first terrorist attack intending to indiscriminately kill civilians while in flight was Cubana Flight 455. It was a Cubana flight from Barbados to Jamaica that was brought down by a terrorist attack on 6 October 1976, killing 73 people, [13].

As a consequence, worldwide airport security systems were installed to screen passengers, luggage and cargo for any kind of weapon and explosives. Figure 4.15 provides a look at the state-of-the-art passenger and hand luggage screening on an airport using X-ray technology.

Further, in the some countries so-called "Sky Marshalls" were introduced to follow flights within the cabin to track potential hijacker.

Since the 9/11 catastrophe aviation security reached a new level of quality, Fig. 4.16.

Passenger, luggage and cargo are checked for explosives, weapons and anything capable of affecting people and aircraft safety was intensified, leading to a review of all security procedures and consequently to more intensive time delays for passengers at the airport.

Further inconveniences due to restrictions on luggage contents led to a decrease in air transport attractiveness (Fig. 4.17).

The main technical and operational tasks to be performed are scanning activities and information exchange and merge. Further, it is necessary to track the individual

Fig. 4.15 Passenger and hand luggage security check at airport

Fig. 4.16 9/11 aircraft crash attacks to World Trade Center and Pentagon

Fig. 4.17 Passenger queues at security check on airport

passenger and cargo from the entrance of the airport terminal or cargo centre up to the final delivery.

For this purpose body X-ray scanner and RFID systems are used as well as sniffer dogs and photographic or radiative analysis methods to detect unwanted elements like weapons, knifes, explosives or poisons.

The effectiveness of aviation security measures is on the one hand vital for secure air transport. On the other hand, these measures hinder more fluent air transportation.

On the aircraft side only a few weeks after the 9/11 event, Airbus introduced new protected cockpit doors, which are strengthened and can be opened only from inside the cockpit. They also include spy glass. By this measure the cockpit crew shall be protected from any direct terroristic threat. On the other hand, the fatal crash of Germanwings flight 4U9525 on 24 March 2015 has shown that even pilots can be a security issue.

Further, automatic flight procedures are discussed, which are launched if an aircraft is hijacked. The latter can be used in case of pilot crew emergency or in conjunction with future single pilot operation systems.

References

1. ICAO: International Civil Aviation Organization. www.icao.int. Accessed 10 Oct 2011
2. FAA: Federal Aviation Authority. http://www.faa.gov. Accessed 10 Oct 2011
3. IATA: International Air Transport Association. www.iata.org. Accessed 02 Mar 2014
4. Air Transport Action Group. http://www.atag.org. Accessed 02 July 2013
5. ATAG: IATA—Technology Roadmap, 4th edn. International Air Transport Association, Genueva. http://www.iata.org/publications/Pages/technology-roadmap.aspx. Accessed 07 Oct 2013
6. Boeing: Statistical summary of civil jet accidents worldwide. www.boeing.com. Accessed 10 Oct 2011
7. Lau, A.: Air Transportation Safety—Bird Strike Impact and Prevention, Institute for Air Transportation Systems, German Aerospace Center, Internal Report IB-328-001-2009. Hamburg, Germany (2009). in German
8. ICAO: Procedures for Air Navigation Services—Rules of the Air and Air Traffic Services, Doc 4444.RAC/501, 13th edn., Amendment 3 (1999)
9. FAA: Master minimum equipment list—A380-800, Federal Aviation Administration, Issue 31.8.2009. http://fsims.faa.gov/PICResults.aspx?mode=Publication&doctype=MME. Accessed 17 July 2013
10. Bachtel, B.: Boeing; ETOPS. Extended Operations and En Route Alternate Airports, FAA/AAAE Conf (2003)
11. N.N.: http://www.biggles-software.com/software/757_tech/boeing/aero_22_etops.htm#1
12. ETOPS certification: http://www.caa.co.uk/default.aspx?catid=1431&pagetype=90&pageid=8193
13. N.N.: Airport security. http://en.wikipedia.org/wiki/Airport_security. Accessed 6 Jan 2012
14. Williams, E., Heckman,S.: Polarity Asymmetry in Lightning Leaders, Journal AerospaceLab, Issue 5, December 2012, aerospacelab-journal.org
15. N.N.: The Get-home-itis syndrome. http://www.bea.aero/etudes/gethomeitis/gethomeitis.htm. Accessed 17 July 2013
16. Wolf, C.: How safe is my airline? www.aerointernational.de (2012). Accessed 6 Jan 2012
17. N.N: Air accident fatalities recorded by ACRO 1918-2009. Aircraft Crashes Research Office, Geneva, Switzerland. http://www.baaa-acro.com. Accessed 12 Jan 2012
18. Cook, A.: European Air Traffic Management. Ashgate Publishing, Farnham (2007)
19. EASA: European aviation safety agency. www.easa.eu. Accessed 10 Oct 2011
20. EASA: Certification specification CS25—large aeroplanes. www.easa.eu. Accessed 13 Oct 2011
21. RTCA: Software considerations in airborne systems and equipment Certification RTCA DO-178. Radio Technical Commission for Aeronautics (RTCA). www.rtca.org. Accessed 10 Oct 2011
22. N.N.: http://de.wikipedia.org/wiki/Extended-range_Twin-engine_Operation_Performance_Standards

Chapter 5
Aircraft Characteristics

Abstract The chapter starts with different ways of aircraft classification systems. The principles of cabin design follow, as the cabin is the important interface for the airline to the customer and where differentiation between business and leisure travellers for different comfort levels can be implemented. The principles of flight are outlined without going too much into technical details. The atmosphere around the earth is characterized and the standard atmosphere as basis for lift and drag calculation is introduced. Flight controls and their function to operate the aircraft are described. The aircraft structure, which is under a constant challenge to be minimized for a given task, and the major aircraft components are defined. The aero jet engine principle is outlined, being still a major component for further amelioration of fuel consumption. Aircraft performance and mission elements like payload range and flight envelope are specified. The Breguet formula, which characterizes the main parameters for an efficient cruise flight, is a simple but very important formula for the aircraft design.

This chapter describes the main technical characteristics of aircraft vehicles and provides some data of actual aircraft.

A lot of simplifications have to be applied in order to keep the high-level view on air transport. The chapter will start with the cabin features, the interface of the airline with the customer, the travelling persons. Some basic flight physical aspects will follow like the basic aerodynamic assumptions, the critical mass aspects and mass definitions, flight mechanics, and flight performance parameters as well as some cost considerations.

5.1 Classification of Flight Vehicles

There are several ways to categorize the flying vehicles.

A very good characterization has been done in Euromart [1], which illustrates four different axes of challenges, the axes of speed, the axes of maneuverability, the axes of efficiency, and the axes of vertical take-off capabilities.

D. Schmitt and V. Gollnick, *Air Transport System*,
DOI 10.1007/978-3-7091-1880-1_5

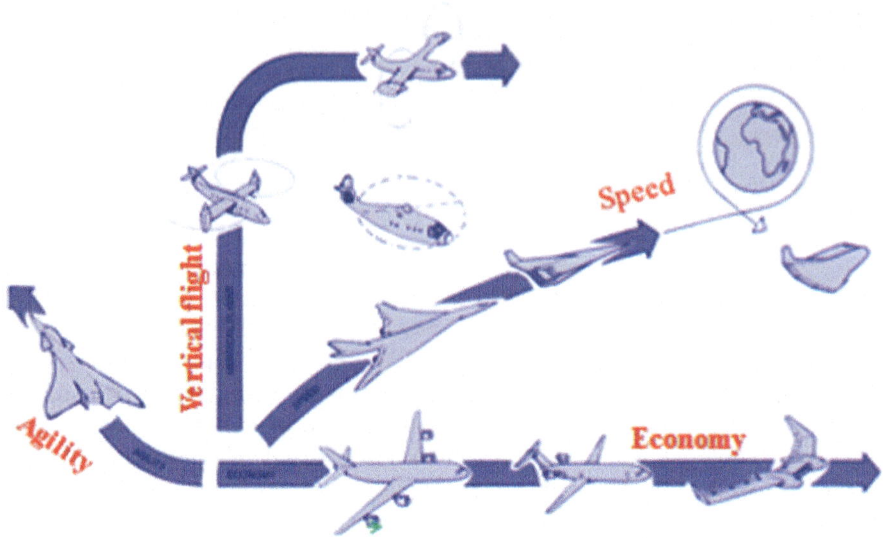

Fig. 5.1 Classification of flight vehicles, as defined by [1]

This book will mainly talk about the commercial air transport, where the efficiency is the relevant driver. Even in 1988 a certain aspect of reintroducing the propeller aircraft and the flying wing concept as a future more efficient aircraft has been already formulated (see Fig. 5.1 the economy axes). This picture is therefore reused, illustrating the long-term vision in civil air transport.

Another way of differentiation is following the different objectives of the users, who may use the aircraft for private, commercial or military purposes.
This book is mainly dealing with commercial transport aircraft, so we have different options to further differentiate the types of commercial transport aircraft by:

- Payload: i.e. passengers, freight and mail/parcels
- Range: short-range SR (2500 nm); medium range MR (<5500 nm) and long range LR (>5500 nm)
- Speed: low subsonic (Ma < 0.5), subsonic (Ma = 0.6−0.9), supersonic (Concorde; Ma = 2.0) or hypersonic transport concepts (Ma > 3.5)
- Size: Air taxi: up to 19 seats for passengers, commuter aircraft (up to 100 seats), airliner aircraft (from ~100 seats to 800+)

As mentioned in the historical review, the main drivers in the past have been payload, range speed and size. As shown in Fig. 6.3 (Airbus—Boeing family concept) certain standardization has been achieved:
Regional Aircraft: Speed: Ma = 0.75–0.78; 70–120 seats; range up to 2500 nm

Examples: Embraer E170-195, Bombardier CSeries, Mitsubishi MRJ, COMAC ARJ, Sukhoi SJ21

Short-range (*SR*) *aircraft*: Speed: Ma 0.76–0.78; 100–200 seats; range up to 3500 nm

Examples: Boeing B737 family, Airbus A320 family

Medium-range (*MR*) *aircraft*: Speed Ma = 0.8–0.85; 200–350 seats; range up to 6500 nm

Examples: Airbus A330, A340, A350 and Boeing B767, B777 and B787

Long-Range (*LR*) *aircraft*: Speed Ma = 0.85; 300–800+ seats; range up to 8500 nm

Examples: Airbus A340, A350, A380 and Boeing B747, B777, B787.

5.2 Cabin Design, Focus for the Airlines

In the aircraft design process, attention is given to the difficult task of finding an optimum compromise between the different disciplines of aerodynamic, structural design, flight mechanics aspects for control and stability and aircraft subsystems. In this optimization process is the cabin size more or less defined via the "payload—range" requirement from the marketing specification. From the point of view of operation the cabin is a very dominating factor. The cabin is defining the volume and space for an airline, where the passengers can be integrated with their seats, their baggage, where toilets and galleys have to be integrated and where additional services during the flight can be provided.

In this respect the cabin is of prime importance for an airline. Here the airline can define and develop their individual "airline brand", specific design concept, look and feel, cultural and regional characteristics, symbols and cabin atmosphere, where the passengers will feel at "home" and very comfortable.

5.2.1 Transportation Task Requires Volume and Space

5.2.1.1 Cabin Requirements

From the aircraft manufacturer's point of view, the main customers are the different airlines, operating the aircraft. They are purchasing the aircraft and they want to earn money with these flying vehicles. They are defining their specific requirements, i.e. size of aircraft, network which has to be flown (max and min range), airport constraints, environmental constraints, etc.

But today there are very often financial institutions, which are buying aircraft and leasing them to operators.

At the end, it is however the passenger who is flying in the aircraft, who wants to feel comfortable and good in the cabin and who will express his opinion about the good atmosphere, ambience and comfort in the cabin or his dissatisfaction!

Table 5.1 Payload elements for the cabin (as seen by Airbus)

Level 1: Products with direct benefit for passenger and airline branding	Level 2: Products of special interest for crew, operations, airline branding
• Layout	• Lining (sidewall, ceiling, supply channel)
• Seat, bed	• Medical + emergency equipment
• In-flight entertainment IFE	• Carpets, non textile floor
• Lavatory, shower, wash room, etc.	• Cabin attendant seat
• Galley	• Galley catering (trolley lift, chiller)
• Hat rack; stowage	• Class divider (curtain,..)
• Lighting (mood light, reading light)	• Video surveillance
• Special cabin, social Area (bar, sales)	• Cargo loading system
• Communication, information system	• Crew rest area
• Humidification	• Cabin communication (attendant panel)
• Lower deck facilities	• Cabin work station
• Auxiliary equipment (magazine, baby)	• Colour and material
• VIP equipment	• Lining (sidewall, ceiling, supply channel)

As can be seen it is not so easy who will finally define the cabin requirements! But at the end it is the airline which defines the cabin to be attractive for their specific customers/passengers!

The following table shows the main cabin elements, which are needed to define good cabin architecture, following the specific request for each airline and their national interests. Table 5.1 is defining all the passenger-related aspects as well as the airline-related aspects like security aspects and intra-airline communication means.

5.2.1.2 Passenger Requirements

The passenger requirement will normally be defined by the different airlines following their cultural and national environment. On the other hand the aircraft has to fit to nearly all international airline requirements, so the aircraft manufacturer is already defining a certain standard of passengers out of statistical data in order to achieve a maximum of acceptance from the customer airlines *afterwards*. The definition of accessibility is a major point of interest.
Generally, two groups of passengers can be defined (see Fig. 5.2):

1. Passengers, who are paying their tickets themselves. These are normally all private persons, who are travelling by private reasons (holidays, visit of family or friends, etc.) and who are optimizing their travel value and are very cost conscious.
2. Passengers, who are travelling on behalf of their company, the "business traveller". He is characterized that somebody else—his company or an organization—is providing him with the ticket and wants him to travel from A to B in order to participate in meetings, discuss scientific or commercial items with customers. In this case the business traveller expects that a reasonable and comfortable travel arrangement will be provided.

Leisure

- Ticket price is of high importance
- 3 hours before check-in are accepted
- Flight is part of holiday adventure
- A lot of baggage (bike, surfboard, ...)
- Entertainment is very important
- Comfort could be better, but ....

Business

- Ticket price is less important
- Quick check-in (last minute)
- Minimise non-working time
- A lot of hand luggage
- Needs communication on-board
- Comfort and service are important

Fig. 5.2 Passenger requirements (two different and opposite views)

Figure 5.2 is defining the main different characteristics for these two different passenger types.

The *private traveller* is very cost conscious, is normally also slightly flexible when ordering a ticket and can shift the forward and return flight slightly if this will have some cost benefits for him. He is accepting to show up at the airport some 3 h in advance, he sometimes needs a lot of baggage (sports or entertainment items) and he is using the flight as part of his travel adventure. Comfort could be much better, but flight cost will normally be his ultimate parameter for optimizing his flight parameters. Therefore, he is accepting a "reduced" comfort level.

The *business traveller* has a different attitude. Somebody else (his company, a customer, an agency) is requesting him to travel. Very often these flight arrangements are defined on short notice and there is not a lot of optimization feasible to get the best price. The urgency of the travel is more important than an optimization of low-cost fares.

But these persons are then expecting a different service on-board. Either they want them during the flight to be connected (internet, email, phone) to their offices or to the outside world or they are using the flight hours to fully relax from all the busy hectic in the office and they just want to use the on-board time to concentrate on some strategic thinking or just relax, detente and sleep. All this has to be considered for these different expectations and the cabin items have to provide such wide range of service functionalities.

5.2.1.3 Reference Passenger

People all over the world are having different personal sizes and geometrical dimensions. Each cabin design has to take into account that people are quite different in size, arm lengths, body dimensions, etc., and that seat arrangements and accessibility to overhead compartments, overhead lights, screens in front, etc., can easily be ensured for nearly everybody. The cabin designers are using therefore standardized persons from different continents and these standardized persons—a very small lady—5 percentile from Japan—and a very tall man—95 percentile of European Nordic men—are then taken for investigations with respect to accessibility, comfort, space volume, etc.

A 5 % percentile women means, that only 5 % of the total population of this country (i.e. Japan) is smaller than the 5 percentile person. The 95 % percentile man characterizes a male person in this country/population (Northern Europe) where only 5 % of this male population is taller than this person.

With the physical data of these two fairly extreme persons, all detailed design studies of accessibility are undertaken to ensure that both persons

(a) do have a chance to see in the overhead compartment and identify that no personal belongings have been left there,
(b) can reach his seat without major disturbances and also be seated comfortably below the overhead compartment without any disturbances,
(c) can see all information panels in front or above them and
(d) can have access to the overhead buttons for light, service demand, etc.

Figure 5.3 is giving an example.

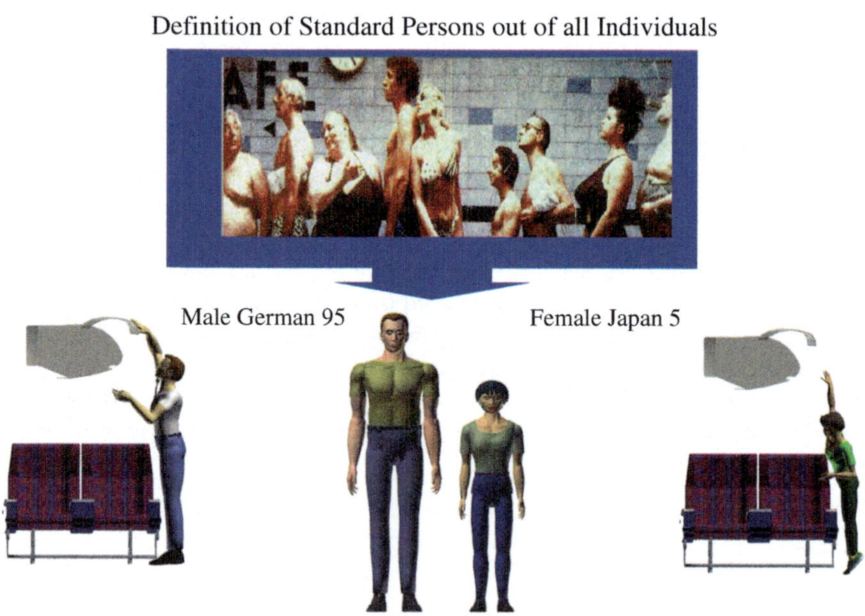

Fig. 5.3 Definition of standard passenger persons for cabin design

5.2.2 Cabin Design

Due to the need of pressurizing the fuselage for flights above 3000 m, the fuselage cross section is normally defined by cylindrical parts or a complete circular cross section (see Fig. 5.4). This leads to some compromises for the cabin interior. There are always some new attempts to define a rectangular cabin cross section, but for all transport aircraft, flying above 10,000 ft, the cabin needs an internal pressurization and this will lead to a circular cross section of the fuselage. Otherwise, the aircraft would become too heavy despite some possible advantages for the cabin design.

The cabin interior is normally designed to make best use of the available cabin floor space.

Besides the seats several other important cabin elements have to be integrated which are also taking floor space as:

- toilets,
- galleys (including all kitchen-related aspects like refrigerators, coffee machines, coolers, pre-packed trolleys, etc.),
- flight attendant seats,
- storage space for cabin baggage, suitcases, coats,
- coat storage (mainly for business and first class compartments),
- door entrance clearances.

For some of these items, the certification requirements demand a certain minimum of equipment, which is highlighted in Table 5.3. But certain line carriers are proposing a better standard to their customers and are exceeding these minimum standards as also can be seen in the same table.

To get the maximum number of seats in the cabin, a lengthy iteration process is required to make optimum use of the available floor space and integrate as much seats as possible. For sure a certain standard of seat pitch, seat width has to be applied. An experienced cabin engineer is required to check the consistency of the final layout with the rules and the feasibility and acceptability of the cabin by the customers and certification authorities.

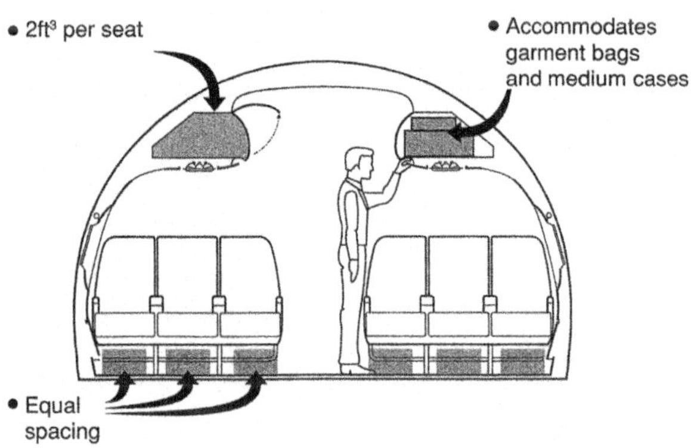

Fig. 5.4 Cabin space for a 6 abreast cross section

Figure 5.4 is showing a typical cross section. This cross section is named as "6 abreast", as there are six seats available per cross section with one aisle. Another nomination for the fuselage cross section is "Single aisle", as it has only 1 aisle. In addition, the storage space for hand luggage is visible, mainly in the "hat rack" above the seats, but also below the seats in front of each passenger. It can be seen that this 6-abreast cross section is providing sufficient spare room for passengers, walking in the cabin. Smaller cross sections like a 3- and 4-abreast cross section (Fig. 5.7) are providing less cabin height in the aisle.

Cabin comfort is a very important item, but always in direct conflict with the economic side of the configuration. The more comfort will be installed, the less seats will be available for a given cabin size. This conflict has to be analysed and several trade-offs between comfort level and possible seat number have to be investigated. Reviewing the tendencies with actual airline configurations will provide a reasonable database which will help to define the standard cabin configuration for a given aircraft type which then will be an important input for defining the fuselage length with all related door and window positions. The comparison with aircraft from the competitor will be another important element in the final choice.

In a more general way, the aircraft manufacturer has to do this optimization task at the beginning of a new aircraft programme, before fixing the basic geometric parameters of the aircraft design.

A lot of parameters are influencing the comfort behaviour (see Table 5.2). Comfort is also a very individual feeling, where it is difficult to generalize the comfort feeling and the well feeling on-board of an aircraft. A farm worker will have a different comfort demand compared to an upper class lady. This strongly depends what you are used to have in your homely environment.
There is first of all the human factor of "comfort": [2–4], which includes:

- Physical, psychological and emotional aspects
- Individual desires, needs, fears

Table 5.2 Comfort influencing parameters

• Individual and public space, pitch, load factor
• Air conditioning, flow, temperature, pressure, humidity
• Noise and vibration level, acoustics
• "Horizontal" cabin attitude
• Comfortable seats (different positions)
• Low gust sensitive aircraft
• Catering, food and drinks, galley, bar
• Cabin service, attendants
• Lavatories, toilets, showers
• Wardrobes, overhead bins for luggage
• Friendly interior design, colours, lighting
• Sun protection, size of the window
• Entertainment (info, music, video, TV, games…)
• Telecommunication (phone, fax, SMS, Email, internet access)

Fig. 5.5 Seat pitch as basic comfort parameter

Fig. 5.6 Cabin design to create a well-feeling atmosphere

Out of the comfort requirements, a technical cabin concept has to be defined which has to include the following elements:

- Ergonomics (see Fig. 5.3, accessibility)
- Safety
- Industrial design

The industrial design will have to include the aspects of:

- clear shapes and design surfaces
- optical clearness—interior space—comfort (Fig. 5.5)
- avoiding the "tube effect"
- lighting design for more spacious effect in the small "tube", mood lighting

Several investigations have clearly identified the most dominant comfort parameter (see Fig. 5.6):

> The most important aspect of comfort is space! [4].

5.2.3 Fuselage Cross Section, Floor Area (2-D Aspects)

The fuselage cross section defines the basic size of an aircraft. It has to be carefully selected at the beginning of a new aircraft design. Depending on the number of passengers and a possible family concept, the fuselage cross section will be carefully chosen. The number of seats, which can be installed in a cross section, is defined as abreast (see Fig. 5.4).

Figure 5.7 is showing several cross sections, a 3-abreast, a 6-abreast, an 8-abreast and a double deck solution. The double deck cross section—chosen for the A380—has on the main deck a 10-abreast seating and on the upper level an 8-abreast seating, leading to an 18-abreast cross section.

The 6-abreast cross section is the best cross section for a single aisle configuration. This 6-abreast fuselage can accommodate between 100 and 200 passengers, depending on the fuselage length and the seat pitch chosen. A 7-abreast configuration needs a 2nd aisle, as the certification rules are not allowing more than three seats in a row or more precisely, the window seat can only have two seats besides him before the aisle can be reached. Otherwise—in case of emergency—it will be too difficult for the passengers at the window seat to escape from the cabin within a reasonable time period. So from a seven abreast seating concept, 2 aisles have to be provided. The Boeing B767 provides such a fuselage with a 7-abreast cross section. By increasing the fuselage diameter further, the next reasonable cross section will be an 8-abreast seating, chosen from Airbus for their A300, A310, A330 and A340 fuselages and from Boeing for their B787. The 8-abreast configuration is well received by passengers and airlines as there is from each seat only one seat to reach the aisle. Sometimes this middle seat is also called a "prisoner seat" or a "single

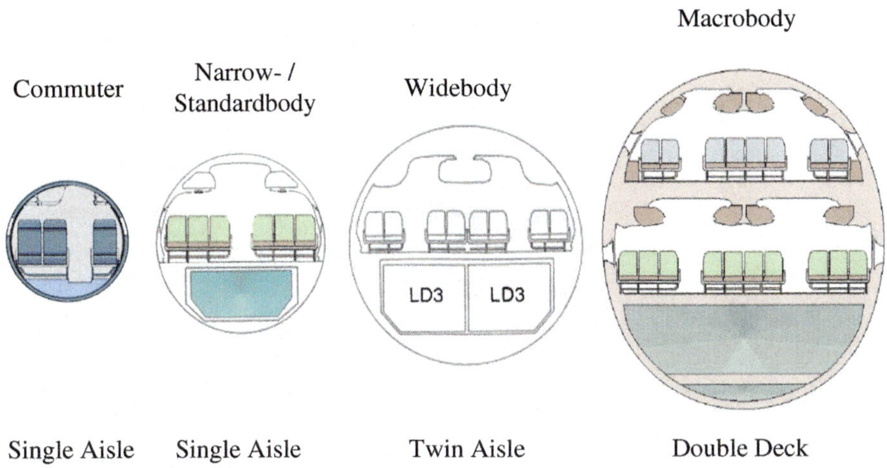

Fig. 5.7 Different aircraft fuselage cross sections and their characteristics

excuse me" seat, because when seated in the middle, you have only once to say "excuse me" to your neighbour in order to reach the next aisle.

The 9-abreast cross section is the next one, chosen for the B777 from Boeing. Here are two choices, either a 3 + 3 + 3 seating possibility or a 2 + 5 + 2 or even an asymmetric seating with 3 + 4 + 2.

The 3 + 3 + 3 seating has 2 window seats, 4 aisle seats and 3 middle seats, which are generally more difficult to sell, if it is not for a family or group. The 2 + 5 + 2 seats has also 2 window seats, 4 aisle seats and 3 seats in the middle bench, which are difficult to sell, unless as unsold seats for medium and long range and passengers can use the bench for sleeping.

The next bigger cross sections are the 10-abreast and 11-abreast cross sections. If the 2 aisle concept is maintained, then there are no additional "good to sell seats" added. So the bigger cross sections are not very popular, neither by the passengers nor by the airlines. Another negative aspect is also linked with the bigger cross sections above 9-abreast: the head space will increase considerably but it may be difficult to use this space. This aspect can be seen in Fig. 5.9, where a 9-abreast cross section like the B777 is shown. This configuration has already a lot of free space in the upper lobe, which is not really usable for cabin comfort. More technically speaking, increasing the cross sections beyond 8 or 9-abreast will increase the outer circular fuselage surface (SF) with

$$\mathrm{SF} = \pi * D * L_f \tag{5.1}$$

Taken 20 inch per seat and aisle, the fuselage diameter will increase for a 10-abreast cross section from 9 seats + 2 aisles ((9 + 2)*20 inch) to 10 seats + 2 aisles ((10 + 2) * 20 inch), i.e. from 220 to 240 inch. In terms of scrubbing drag

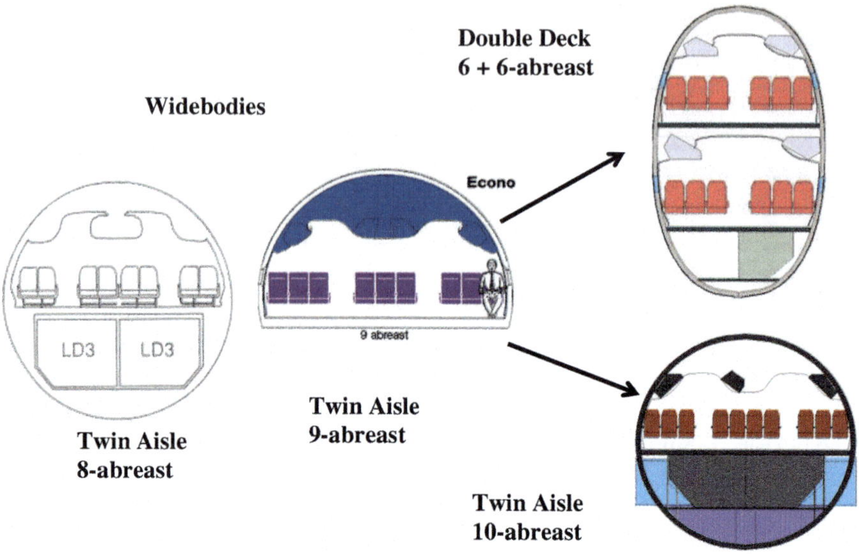

Fig. 5.8 Increase in fuselage cross section and related usage of volume

(aerodynamic friction drag from the outer skin), the outer surface increases by roughly 10 % for only adding one seat per row on the main deck and this seat is not easily to be sold. This shows clearly that a further increase in fuselage diameter has to be done in a different manner, i.e. adding a second deck.

So there is a point where the additional increase in fuselage diameter has to be done in a way that an additional deck will be created by putting different circular cross sections between the two main decks (Fig. 5.8). This increases the volume in such a way that two decks can be used for a comfortable seating by still increasing the outer surface in a reasonable manner (Fig. 5.8).

But there is also another important point which has to be considered.

Up to now only normal seating arrangements are mentioned, which means the classical economy class. There are, however, also travelling persons, who would like to have a more comfortable seat, which means a wider seat and a seat with an increased pitch.

Generally, there are normally three different classes with different seats. In Table 5.3 typical differences in seat width and seat pitch for different classes (FC First Class, BC Business Class, YC Economy Class) are shown with their normal standard dimensions. But again it has to be mentioned that each airline can choose their seat pitch and seat width value as they like. But for the given cabin cross sections, there are more or less certain standards fixed by the fuselage cross section and length.

There is no unique definition of seat standards for economy, business and first class seats. Each airline has their own standard which they have developed, depending on the typical size of their own population (Asiatic airlines normally are

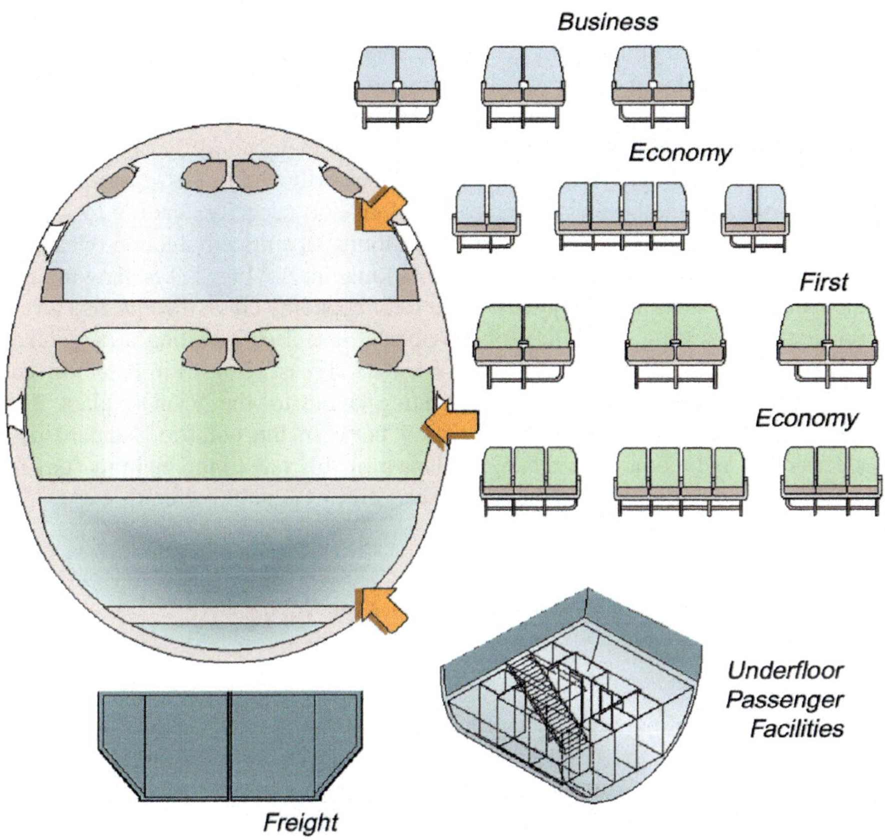

Fig. 5.9 Compromise of different class seats for the 2-deck configuration (A380)

Table 5.3 Definition of cabin standards for different design ranges

		Short range (SR) SR ≤ 3000rm	Medium range (MR) 3000 nm < MR < 5500 nm		Long range (LR) LR ≥ 5500 nm		
		YC	FC	YC	FC	BC	YC
Seats	Seats in %	100	8-10	90-92	5-7	18-20	73-77
	Seat pitch (inch)	32	40	32	60	38	32
	Seat decline (inch)	5	7.5	5	15	7	5
	Seat width (two-man bench)	40	48	40	53	50	40
Cabin attendants per pax		1/45	1/8	1/35	1/8	1/20	1/35
Lavatory per pax		1/60	1/14	1/45	1/14	1/25	1/45
Trays per pax		1.7	9	2.3	9	7	2.7
Coat: stowage (inch/pax)		No	1.5	No	1.5	1.5	No

offering a slightly reduced seat pitch compared to European or US airlines.) seating standards are also changing with the local competition. It is now visible in Europe that the big national flag carriers are intending to reduce their seat pitch in the short range aircraft as the competition from the Low Cost Carriers (Easyjet, Ryan Air; Air Berlin) are also offering only a very low seat pitch in their cabin! By offering a poor standard with a pitch reduction from 32 to 29 inch will lead to 10 % more seats in the cabin, offering the airlines an increase in transport capacity, leading to a 10 % better efficiency which can be given to the customer as a price reduction (Chap. 8).

Coming back to the double-deck seating arrangements, Fig. 5.9 is showing that at the main deck with their 10-abreast seating for economy class, there is also a very good first class seating capability for a 6-abreast in a 2–2–2 seating arrangement. On the other hand, the upper deck seating with 8-abreast (2–4–2) in economy can also be used for a 6-abreast in a 2–2–2 seating layout for the business class. This allows also a clear and good differentiation between the comfort standard of a business class and a first class seating. This is probably one of the main reasons for Airbus to have finally selected this cross section for the A380 aircraft.

5.2.4 Cabin Layout for Several Comfort Standards (3-D Cabin)

After the selection of the best fuselage cross section (the 2-D aspect), the total cabin has to be considered and configured. The front fuselage part is reserved for the cockpit. The normal cockpit has two seats for pilot and co-pilot. In addition, there are normally also 1 or 2 additional seats for a supervisor or training pilot. So the cabin starts only after the cockpit.

The cabin has normally one floor level without any obstructions on the bottom. Some aircraft—mainly with high wing layout—will have in the middle section some compromise for cabin height and overhead storage compartments. Figure 5.10 shows some typical seat layouts for a long-range aircraft, here A340-300.

Besides the seats, also a certain number of toilets, galleys, coat stowage, cabin attendant seats have to be installed in the available cabin space (see Table 5.3).

The door arrangement is giving some restrictions, but doors are arranged due to emergency evacuation considerations and are also fairly evenly distributed between front and rear. Door layout aspects can be seen in [5–9]. For each aircraft the door arrangement is a certification item and is fixed per aircraft type.

Doors are providing a specific chance, especially for twin aisle configuration, to provide a connection between the 2 aisles for passengers during the flight. This connection which is needed at the 2 front doors (see Fig. 5.10) for boarding and deboarding, is also an important element for long-range flights to provide these connection between the 2 aisles. So the door area is a natural placement for installation of toilet blocks and/or galley areas. It could also be used as a natural

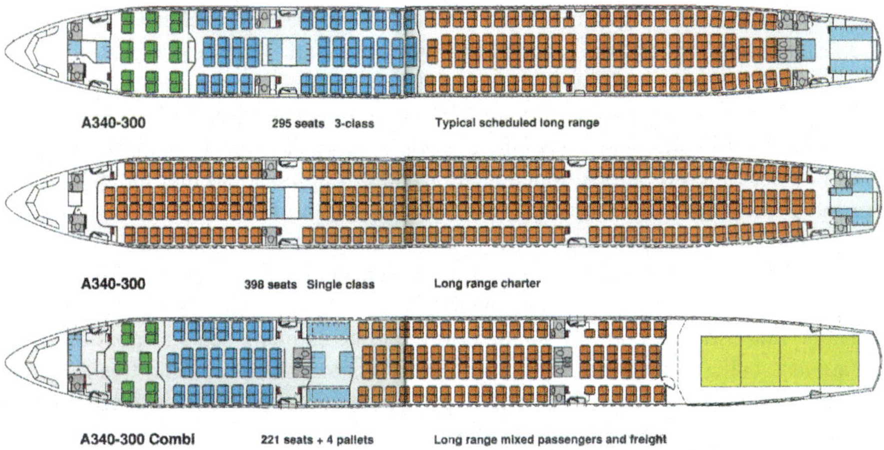

Fig. 5.10 Standard cabin layout for a typical long-range aircraft (A340-300)

separation between cabin classes (door 2 area as separation between business and economy class).

There is a certain minimum standard for the location of doors as defined in the Certification Requirements [10, 11] for the number of toilets per passenger. However, the airlines are normally offering a better standard. Reasonable numbers are given in Table 5.3, differentiated for short range missions, medium range missions and long range missions. Taking those numbers allows developing a cabin layout as shown in Fig. 5.10 for a typical long-range aircraft. There is, however, one remark to be done: the layout in Fig. 5.10 shows in the business class a 7-abreast seating, which was proposed at the beginning of the 1990s, but is today no longer accepted by the airlines. As mentioned before, a business class can have no middle bench with three seats. The passenger, paying the price for a business ticket, will not like to be seated in business class between two other persons.

The 3-class layout shows in green the First Class with 18 seats. This corresponds to 6 % of the total number of seats. It would be ideal to have 5 % from 295 seats in First Class. However, the first class compartment has to present itself as an integrated unit. Two rows only would be a fairly small space. There is only the option to have 2 or 3 seat rows, so the 18 seats are from the cabin design the only reasonable answer and solution.

25 % of all seats should be in the business class (see Table 5.3). The 3-class layout in Fig. 5.10 shows 81 business class seats and 196 seats in the economy class. The partition between classes should always be done between complete rows to allow a good cabin separation and feeling.

Each class has to have its own block of toilets and galleys. In our example in Fig. 5.10 has the first class two toilets installed, also one more than a standard solution would require. But for 18 passengers in first class, 2 toilets are mandatory to have always a free toilet. Also the fairly big galley block in the first class may be surprising. But for a long-range flight, 3 meals will be served, each meal will have 3 courses, each presented on a separate tray. All in all 9 trays per passenger are needed, which leads to $18 \times 9 = 162$ trays storage space only for the first class, with extra space for drinks, snacks, coffee and ovens for heating part of the meals.

For each cabin attendant a specific *cabin attendant seat* has to be provided. This attendant seat should be positioned close to a door and also so that the cabin crew can have a free look into the cabin during take-off and landing.

Specific *coat stowage* compartments are needed for business and first class areas.

The cabin configuration for Low Cost Carriers is slightly different. The need for galley space is reduced. Toilet standard is at the minimum; seats as much as possible with reduced comfort standard to integrate as many seats as possible. This is then also called a high-density layout. But it has to be kept in mind that there is a maximum number of seats per aircraft type, which is part of the aircraft certification procedure and which cannot be exceeded.

The passenger is also expecting a certain entertainment service. Besides info's to the flight route, weather conditions and time till arrival, there are also offers of several audio and video channels. Videos should also have the choice for selecting at least 2 languages, English and the national flag carriers language with a third option of the language of the destination. The entertainment market is developing enormously and cannot be treated here in detail. Live TV, direct internet access and several services via internet are already today feasible and will be further developed. A more difficult subject will be the mobile phone connection. This is not so much a technical feature but more a social problem. If it will be allowed in a high-density cabin and everybody could communicate directly from his seat via his mobile phone, this will then lead to conflicts between persons, wanting to rest and sleep and others, being always active and difficult to control in their loudness. The actual banning of all cell phone calls is therefore a solution, which has a great chance to be kept as standard.

5.2.5 Aircraft Cabin Systems

In addition to a good cabin arrangement with seats, toilets galleys, etc., there is also a need to provide a comfortable cabin environment.

The well-being in a cabin is depending—besides the space—also from several other aspects linked to all the human senses:

- Air conditioning—individual temperature selection
- Air pressure in the cabin
- Lightning environment including individual reading lights
- Low noise
- No abnormal smells
- Harmonic colours for seats and cabin walls and ceiling

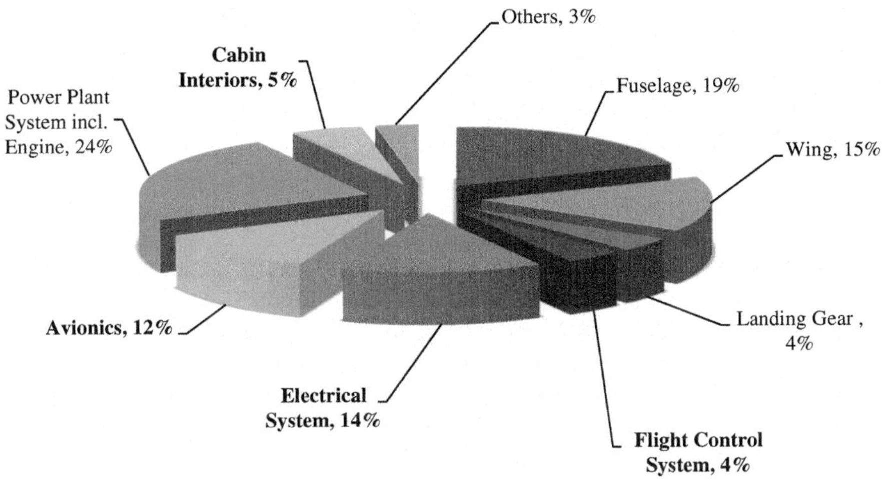

Fig. 5.11 Value of avionics and electrical systems for modern wide-body aircraft [25]

The value of air cabin systems plays only a minor part (5 %) in the context of all avionics and electrical systems of an aircraft today, as shown in Fig. 5.11.

The relevance of electronic systems mainly based on computing devices and software will further increase in the future, while mechanical and aerodynamic systems will rest at their maximum current maturity level. From this perspective also cabin interiors mainly cabin systems control and passenger communication and entertainment systems will get higher value and relevance in future aircraft developments, especially for long-range aircraft. On short-range aircraft, which operate more than 85 % below 500 nm, simplicity and attractive pricing is more relevant.

5.2.5.1 Air Conditioning System

In high cruising altitudes the *pressurization of the cabin* is limited in such a way that the internal cabin pressure corresponds at least to a pressure level corresponding to the pressure in 8000 ft altitude. The differential pressure between outside and cabin side is depending on the flight altitude. But this cabin pressure level is never to decrease below the 8000 ft level. In case of a depressurization, the pilot has to start a descend maneuver and go to a lower flight level (8000 ft—typically, 10,000 ft in abnormal cases), where the cabin pressure can be maintained.

There is a lot of experience that for all normal people, this pressure level is not causing any health problems. Normally in a cabin there will be no physical exercises to be done by the passengers, they are normally seated and fairly calm in their physical activities, which supports the acceptance of the today's requirements. Several researches have been done by the US [12, 13] and from European Research

Centers [14, 15], which are normally confirming the standards of today. There is, however, a new attempt in the competition between Boeing and Airbus, where the B787 will be designed to have a higher cabin pressure (6000 ft instead of 8000 ft). New fuselage materials are allowing this pressure increase which corresponds to a small weight penalty.

The other parameter, which goes parallel to the cabin pressure, is the *air humidity*.

The humidity in aircraft cabins is normally at a very low level (less than 20 % relative humidity), which is felt as uncomfortable by most people [15]. There is however the fact that due to the very cold outside temperature of – 50 to −60 °C, all air humidity in contact with the cold outside wall will accumulate to water droplets at the fuselage, leading to higher corrosion risks of the metallic fuselage and weight penalties inside the insulation packs. This is a design point of the aircraft, which needs improvement in the longer term. Human beings feel better and report less dryness symptoms if the relative humidity is higher than 30 %. Specific humidity provision elements are installed in some specific cabins to improve the comfort. No clear guidelines do exist, but there is a chance in the future with the newly proposed composite fuselages that this will allow to improve and increase the humidity in the cabin and thus improving the well-being of the passengers.

The best recommendation for all passengers is to drink as much as possible during the flight phase to avoid dehydration and related health problems.

Another element is the *oxygen content of the air*. With the existing environmental system installed already on-board the aircraft, there is also the possibility to influence the composition of the air in the cabin (oxygen, nitrogen, carbon dioxide CO_2, microbiological components, water content, and all other elements) besides temperature, pressure and humidity.

An increase in oxygen content can also be a mean to improve the quality of the cabin air. It is state of the art today that part of the cabin air is sucked off the cabin, filtered and partly reused after mixing with fresh air from the atmosphere. The air filtering systems, which are installed today are of very high quality and ensure that even with a recirculation of nearly 50 % of the air, there is no risk that infection may develop or increase.

Several research projects have analysed the recirculated air with respect to microbiological content, dust, bacteria, infectious elements, etc. In summary the air quality is extremely good. The major critical point is still the problematic of the very dry air for long-range flights and the low pressure during cruise.

5.2.5.2 Other Cabin System Aspects

New Lightning technologies (LED; OLED and other elements) are allowing a lot of new design features to create specific atmospheric illusions in the cabin. A moon lightning capability is offered at the A380 aircraft and will be offered by the new B787 and A350 versions [16].

Cabin noise has been a major research issue over a very long period. There are two major noise sources [5, 8, 17]:

- the engine noise with the elements fan noise and exhaust noise and
- the aerodynamic noise, generated either by the boundary layer of the fuselage or by the vortices, being transformed into vibrations at all outer surfaces and being transferred through the structure to initiate vibrations of the fuselage.

With the modern design tools, the reduced noise generation from the engine source and the good aerodynamic design, the modern commercial aircraft have no longer a cabin noise problem. The insulation material close to the fuselage skin has also been improved. The aircraft with propeller engines are still today suffering often from fairly high noise levels in the cabin and modern propeller-driven aircraft have integrated several technologies to improve this phenomenon.

Some former aircraft are still suffering from some specific noise frequencies (Buzz saw noise) at the front fuselage part, generated from the engine inlet in some SA aircraft types. This was specifically painful for some airlines as this front part is reserved for the business class passengers and major efforts were needed to reduce the noise levels in this area.

A specific technique, investigated for several years, is the so-called active anti-noise technology. This technology generates noise in such a way that the noise peaks are counterbalanced by a counterwave where all peaks are equalized. Several research studies have shown the benefits of such a system, but it is only effective if there is a very clear singular tone, which can be eliminated or better reduced. This is seldom the case. In most of the actual noise cases, there is a broad band noise, where anti-noise measures cannot really improve the overall noise level [17, 18].

Another major element in cabin design is the *entertainment system* with all the possibilities of *connectivity* to earth stations. Access to emails, life television, radio, Internet, etc., is under research and several options are already available in modern aircraft [4]. This is a very wide and complex subject, which will not be further treated here, as it is a specific topic and under constant development.

5.3 Basics of Flight Physics

The following chapter will cover the basic principles of flight physics. It will start with the atmosphere as basis for all flight activities. Then the main four acting forces on the aircraft like:

- aerodynamic lift,
- aerodynamic drag,
- weight and weight breakdown of the aircraft,
- thrust requirements as propulsion forces during take-off and in cruise.

will be shortly described to provide an understanding of the basics of flight performances. The basic principles and references for stability and controllability of the aircraft with the trimming of an aircraft will be treated. The importance of centre of gravity for load ability will be mentioned. The basics of flight performance will be an important chapter. Some cost aspects will conclude the chapter.

It should be mentioned again that only the basics can be covered here, as far as they are important to understand the efficiency of flight vehicles in comparison to other transport modes. But there are several good standard books on the market like [5–9, 19, 20].

5.3.1 ICAO Standard Atmosphere

The aircraft is flying in the atmosphere at different flight altitudes. It is therefore important to define the physical conditions of the atmosphere, their basic characteristics like temperature, pressure, density as a function of the altitude.

The atmosphere has a fairly complex structure. With higher altitudes the density of the air is diminishing. But this is not happening in a constant and linear way. The air density varies with meteorological, seasonal and even local topographical changes. But it is important to define a mathematical model which will allow calculating in a fairly general way the basic characteristics of the air with variation of altitude. The ICAO has accepted the ISA Standard Atmosphere (ISA), which has been defined in the year 1975 by the ISO [21] and is mainly developed and valid at the northern hemisphere in the range of 40–50° latitude. The measured and average values in this region have been chosen. ISA defines a linear decrease of the air temperature up to an altitude of 11 km.

At 11,000 m the tropopause is located and fixed. The tropopause is an interlayer between the troposphere and the stratosphere. In reality the tropopause is not a fixed value at 11 km but changing the altitude as a function of the earth latitude (Fig. 5.12).

At the North and South Pole the tropopause is located at only 8000 m and at the equator the tropopause is located at around 17,000 m. The tropopause is not fixed at a constant altitude; it varies with the earth latitude. However, the ISA standard is assuming a constant point at 11,000 m.

Above the 11 km threshold starts the stratosphere. The stratosphere is basically defined that the temperature stays constant above 11 km. Figure 5.13 shows the temperature gradient versus altitude as defined by ISA [21].
The troposphere is characterized by:

- temperature is constantly decreasing with altitude up to the tropopause
- High vertical exchange of air due to weather phenomena
- Diminishing duration of local emissions at a certain position of input.

The stratosphere is characterized by:

- Constant temperature with altitude above 11 km
- Limited vertical movement of air and low exchange rates

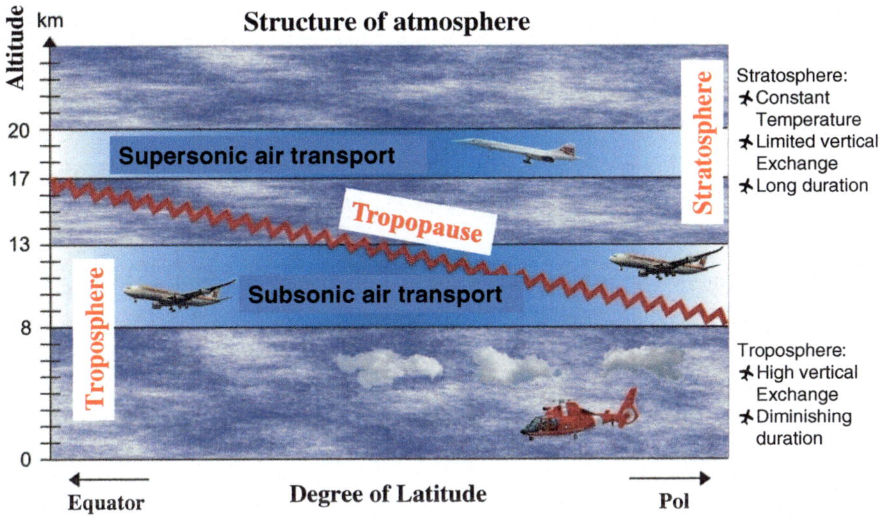

Fig. 5.12 Schematic view of troposphere and stratosphere versus altitude and earth latitude

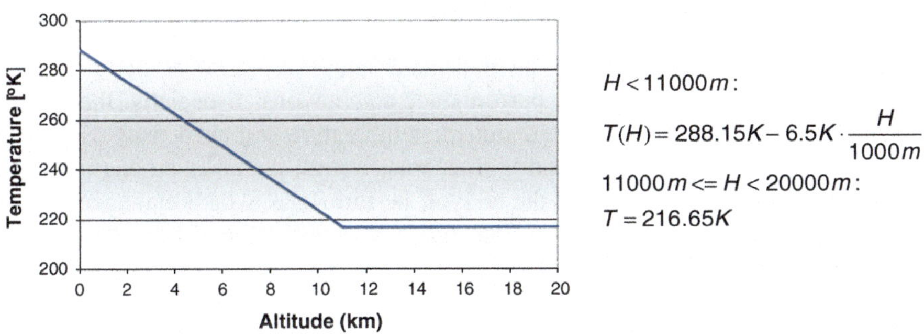

Fig. 5.13 Air temperature versus altitude, as defined by [21]

- Long duration of emission parameters in this region.

The temperature at sea level is defined with:

$$T_0 = 288.15\,\mathrm{K}\,\mathrm{or}\,15\,°\mathrm{C} \tag{5.2}$$

The air pressure at sea level is defined with

$$P_0 : 1013.25\,\mathrm{hPa} \quad \mathrm{or}\ 101325\,\mathrm{N/m^2} \tag{5.3}$$

Air pressure and air density are calculated as a function of temperature (Fig. 5.14). It is obvious that the density at altitude sea level (0 m) is around five times higher than the density at flight altitude ($\sim$11,000 m). But it also has to be

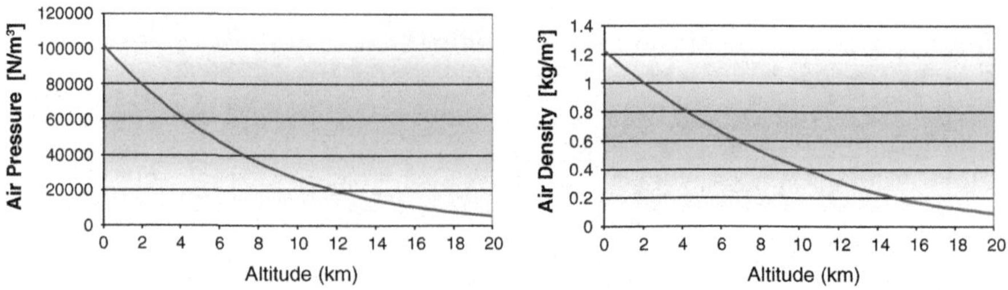

Fig. 5.14 Air characteristics pressure and density as a function of altitude (ISA) [21]

mentioned that the speed is changing with temperature and Mach number is slightly decreasing with altitude. For details see [21] or [19].

As mentioned above, the ISA atmosphere is an idealized and simplified definition. Real temperatures in the atmosphere are easily handled by a delta temperature to ISA.

If an outside temperature of 25° is measured, it can be transferred to a temperature ISA + 10, i.e. 10° above ISA at sea level. With this model, all temperature as function of altitude can be calculated with reference to ISA. Very often in specifications, airports with high temperatures and high altitudes (so-called hot and high airports) are referenced where the outside conditions can be, for example 40 °C and 2500 ft altitude. These data can be translated into ISA + 25° at 2500 ft. This information is very important for performance calculations. Especially, the engine thrust at take-off is a function of the outside temperature and air density. So it can happen that at airports with hot and high characteristics, the installed engine thrust may lead to some degradation for the take-off performance, which may mean at the end that the aircraft cannot take-off with full passenger load and full fuel tanks at certain outside conditions. This may lead in practice to a reduced payload or reduced fuel volume (reduced range capability) for the aircraft.

5.3.2 *Aircraft Forces: Lift, Weight, Drag, Thrust*

To keep an aircraft flying in the atmosphere, a certain aerodynamic upward force is needed, the *aerodynamic lift force*, which is generated mainly by the wing of an aircraft. In addition a forward force to push the aircraft through the atmosphere is needed, which is called *thrust*, produced by the engine(s).
The main forces acting on the aircraft can be described by the following four forces:

- Lift generated by the wings of the aircraft
- Weight of the total aircraft including the aircraft empty weight plus payload and fuel
- Aerodynamic drag
- Forward thrust of the engines

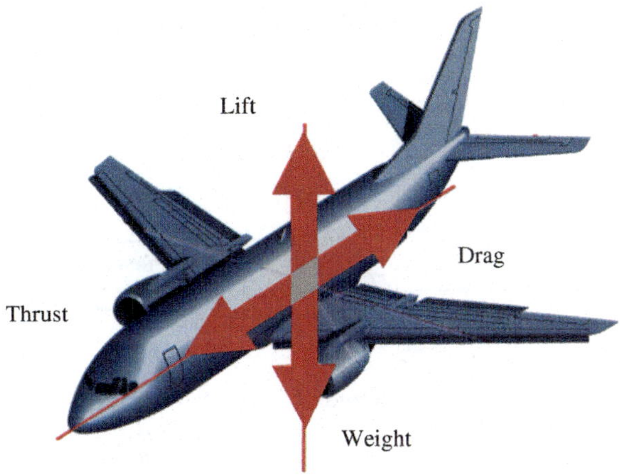

Fig. 5.15 Aircraft forces in horizontal and vertical axis

If we consider a steady flight of an aircraft during its cruise phase, i.e. speed is constant and the flight altitude is constant, then the 4 main forces acting on the aircraft have to be in an equilibrium.

In a simplified way, the lifting force of the aircraft is needed to balance the aircraft weight.

$$L = W, \text{ where } W = m * g \text{ (aircraft mass)} \tag{5.4}$$

And the thrust has to be equal to the aircraft drag in order to fly at constant speed

$$T = D \tag{5.5}$$

For these conditions, the equilibrium is given and the aircraft flies in a given altitude with a given speed (Fig. 5.15).

5.3.3 *Lift*

Lift will be generated mainly by the wings of an aircraft.

The wing flying at a certain speed and certain altitude is generating an aerodynamic force R which will then be split into a lift component L, acting perpendicular to the speed vector and a drag force acting in the level of the speed vector (Fig. 5.16).

The lift of the wing is mainly generated by the shape of the wing profile (camber of profile) and the angle of attack of the wing, relative to the wind vector V.
In different words the Lift L is dependent on two basic elements:

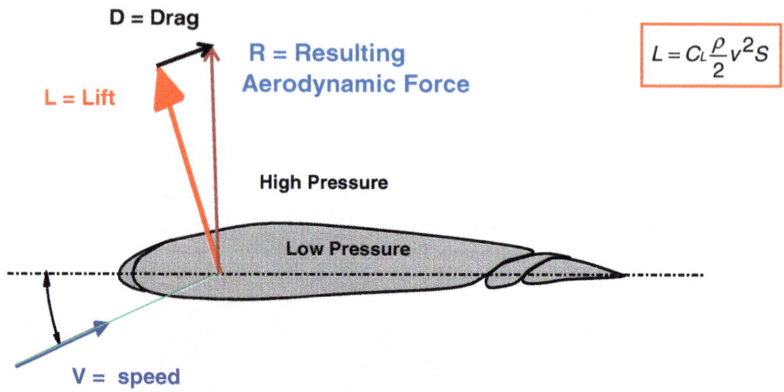

Fig. 5.16 Lift force as resulting force, perpendicular to the speed vector

1. The Flight condition, which includes
   - Flight Speed V
   - Air density $\rho = f(H)$
   - Angle of attack α
2. The aircraft configuration
   - Size of the lifting surface (Wing) S
   - Geometry of the lift generating surface (i.e. the wing shape with its profiles, the profile camber, profile twist along the span and its particular flap and slat deflection during take-off and landing). The wing geometry is the result of a very sophisticated aerodynamic design process and is not treated here [5, 6].

For a given aircraft type (wing condition and wing area fixed) the flight condition parameters can be varied to control the lift force. When flying in a cruise condition (a given altitude and a given speed) a certain angle of attack α is needed to provide the necessary lift force to be equivalent to the actual mass of the aircraft. The only means to control the lift force is by changing the angle of attack α or by changing the speed. At a given speed, the only parameter for control of lift is α! The trimming of the aircraft is the mean to control the cruise condition (Sect. 5.3.5)

In cruise flight (this means, all high-lift devices (slats and flaps) are not extended and the wing profile is "clean") the angle of attack varies linearly with the lift in the normal flight domain. Figure 5.17 is showing this linear relationship between the angle of attack α and the lift coefficient C_L. If the angle of attack becomes too high, a certain flow separation will start and a nonlinear behaviour can be seen. This is a critical flight condition which has to be avoided in any case during all flight phases. Therefore, a certain limit of maximum angle of attack α has to be considered and the flight control system is indicating this limit to the pilot as never to exceed.

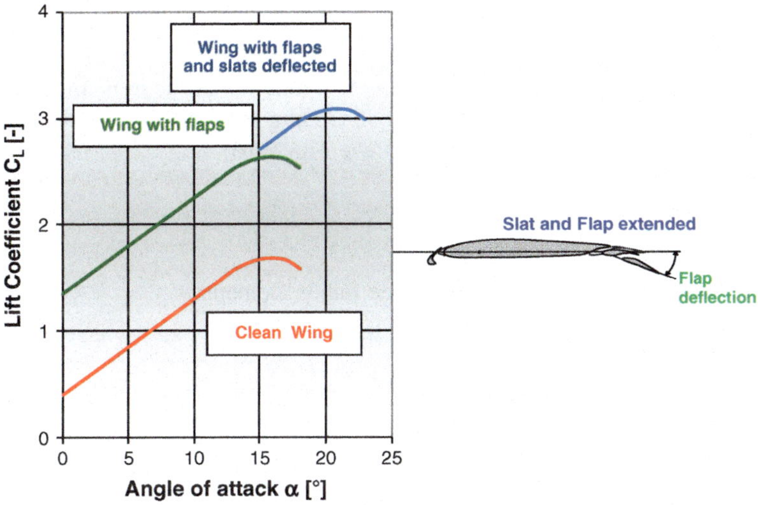

Fig. 5.17 Wing lift as a function of α and slat/flap position

The aircraft is normally designed to fly with a given speed (Ma = 0.78 for short-range aircraft) and at a given flight altitude. The typical lift coefficient is about 0.5.

If the aircraft has to land on ground, with this given lift coefficient, the air density on ground will be increasing and the speed can be reduced. Assuming the air density between flight level (33,000 ft) and the ground level (100 ft) will increase by a factor of 5, then the landing speed V_L can be calculated in relation to V_{Cr} as indicated in the formula:

$$V_L = \sqrt{\frac{\rho}{\rho_0}} * V_{\mathrm{Cr}} \tag{5.6}$$

This leads to a reduction in speed by a factor of 2.24 or a reduction of speed from Ma = 0.8 down to Ma = 0.36. This Ma = 0.36 is still far too high to land an aircraft. Some additional aerodynamic means are required to increase the lift coefficient during the take-off and landing phases. The increase of lift is possible by the introduction of slats on the leading edge and flaps on the trailing edge of the wing. In simplified terms, the flap extension will shift the lift curve to higher C_L—values and the slat extension will further enlarge the operating range to higher angle of attacks α (Fig. 5.17).

Modern aircraft can increase their lift capabilities with the extension of slats and flaps by a factor of 3–4 (the C_L can go up to 1.5–2.0 for landing and 1.5 for take-off), which allows the aircraft to be landed with a much lower speed (Ma < 0.2). This is a reasonable approach speed and manageable for all pilots when properly trained.

5.3.4 Drag

The Drag force—similar like the lift force—is a function of the dynamic pressure, a reference surface and the drag coefficient. The Drag force acts perpendicular to the lift force and in line of the speed vector (see Fig. 5.16)

$$D = C_D * S * \frac{\rho}{2} * V^2 \tag{5.8}$$

The drag coefficient C_D contains the three major elements:

- drag at zero lift (mainly skin friction drag) C_{DO}
- induced drag or lift dependent drag $C_D^i = f(C_L^2)$
- compressibility drag or Mach-dependent drag $C_{DM} = f(\text{Ma})$

Figure 5.18 illustrates the different drag terms in a simple way. The blue curve represents the parabolic curve

$$C_D = C_{Do} + k * C_L^2 + C_{DMa} \tag{5.9}$$

The blue theoretical curve is fairly close to the real measured curve at Ma = 0.3 (low speed)

The difference in drag between the red and yellow curve is fairly small, indicating there is not a lot of change in drag in the range between Ma = 0.3 and Ma = 0.7. But with further increase in terms of speed resp. Ma number up to Ma = 0.82 (the green line) the total drag is increasing due to the compressibility drag C_{DMa}.

This rough introduction to drag is sufficient in this context. There are several references, where more details can be found like [5–9].

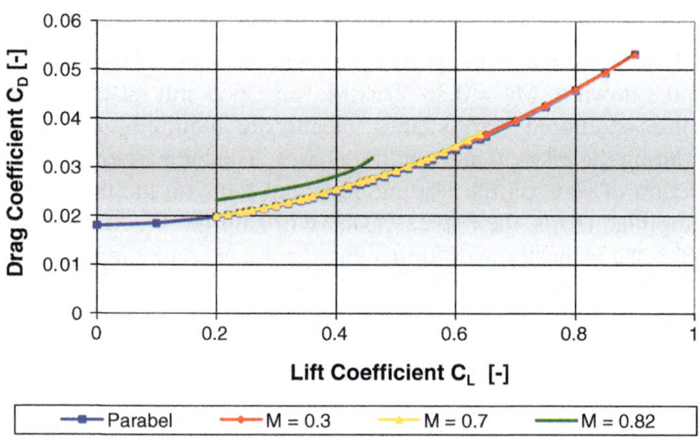

Fig. 5.18 Aerodynamic drag as a function of aerodynamic lift

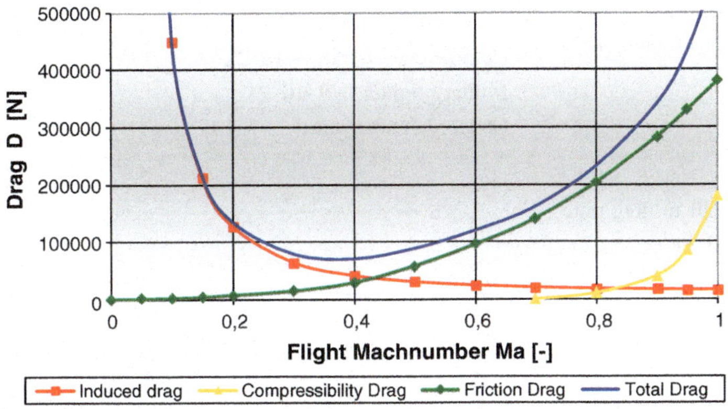

Fig. 5.19 Drag as a function of speed

Another interesting presentation is the dependency of Drag as function of speed. Taking Eq. 5.8 and isolating the induced drag D_I, this leads to

$$D_i = k * C_{Di} * q * S = k * C_L^2 * q * S = W^2 * k/q * S \tag{5.10}$$

Assuming, that the aircraft is well-defined and all aircraft parameters like wing area S and factor k are fixed, then

$$D_i \sim K_1/q = 2K_1/\text{rho} * V^2$$

where q defines the dynamic pressure $q = \text{rho}/2 * V^2$

This means, that the induced drag goes to infinity when speed is zero and decreases with increasing speed (Fig. 5.19).

The other parameter D_0 (skin friction drag + parasitic drag) is, however, increasing with speed due to the definition of

$$D_0 = C_{Do} * q * S = C_{Do} * S * \rho/2 * V^2 \tag{5.11}$$

5.3.5 *Aerodynamic Efficiency*

There is another parameter, which is of interest in this context; this is the aerodynamic quality of an aircraft, defined by the ratio Lift /Drag. In the sailplane world, all sailplanes are characterized by the glide slope ε, which is characterized by

$$\tan\gamma = -\varepsilon = -\frac{W}{A} \tag{5.12}$$

Table 5.4 Aerodynamic efficiency for different aircraft types

Sailplane ∼40–60
Commercial aircraft ∼18–22
Military fighter ∼9–10
Concorde (supersonic) ∼7–10

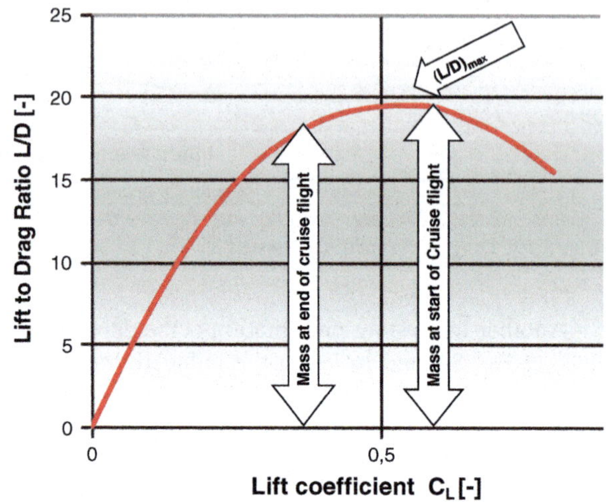

Fig. 5.20 Lift to drag ratio for optimum flight

The glide slope is optimum when there is only little loss in altitude with a maximum of horizontal distance achieved. If an aircraft achieves an aerodynamic efficiency $L/D = 20$, it means that for 1 N in drag 20 N in lift will be achieved, which is a very good quality for transport aircraft. The best commercial aircraft today have an L/D of more than 20 at a given cruise speed. At lower Ma numbers they are even better (see Table 5.4).

During the different phases of flight (take-off, climb, cruise, descend) the aircraft is burning fuel and the mass is diminishing. This means, that the needed aerodynamic lifting capability is decreasing between climb, cruise and descend. Especially for a long-range flight, there is quite a big difference between the lift needed at the beginning of the cruise phase and at the end of the cruise phase. This is shown in Fig. 5.20. It also shows that the aircraft characteristics like L/D has to be selected that it is optimal for a wide range of C_L.

5.3.6 Aircraft Mass Breakdown

The aircraft total operating mass consists of four major blocks, which are important for the aircraft operation. Figure 5.21 is showing these different mass terms. These terms are correctly defined as masses of the aircraft. But a lot of handbooks and also the operational side are still using the terms weight instead of mass. We will use

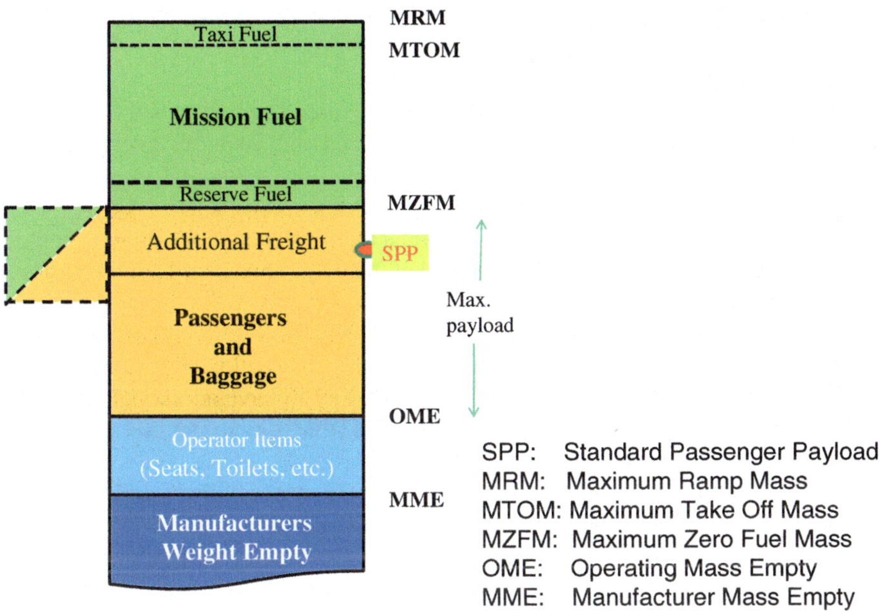

Fig. 5.21 Mass breakdown for a transport aircraft

mainly the term mass in this book, but recognizing and indicating, that the terms MTOW and OWE are used very frequently instead of MTOM and OME.
The most important aircraft mass terms are the following:

1. *MME*: *Manufacturer Mass Empty*
   The MME is defined by the aircraft manufacturer. It contains all elements of an aircraft, which are required by the authorities to operate an aircraft, including all masses for safety elements the two pilots, but without specific cabin arrangements.
2. *OME*: *Operating Mass Empty*
   The OME is defined by the MME plus all elements an airline (the operator) will define to provide a very comfortable cabin including seats, toilets, galleys, entertainment systems, etc., for their passengers. The OME includes also the cabin items. There is a certain minimum of cabin staff requested by the authorities. However, the airline can choose a much higher standard corresponding to the airline image. The OME is therefore defined by the individual airline and its comfort standard and differs from airline to airline.
3. *MZFM*: *Maximum Zero Fuel Mass*
   The MZFWM is based on the OME plus the maximum payload mass, which is allowed. The MZFM is defined by the aircraft manufacturer and is an important figure for the dimensioning of certain fuselage structural elements. The MZFM is fixed by the aircraft manufacturer and defines the maximum payload mass, which can be used for any operation. MZFM is defined by the MME plus the

maximum payload mass including all operator items or in other terms, MZFM is the maximum mass without fuel.

4. *MTOM*: *Maximum Take-off Mass*
   The MTOW is defined by the aircraft manufacturer and indicates the maximum weight for the aircraft, just before take-off and this includes everything, the payload, the fuel and the MME.
5. *MRM*: *Maximum Ramp Mass*
   The MRM is fairly close to the MTOM, but it allows an operator to have a bit more fuel in his tanks, while the aircraft is still at the gate. This additional fuel is just a small amount of fuel, which will be needed to taxi the aircraft from the gate to the take-off position. At big hubs, sometimes the aircraft has to wait quite some time in a long queue before arriving finally at the take-off position. To compensate for this fuel needed for taxiing, the aircraft is allowed to have additional fuel in the tank, the MRW. At the take-off position the MTOM should not be exceeded.
6. *SPP*: *Standard Passenger Payload*
   The SPP defines a mean passenger payload, which normally includes different classes in the cabin. It is taken from airline statistics and is a mixture of several airline standards. For this standard payload, a certain flight range can be defined, the standard range, given in the aircraft brochures. (Sect. 5.4.4).

As can be seen in Fig. 5.22 payload and fuel are normally taking 20 % each of the total aircraft weight. For long-range aircraft, the fuel proportion increases to 30 %.

Another typical value is given by the "Wing loading"

$$\textit{Wing loading} \text{ is defined as ratio : } \mathrm{MTOM}/S \tag{5.13}$$

where S is the reference wing surface of the aircraft. The wing loading describes, how many kg of weight will be lifted by 1 m^2 of wing surface.

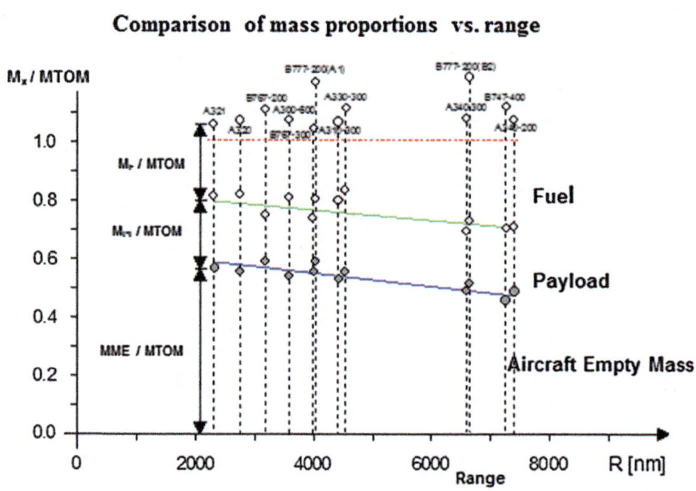

Fig. 5.22 Typical payload and fuel proportions of weight for different aircraft

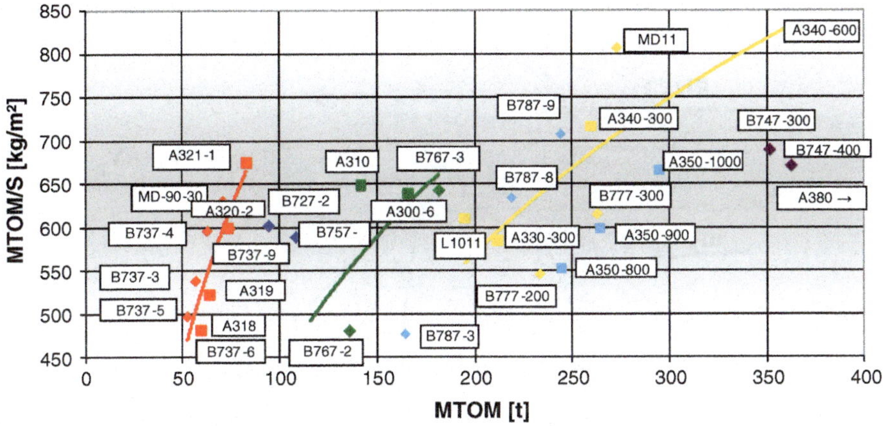

Fig. 5.23 Wing loading for several transport aircraft

A typical value for wing-loading MTOM/S for all transport aircraft is in the order of 500–700 kg/m^2 (see Fig. 5.23).

5.3.7 Thrust Requirements

Following Fig. 5.15, a propulsive force, called thrust is needed to overcome the aerodynamic drag in cruise flight. The thrust should at least be as large as the drag, even a bit bigger in order to have some excess power, needed for acceleration and for maneuvering and control.

There are a lot of interesting books about the functioning of jet engines [5, 22, 23], the requirements for effective engine design and a catalogue from existing jet engines, where all the necessary data and calculation methodologies for jet engines are provided.

It seems sufficient here to just outline the basic principle of a turbofan engine. Figure 5.24 is indicating the mass flow of air which is sucked into the engine during its operation. A part of this air mass m_1 is used in the inner hot section, will be further compressed in different axial compressor stages, will then be mixed in the combustion chamber with fuel, will be ignited and the hot air will pass through the axial turbine stages, will be further accelerated in the exit nozzle and deliver one part of the thrust (5.12). This inner hot part is also called the gas generator, providing sufficient power to drive the big fan.

The other part of the air mass m_2 will be accelerated in the fan and will deliver the other part of the thrust. This is the cold flow, producing the major part of the thrust in modern engines.

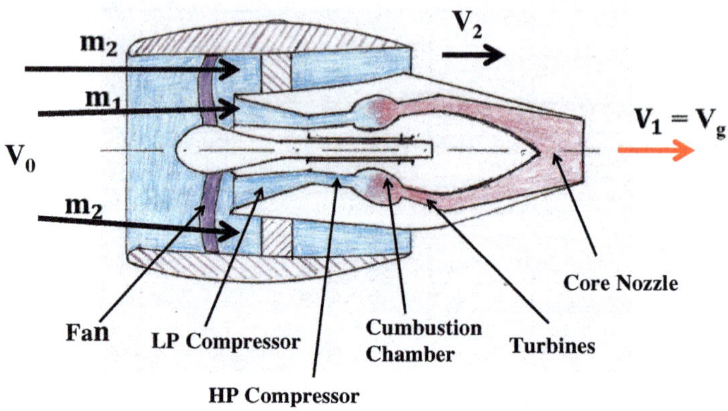

Fig. 5.24 Schematic representation of a turbofan engine

The thrust—in an idealized form—can be expressed by

$$T_i = \dot{m}(V_i - V_0) \tag{5.14}$$

where V_i represents the exhaust speed of the air flow and
V_0 the entrance speed of the air flow $\dot{m}$
$dm/dt = \dot{m}$ represents the mass flow (the inner or outer)

So the total thrust as sum of inner and outer thrust (hot and cold flow) can be defined as:

$$T = \dot{m}_1(V_1 - V_0) + \dot{m}_2(V_2 - V_0) \tag{5.15}$$

So the engine thrust can either be achieved by increasing the mass flow, increasing the difference between exhaust and entrance speed or increasing the bypass ratio.

The ratio between the cold outside flow (fan flow) and the inner hot flow (gas generator) is called the bypass ratio:

$$\text{Bypass} - \text{Ratio BPR} = \frac{\dot{m}f}{\dot{m}g} = \frac{\dot{m}_1}{\dot{m}_2} \tag{5.16}$$

The jet engine development is still continuing at very high and complex level. The main routes for further improvement is the increase in pressure ratio, the increase in temperature ratio and the increase in bypass ratio. Temperature and pressure increases are the themes for engine specialists. The bypass ratio is an interesting engine design parameter, which has been constantly increased over the last 50 years as shown in Fig. 5.25.

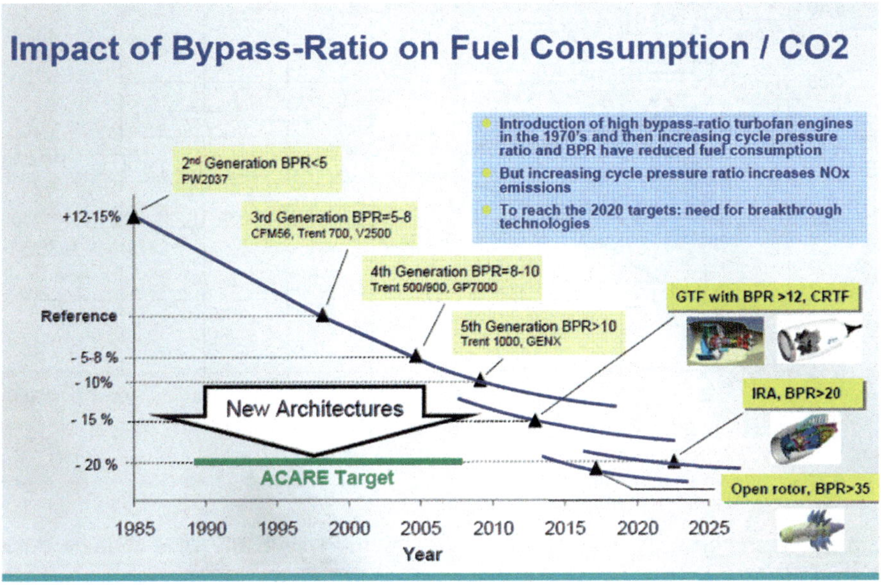

Fig. 5.25 Influence bypass ratio (BPR) versus fuel burn (SFC) (*Source* Rolls-Royce)

The efficiency of a modern jet engine is measured by its "specific fuel consumption (sfc)". The unit is given in kg fuel/N thrust and per time (sec). The smaller the sfc value, the better the engine efficiency. Figure 5.25 is showing the historical development of technology, where a constant increase in engine efficiency (sfc) has been obtained over the last 50 years by increasing the bypass ratio from 1 to about 10.

Today, there is a very detailed discussion between all stakeholders, which technology trend should be further developed, either to continue the fuel reduction by a further increase in BPR or whether a real step change should be made by taking off the engine cowling and provide a modern high-speed propeller engine. The latter will have the best fuel consumption, but may also have some problems with noise issues, as the shielding effect from the cowling will no longer be available.

The pressure for more green aircraft concepts will push the engine and aircraft manufacturers to prepare the next big step in engine design and prepare the technology for a new generation of Open Rotor engines (single or contra-rotating)

A lot of research is done at the moment in the USA and in Europe in order to investigate and prepare the "Open Rotor" technology in two different concepts (Single open Rotor concept with 1 propeller, versus 2-Contra-rotating propeller concept, where the 2 propellers are rotating in a different sense in order to reduce the wake vortex and the swirl effect behind the propeller. The contra-rotating Open Rotor is a top technology item today in all R&T research portfolios and if this technology will have achieved a certain maturity, the Open Rotor Concept may be

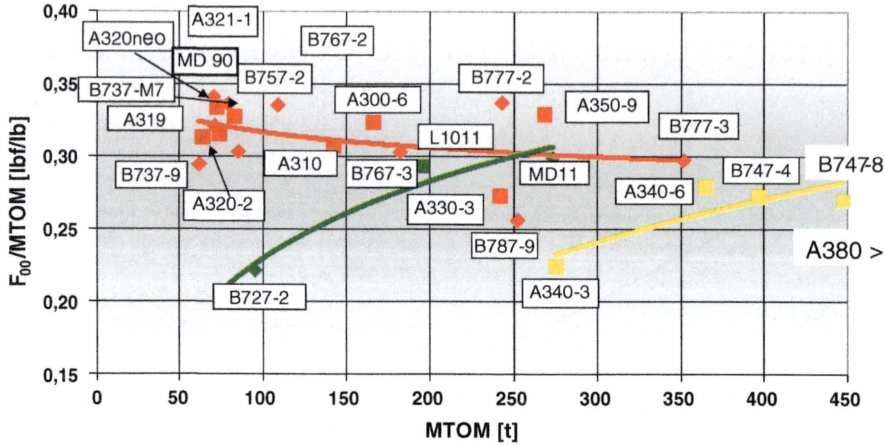

Fig. 5.26 Thrust to weight data of flying aircraft

introduced in the next generation of aircraft, most probably in a "Single Aisle Aircraft"!

An interesting design parameter in the aircraft design process is the thrust to weight ratio. This ratio is expressed as the engine thrust installed (T_{∞}) at sea level and at full power setting related to the maximum take-off mass MTOM. Figure 5.26 summarizes actual data from built and delivered aircraft. It can be seen, that most of the actual designed aircraft have a Thrust to Mass ratio of ~0.3. This means that 30 % of the MTOM has to be installed as thrust force (first estimate!) to be able to take-off in critical weather conditions.

Figure 5.27 shows that for a given aircraft design and a given speed range, the thrust is always to be higher compared to the total drag situation. But with decreasing speed, the Drag is increasing (Fig. 5.24) and there is not sufficient thrust available to equilibrate the aircraft. A similar situation is happening at higher speeds, where the skin friction drag is further increasing so that the available thrust cannot compensate and no steady speed flight is possible.

Fig. 5.27 Relation of thrust to drag in different altitudes

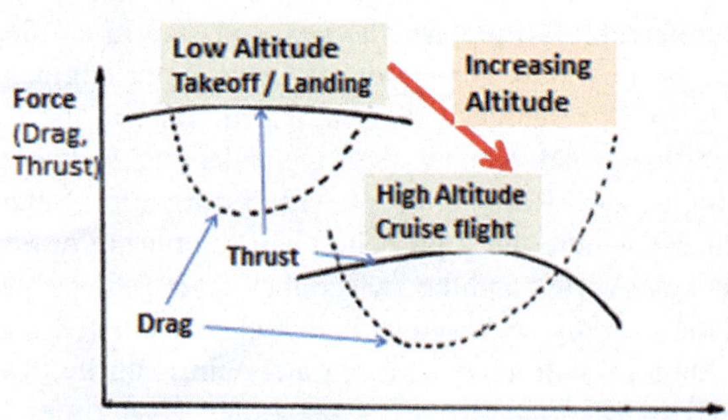

There is a surprisingly good arrangement that the thrust in altitude is far less than the thrust at low level conditions. But fortunately, the aerodynamic drag is also decreasing with altitude (The air density is decreasing with altitude! (Figure 5.16) so that a good match of thrust and drag versus altitude and speed is a main characteristic of jet aircraft design.

5.3.8 *Aircraft Stability and Control*

Figure 5.15 is showing the four main forces Lift, Drag, Weight and Thrust and the main characteristics of these forces have been explained in a simplified manner before. The aircraft has in general 6 degrees of freedom, 3 axial degrees and 3 rotational degrees (Fig. 5.28). In the chapter before, the forces in longitudinal (thrust and drag) and vertical axes (Lift and weight) have been described. The forces along the lateral axes are needed for control of the aircraft. In normal flight no lateral force should be acting!

In addition to the movement along the three axes, there is also the possibility to use the rotational degrees of freedom and turn the aircraft around all 3 axes. The movement around the lateral axis is called the "pitch movement", necessary for take-off and landing, when the aircraft needs to be rotated around the lateral axis.

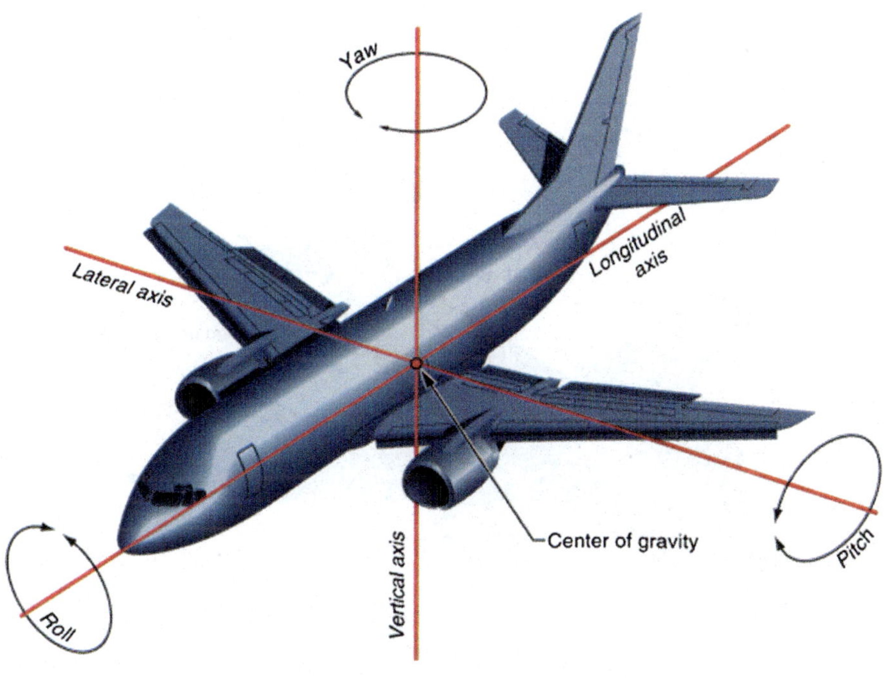

Fig. 5.28 Six degrees of aircraft movement

The rotation around the longitudinal axis is called the roll movement. Roll movements are needed to change flight directions and start a roll maneuver.

The rotation around the vertical axis is called "Yaw movement". This movement is also used—in combination with the roll movement—to change direction of flight in all altitudes. The specific yaw movement and control of the aircraft is needed, for example for landing, especially when there is a strong crosswind during landing and the aircraft has to be turned shortly before landing from the wind axis into the runway direction.

In order to produce control forces, the aircraft needs specific control surfaces, which—when deployed during flight—will provide a lift force at the control surface and this force will then act as a force or moment around the centre of gravity of the aircraft and allows the aircraft to be maneuvered in the air space.

Figure 5.29 shows the classical control surfaces of civil transport aircraft. It can be differentiated between primary and secondary control surfaces.

Primary control surfaces are the elevator for pitch control, the ailerons for roll control and the rudder for yaw control.

Pitch control is provided by the elevator (two control surfaces at the horizontal tailplane). Roll control is provided by the ailerons at the tip of each wing. Both ailerons are deflected always asymmetrically. To enter into a right turn, the left aileron has to be deflected downwards in order to increase the lift at the left wingtip. The right aileron is deflected upwards in order to reduce the lift at the right wingtip. This will increase the lift of the left wing and decrease the lift of the right wing,

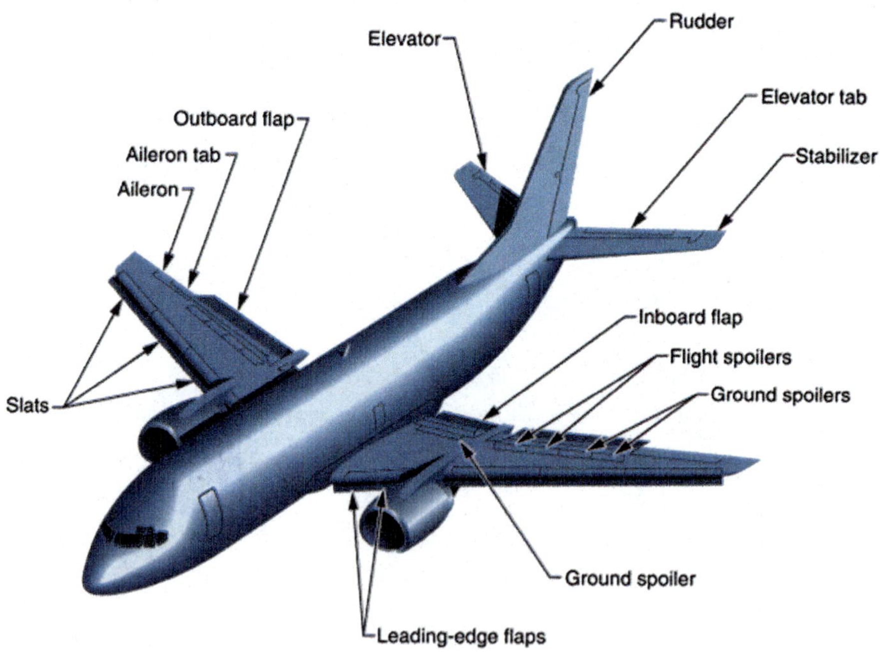

Fig. 5.29 Primary and secondary control surfaces for aircraft control

therefore introducing a right turn of the aircraft. Normally these aileron deflections are supported by a rudder input to compensate for the negative yawing moment which is produced by the slightly higher drag of the left wing due to the increased lift on this wing side. But details about the flight mechanics and flight control aspects can be found in all relevant flight mechanics and aircraft design books [24–19]. Secondary flight controls are the

- high-lift surfaces, i.e. slats and flaps
- the horizontal tailplane, which is movable and used for trimming the aircraft
- the airbrakes and spoilers.

Slats and flaps are needed—as already shown in Sect. 5.3.3—to increase the lift of the aircraft for take-off and landing. The aerodynamic characteristics of flaps and slats have been briefly described in Fig. 5.17, the operational aspect is shown when explaining the flight envelope in Sect. 5.5.1.

The spoilers are movable surfaces on the upper side of the wing and can be deployed upwards, which then will reduce the lift of this wing area. By deflecting all spoilers symmetrically, the aircraft lift will be reduced, the aircraft drag will be increased and the aircraft can be decelerated. This function of the spoilers is called the "airbrake" function, either in cruise, when there is a need for a rapid descend after a loss of cabin pressure or on ground, when the aircraft wheels have just touched ground and the airbrakes are used to support the braking function of the aircraft.

Asymmetry of the aircraft is appearing when for example an engine failure occurs and the asymmetric thrust of the aircraft needs to be compensated by a yawing moment of opposite direction. This is then provided by the full deflection of the rudder. This one-engine inoperative case is a dimensioning fact for the sizing of the fin and the rudder (see [5–9]).

5.4 Structure, Mass and Balance

5.4.1 Structural Components

Each aircraft structure consists of six major elements (see Fig. 5.30):

- The wing for generating the lift
- The fuselage to integrate the payload (passengers and cargo)
- The tailplane to control the aircraft during all flight phases
- The engines to provide sufficient thrust during all flight phases
- The undercarriage, to allow the aircraft to taxi, take-off and land on ground
- The cockpit to provide the pilot with all necessary data and allow the control of the aircraft, which is normally put in the front fuselage to provide sufficient pilot view.

In addition, a lot of support systems are needed to control the aircraft, keep the passengers in a comfortable environment (Sect. 5.2.5) and provide the necessary functionality during all flight phases.

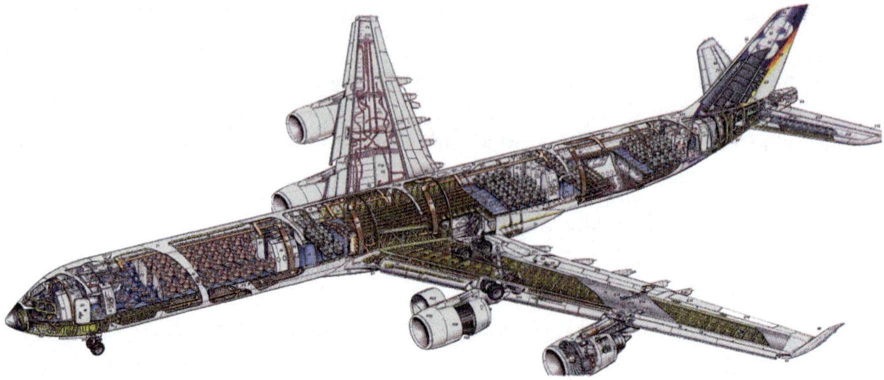

Fig. 5.30 Structural aircraft layout

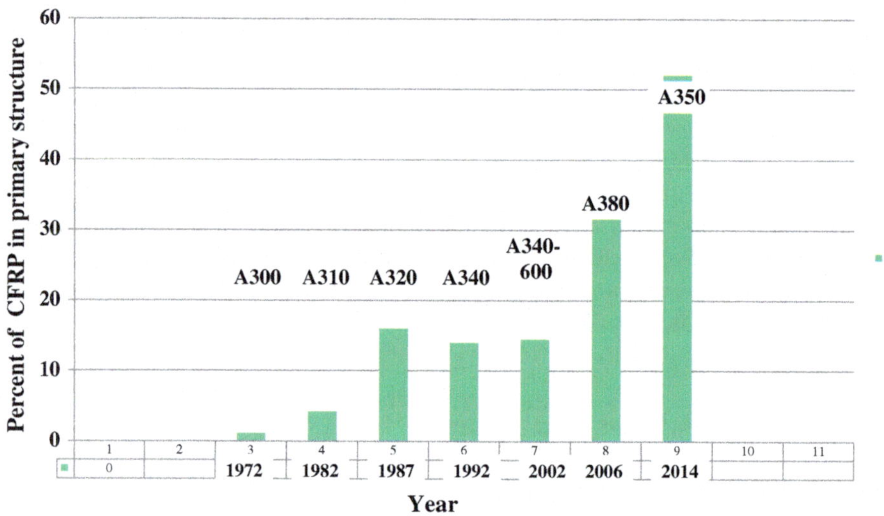

Fig. 5.31 Evolution of composite structure due to time

Light weight structures are an important element in the aircraft design. As has been indicated in the Lift chapter, the less aircraft mass will be needed, the less lift is necessary and less drag will be generated, reducing the engine thrust and thus improving the fuel burn.

Light weight design is a basic discipline which has been developed specifically for the aircraft design. Further details can be found in the literature [5, 8, 9].

Specific lightweight materials have been developed over the last 50 years of commercial transport aircraft design. The basic primary structure consists of aluminum alloys, where different and specific alloys are used for wing surfaces, fuselage primary structure and tailplane structures. During the last 30 years tremendous efforts have been undertaken to reduce the aircraft weight by use of new materials besides the classical aluminum alloys. Very promising classes of material are Carbon Fiber Reinforced Plastics (CFRP), which are not only used in the secondary structure but also in the primary structure. The amount of CFRP in the aircraft structure has increased from 5 % in 1985 (A310) till today (B787 and A350) to over 50 % (Fig. 5.31). However, there are also major risks involved to prepare the CFRP technology to such a mature level that the automated manufacturing will provide all the cost benefits expected. The tremendous delay in the certification of the B787 (3 years) is mainly due to the production refinement with CFRP material for fuselage and wing construction.

5.4.2 Mass Breakdown

The aircraft weight consists of mainly four major elements as outlined in Sect. 5.3.6.

Figure 5.22 shows a graph with all major civil aircraft flying today. The statistics identify clearly that for short range aircraft (Range up to 3000 nm) 60 % of the weight consists of the aircraft empty mass (MME). 20 % is linked to the payload and the other 20 % are fuel. For the long-range aircraft, fuel weight increases up to 30 % and the relative part of MME reduces to 50 %. These are just weight proportions. It is clear that the total aircraft weight increases considerably for long-range aircraft. 30 % of fuel for a long-range aircraft like A380 means that there are nearly 200 t of fuel possible.

5.4.3 Payload—Range Diagram

One of the most important aircraft characteristics is the payload—range diagram. It describes the capability of an aircraft, which payload it can transport across which range. It is a basic aircraft design parameter, and is fixed—out of market studies—at the beginning of an aircraft programme. The wing size is an important factor, as the wing box is usually the natural fuel reservoir. The bigger the wing the more fuel volume can be stored. As shown in Fig. 6.3, the bigger aircraft have naturally more range due to their larger wings. As mentioned in the Sect. 5.3.3 (aerodynamic lift), the lift capability increases with the wing surface. The fuel volume, however, increases with $S^{3/2,}$ which allows to store in big wings disproportionally more fuel. The other effect visible today, is the further development of the engine technology. With increasing fuel efficiency of modern engines, there is also a tendency that the aircraft range is increasing. A good example is the A320 neo, which has as main effect besides the improved fuel efficiency (~ -15 %), also a range increase from about 15 % or 400 nm range.

The typical payload—range diagram has three characteristic borders: There is a maximum payload border, which is defined by the aircraft structural design. With all the fuel, which can be put as a delta between MTOM and MZFM, the aircraft can fly a certain range. For safety reasons, the aircraft has always to load a defined quantity of reserve fuel (Sect. 5.4), which in all normal situations has still to be in the fuel tank when landing. So for the useable fuel the aircraft can fly with max. payload a range, defined as "Max. payload range". But all aircraft normally are designed in a way, that full payload and maximum fuel volume are exceeding the certified MTOM. So with full payload the aircraft cannot use all the fuel volume capability. But the aircraft operator can choose to either fly the aircraft with full payload over a shorter distance or with less payload over a longer distance. Figure 5.32 shows the second border—named the MTOM limit in the payload range diagram—where a longer range can be flown with reduced payload. This goes up to a limit, where the full fuel volume will be used and only a limited payload can be transported. The aircraft can still be operated for higher ranges, by further reducing the payload. As there is less lift needed, the induced drag is also reduced, less engine thrust needed, which will reduce fuel consumption and will further increase the aircraft range. This third border of the payload range aircraft is called the "maximum fuel limit". This does not make real sense for commercial flights, but for specific events—a flight with just some journalists for a specific long range mission or for a ferry or transfer flight of an aircraft—the aircraft can fly with no payload still further. The additional range is relatively small and is achieved as the aircraft mass will then be OME + M_{fuel}, which is less than MTOM.

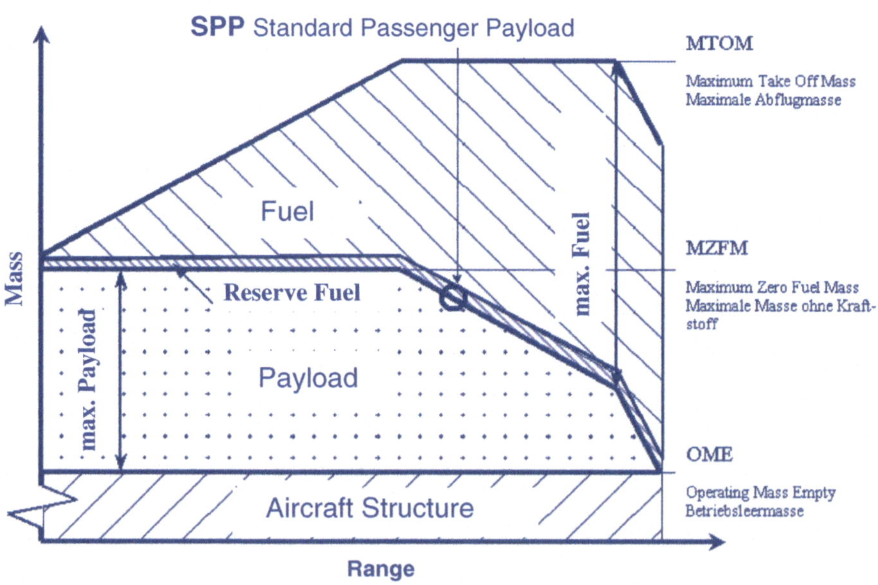

Fig. 5.32 Typical payload—range diagram for an aircraft

The payload range diagram is defined for each transport aircraft. It is defined without wind. For a realistic flight, the pilot or the airline will calculate the required fuel by defining the flight trajectory, defining the available payload, using actual wind conditions, defining some safety margins and using the performance data of the aircraft. In former times these calculations had to be done by the pilot prior to each flight. Today specific computer programmes exist which help the pilot to do this calculation.

5.4.4 Weight and Balance

Another important feature for the aircraft operation is the calculation of the *Centre of gravity* (*CG*) for each flight mission. For each aircraft, the CG boundaries are defined. There exists a limit for the rear CG location, which is called the stability limit. This rear CG position is close or slightly before the "Neutral point" of the aircraft. ("neutral point" is explained as point in the aircraft centre line, where the aircraft lift is acting as integral force). More details are given in all aerodynamic or flight-mechanics literature [5, 8, 9].

The forward CG limit is defined by the controllability of the aircraft. The CG boundaries of an aircraft are fixed so that there is sufficient CG margin for all reasonable loading cases for an aircraft. But if these are chosen very widely, this will lead to bigger control surfaces and reduce the aircraft performance parameters. So again a compromise between good overall performance and sufficient and reasonable flexibility for operational loading with also some restrictions has to be defined and accepted.

For some special loading case—passengers mainly in the rear cabin (no business class passengers, but economy class is full!) all cargo also stored in the back—the aircraft overall CG position may be pushed so far back, that this could lead to a very rear loaded aircraft, which may cause violations of the boundaries for loading and will not be allowed. But there is an easy solution to first fill the forward cargo hold and put the rest of the cargo in the rear cargo part. As can be seen, the CG loading capabilities allow fairly flexible solutions. But there are always some cases—full front or full rear loading, where the critical limits can be achieved and restrictions may arise.

5.5 Flight Performance and Mission

5.5.1 Flight Envelope

The flight envelope is a diagram, which defines the flying envelope of the aircraft with respect to speed and altitude. The basic information here is that the aircraft should never fly too slowly or too fast. The proper speed as a function of altitude is essential.

There is a lower limit in the flight envelope (the left side), which is defining the aerodynamic limit. If the aircraft has not enough speed (Sect. 5.3.3) there will be not enough lift to keep the aircraft weight in balance for flying. This aerodynamic limit is also linked to the stall characteristics of the wing, where the wing cannot generate more lift even with higher angle of attack. This aerodynamic limit is a clear border not to fly too slowly and enter into a dangerous situation.

The other extreme is the right border of the flight envelope. Here the aircraft is not allowed to fly faster or increase the maximum speed for a given altitude, as the aircraft structure will reach its design limits. The right hand side has two different limitations, but they are fairly similar in their importance. At lower flight levels (in Fig. 5.33 up to 25,000 ft altitude) there is a speed limit defined by V_{TAS} (true air speed) expressed in knots (kts). At higher altitudes the speed limit is expressed as "maximum Mach number" which never should be exceeded.

There is also a design limit in altitude for each aircraft. This can be either an engine limit or it can be defined by the internal cabin pressure. As shown in Sect. 5.2.5, the cabin pressure cannot be less than 8000 ft of ISA . In the aircraft design a maximum altitude has to be fixed similar like the minimum cabin pressure. The delta between the maximum altitude and the minimum cabin pressure is an important design parameter during the aircraft optimization process.

The flight envelope is defining the operating limits in altitude and speed, where the aircraft can be safely operated.

Figure 5.33 is also showing an area (rose colour), where the high-lift system can be operated. Without any high-lift devices the aircraft could only land at airspeeds

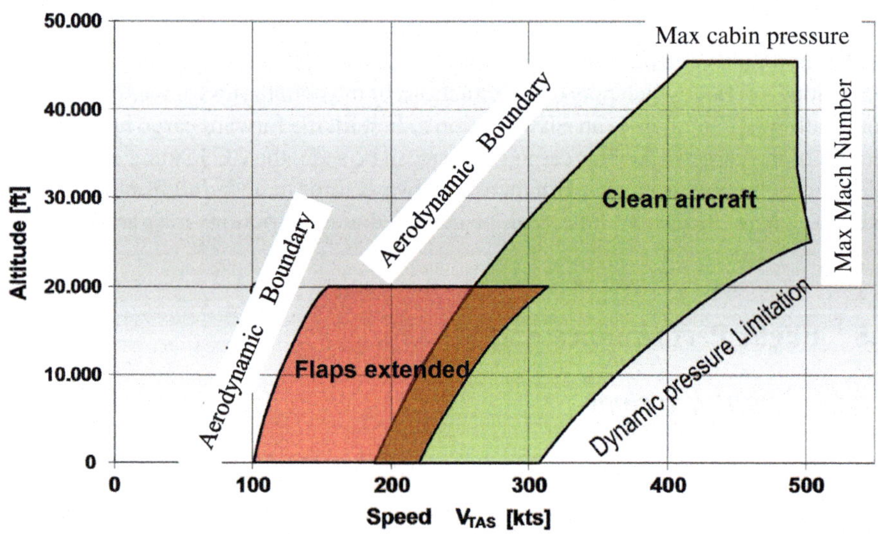

Fig. 5.33 Flight envelope of a transport aircraft

of 200 kts and more, where it is very difficult for the pilots to handle the landing process. With the flaps and slats operative the aircraft can take-off and land at nearly half the speeds compared to the clean aircraft configuration. The operation of flap setting or retracting is therefore limited to a small corridor of speed and altitude and clear procedures are defined for each aircraft, at which speeds and altitudes these transitions have to be done. These data are given in the Flight Crew Operating Manual (FCOM).

5.5.2 *Definition of Speed*

The definition of speed is slightly complex, as there are very different speeds for the aircraft operation. The aircraft forces lift and drag and all the aerodynamic characteristics are very much depending on the speed relative to the vehicle, the airspeed is the important speed for the aircraft operation. It is therefore mandatory to have a good indication of the *True airspeed* (TAS).

This is different for the passenger, who is interested in the ground speed, the aircraft speed relative to the earth. The passenger wants to travel from an origin A to a destination B, independent from any wind and weather turbulences. The flight plan is giving him a departure time and a time of arrival and he expects that the operator—the airline—will manage to bring him in time to his destination!

For the correct and safe aircraft operation, a very good and redundant airspeed indicating system is mandatory.

The relation of true air speed TAS, wind speed and ground speed is given in Fig. 5.34.

The airspeed sensing on-board of an aircraft is done by a pitot system, which has the capability to measure the static pressure and the total pressure and by providing this information to the air speed indicator, the airspeed can then be calculated.

The barometric measurement of

The measurement of speed is done by the aircraft systems in particular by a pitot tube, fixed on the aircraft which is sensing the oncoming flow and its total (stagnation) pressure. In parallel, the static pressure is measured at the static board and the difference is giving the measured "Indicated Airspeed" named IAS. The directly

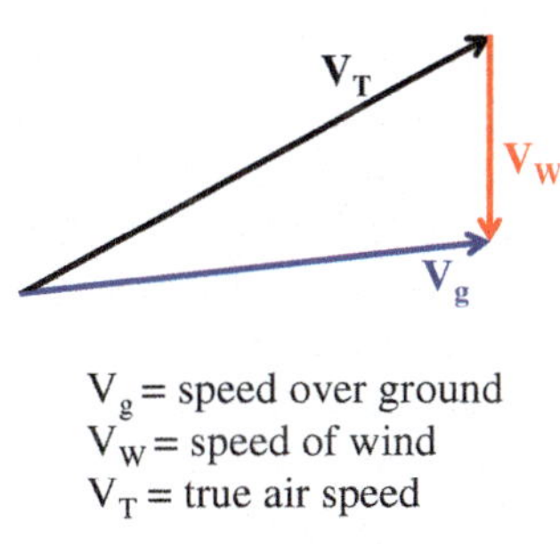

Fig. 5.34 Wind and flight direction

measured airspeed needs some corrections before it is accurate enough to be used in the aircraft control systems. The pitot tube is normally too closely linked to the front fuselage of an aircraft and a correction factor is needed to take into account the ratio of measured speed at the pitot tube compared to the free stream speed. This correction is normally not very large, as the aircraft manufacturer tries to place the pitot tube at a position which requires only small installation corrections. The other corrections needed to achieve the true airspeed or the speed overground are the following:

- Indicated Airspeed IAS, directly measured airspeed
- Calibrated Airspeed CAS, i.e. IAS corrected for compressibility effects
- Equivalent airspeed EAS, i.e. CAS corrected for pressure at sea level
- True airspeed TAS, i.e. the reference for a safe operation
- Ground speed GS, i.e. correct the TAS with the wind effects

TAS is the final true airspeed, used in all performance calculation and all flight planning and mission monitoring exercises.

More details about the formulas to calculate all the speeds are given in [5–9].

Another method of obtaining speed and position are GPS sensors and the Inertial Navigation System (INS). The INS system measures all accelerations and decelerations of the aircraft using gyroscopes and linear accelerometers. This information can then be integrated in time to obtain speed and position. Today's INS or IRS (Inertial Reference Systems) are very precise and have only small deviations over a relatively long period (i.e. 1 nm for 1 h of flight) as long as the INS was properly calibrated before departure. INS is present in civil aviation since some decades and is mostly used in nearly all transport aircraft with more than 50 passengers. It is an additional speed reference system which can help to stabilize the barometric and GPS systems. (see also Chap. 9)

5.5.3 Flight Mission

The flight mission of an aircraft contains all steps from leaving the gate at the airport of departure till arriving at the gate of the destination airport. Figure 5.35 shows such a typical mission profile and all the normal steps in such an aircraft mission.

For each of these phases, it is possible to define the given speed, which allows deriving the time at the end of this phase and the distance travelled in this phase. In addition the necessary thrust setting can be derived from the aircraft performance handbook and with the thrust setting, the fuel flow for each element can be calculated. As shown on top of Fig. 5.35, there are time, distance and fuel consumption for each phase calculated and the sum leads to the total distance, total mission time (block time) and the block fuel needed for this mission.

All calculation is done for a situation without wind. Wind can be favourable (tail wind) or unfavourable as head wind and may change these given numbers. For each

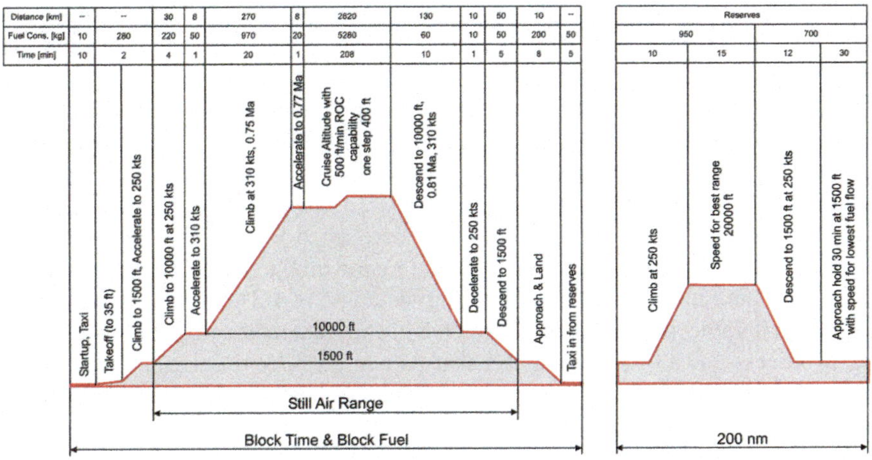

Fig. 5.35 Typical mission profile of a civil transport aircraft

realistic mission, the pilot will calculate these data, using the actual meteorological forecast. He also has to add the reserve fuel as shown in Fig. 5.35 right side. Reserve fuel has to be calculated for the unexpected situation, that the runway of the destination airport is blocked and the aircraft has to divert to the next other airport in the vicinity. In the example, a distance of 200 nm has been assumed. But for each real flight planning the nearest alternate airport has to be chosen and the calculated fuel has to be taken on board in addition. This should ensure that the aircraft can always land safely. The pilot may also add some fuel margin for expected holding patterns, depending on the normal situation of the destination airport.

There is a classical conflict between the pilot—who has the sole and unique authority for the fuel management—and the financial controllers of the airlines, who will push the pilots to use a minimum reserve policy, so that the additional fuel will not be excessive and lead to higher consumption during the whole mission.

Safety however has to be the overruling argument!

5.5.4 *Take-off and Landing*

Take-off and landing are the two most critical elements in the aircraft operation. Here most of the accidents are happening (see Chap. 4) and in these phases the whole attention of the pilot is needed.

The aircraft manufacturer are providing in their FCOM and performance manuals all necessary data to calculate take-off and landing distances. There are several rules defined in the certification regulations from FAR and JAR, how these minimum take-off and landing distances have to be calculated. These calculation rules can also be found in [10, 11] and are too detailed to be discussed here. Some rules and some basics, important to understand the safety aspects, should however be shown:

The aircraft could have an engine failure during the take-off acceleration. If this engine failure occurs before a critical speed V_1, the pilot has to decelerate and stop the aircraft on the remaining length of the runway. The breaking power has to be sufficient to do so and this has to be demonstrated during the certification process. In case, the engine failure occurs after the critical speed V_1, the pilot has to continue to take-off and the full thrust of the remaining good engine(s) has to be sufficient to take off the aircraft, climb at a minimum glide angle of 1.2° and return back to the airport for landing if needed. Also this has to be demonstrated during the certification process and this is also a constant part in all pilot training exercises. This process requires also a sufficient speed of the aircraft called V_{MC}, to ensure that the aerodynamic force from the rudder with full deflection is sufficient to control the aircraft with only one engine running at full thrust.

When the altitude of 1500 ft is achieved, the engine thrust will be reduced from max. Take-off setting to max Continuous thrust, the flaps will be retracted and the "clean" cruise configuration for the further climb phase will be cleared.

Today most of the normal airports are providing runways with sufficient length (3000–4000 m), where even the very big aircraft can be safely operated with full payload and MTOM. But there are still a lot of smaller and specific City airports where there are several constraints for aircraft to take off with full payload.

Critical are also airports, which are situated in regions with a hot climate and at high altitudes. These airports—known as "hot and high"—may also cause restrictions for the airlines, as the engines will not provide full take-off thrust under these conditions and this may lead to restrictions in the maximum payload or in the fuel load needed for long ranges.

5.5.5 *Cruise Performance*

The important parameter at the end is the cruise performance of an aircraft.

The airline is looking at the end, how much positive income can be generated with a certain type of aircraft in a particular network, which the airline is serving, during a certain period (week, month or year!).

It is difficult, to generalize these questions and give a final formula, how to manage this question and provide a final answer! The airline has a certain strategy, has a given network of operation and has a plan, which aircraft type is needed to serve this network best. But it is not only one dedicated network, each aircraft is an element in the overall fleet planning, maintenance aspects have to be considered, a standardization of the fleet is an important element and a single optimization will not lead to an overall benefit.

5.5.5.1 Specific Air Range

An important measure for the comparison of different aircraft in their performance is the specific air range. Specific air range (SAR or SR) is defined in nm/kg fuel. It is a measure to compare cruise performances of different aircraft types and for different aircraft flight conditions. SAR specifies how many nm the aircraft can fly with 1 kg of fuel.

$$\mathrm{SR} = \frac{V}{\mathrm{sfc} \cdot T} \tag{5.17}$$

where
sr defines the specific range (nm/kg of fuel burnt)
T defines the thrust level
sfc defines the specific fuel consumption (given by the engine manufacturers)
V defines the actual cruise speed

5.5.5.2 Breguet Formula

The Breguet formula is very often used in aircraft design and aircraft performance comparison. It is a simple equation, which can easily be derived in the following steps:

$$\bar{\mathrm{d}}m_{\mathrm{fuel}} = \mathrm{sfc} * T * \mathrm{d}t \qquad v = \frac{\mathrm{d}x}{\mathrm{d}t} \Rightarrow \mathrm{d}t = \frac{1}{v} \cdot \mathrm{d}x \tag{5.18}$$

$$\mathrm{d}m_{\mathrm{fuel}} = \mathrm{sfc}\frac{m * g}{L/D} * \frac{1}{V} * \mathrm{d}x \; \frac{L}{D} = \frac{1}{D/L} D = \frac{m * g}{L/D} = T \tag{5.19}$$

$$\mathrm{d}x = \frac{L/D}{\mathrm{sfc}} * \frac{V}{g} * m * \mathrm{d}m_{\mathrm{fuel}} \tag{5.20}$$

Starting from the sfc as Δ fuel flow per Δ time, multiplied with the actual thrust T;
expressing the term thrust T as a function of mass times g divided by L/D and introducing the term dt by dx/V leads to Eq. 5.20

The integration of the term dx leads to Eq. 5.21 where the distance R is the aircraft range!

With the following parameters:
R aircraft range
M aircraft mass in kg
m_1 aircraft mass at the beginning of the cruise flight
$m2$ aircraft mass at the end of the cruise flight

m_{fuel} mass of fuel used in cruise in kg
V aircraft cruise speed
L/D Lift to Drag ratio (see Fig. 5.20)
sfc specific fuel consumption.

The Breguet Formula is defined in 5.21:

$$R = -\frac{L/D \cdot V}{\text{sfc} \cdot g} \cdot \ln\left(1 - \frac{m_{\text{Fuel}}}{m_1}\right) \quad (5.21)$$

The Breguet formula allows to calculate the range of an aircraft (neglecting take-off, climb, descend and landing) by just concentrating on the cruise phase. It shows the three main parameters, which are important for achieving an excellent long range aircraft:

1. The aerodynamic efficiency L/D
2. The engine characteristics culminated in the sfc (specific fuel consumption)
3. The structural efficiency, expressed in the ln—function with mass elements

To achieve the best aircraft range,

- the aerodynamic efficiency L/D has to be very high
- the sfc of the engine has to be very low
- the aircraft structure (OME or MME) have to be very small compared to the payload and fuel part.

This looks very simple, but these are the constant challenges in the aircraft design optimization process!
The Breguet formula is unique as it shows the main driving technologies to further improve the aircraft efficiency. As each new aircraft on the market has to be better in the fuel consumption compared to the old designs, it is very clear, where the improvement has to come from:

1. wingtip devices, better aspect ratio or other aerodynamic features like the aerodynamic efficiency L/D has to be improved (either by new boundary layer control
2. the engine fuel consumption has to be further improved (higher bypass ratio, better thermal or pressure efficiency in any component, etc.
3. the structural efficiency has to be improved by use of new materials (CFRP instead of metal alloys in the primary structure)

The Breguet formula is just referring to technical terms and does not include any cost aspects, which could be generated by new production technologies!

References

1. EUROMART Study Report: Executive Summary (1988)
2. Oborne, D., Levis, J.: Human Factors in Transport Research. Academic Press, New York (1980). ISBN 0125238010
3. Wittmann, R.: Methodology for the evaluation of aircraft cabins, DGLR JT—265, Friedrichshafen (2005)
4. Hiesener, S.: Cabin design, ECATA ABI-course (2003)
5. Torenbeek, E.: Synthesis of Subsonic Aircraft Design. Delft University Press, Amsterdam (1986)
6. Torenbeek, E.: Advanced Aircraft Design. Wiley, New York (2013). ISBN 978 111856811 8
7. Obert, E. (ed.): Aerodynamic Design of Transport Aircraft. Delft University of Technology, Delft; IOS Press, Amsterdam (2009)
8. Fielding, J.: Introduction to Aircraft design. Cambridge University Press, Cambridge (1999)
9. Roskam, J.: Airplane Design, vol. 1– 8. DARcorporation, Lawrence
10. Joint Airworthiness Requirements JAR, Ch. 25: http://www.easa.eu.int/certification/. Accessed 1 Dec 2014
11. FAR regulations for aircraft certification
12. Committee on Airliner Cabin Air Quality: The Airliner Cabin Environment—Air Quality and Safety. National Academic Press, Washington (1986)
13. Cummin, A.C.R., Nicholson, A.N.: Aviation Medicine and the Airliner Passenger. Arnold, London (2002)
14. Mayer, E., et al.: The new pressurised fraunhofer flight test facility offered to the scientific cabin environment network. In: CEAS-2007-468, 1st CEAS conference, Berlin (2007)
15. Grün, G., et al.: Impact of Cabin Pressure on Aspects of the Well-being of Aircraft Passengers, ICAS 2008-6.3.2, ICAS, Anchorage (2008)
16. Diehl Aerosystems: Cabin lighting systems. http://www.diehl.com/en/diehl-aerosystems/aircraft-systems/interior-lighting-systems.html. Accessed 1 Dec 2014
17. Schmitt, D.: Integrated Design—Passenger & Payload. ECATA ABI-course (2008)
18. Isermann U., Schmid R.: Bewertung und Berechnung von Fluglärm, DLR Forschungsbericht 2000–20 (2000). ISSN 1434-8454 (in German, but excellent!)
19. Torenbeek & Wittenberg: Flight Physics, Springer edition. (2009). ISBN 978-1-4020-8663-2
20. Jenkinson, L.R., Simpkin, P.: Civil Jet Aircraft Design. ISBN 978-0340741528
21. International Organization for Standardization: Standard Atmosphere, ISO 2533:1975. http://www.iso.org/iso/catalogue_detail?csnumber=7472. Accessed 1 Dec 2014
22. Mattingly, J. et al.: Aircraft Engine Design. AIAA Education Series (2002). ISBN: 1-56347-538-3
23. Bose, T.: Airbreathing Propulsion. Springer (2012). ISBN 978-1-4614-3532-7
24. Perera, E.: Innovative approach to improve air quality in aircraft cabins. In: Paper presented in session E3 at EC Aeronautics Days 2001, Hamburg (2001)
25. Frost & Sullivan: Aircraft Electrical Power Systems—Charged with Opportunities. IAG. www.iag-inc.com/articles/aeps.pdf. Accessed June 2012
26. Grün, G. et al.: Interrelations of comfort parameters in a simulated aircraft cabin. In: 11th conference on indoor air quality and climate, Copenhagen (2008)
27. Hiesener, S.: Cabin Design. ECATA ABI Course (2003)
28. Marenco, A., et al.: Measurement of ozone and water vapor by Airbus in-service aircraft; The MOZAIC airborne program, an overview. J. Geophys. Res. **103**, D19, 25631–25642 (1998)
29. Vincendon, M.: TANGO: Low Cost Light Weight Structure, vol. 3, No3/4. Air&Space Europe (2001). ISSN 1290-0958

Chapter 6
Aircraft Manufacturer

Abstract The chapter starts with the history of mergers of aircraft manufacturers in US and Europe, leading to the duopoly of Airbus and Boeing as leading players worldwide. Their product portfolio seems to be fairly identical, which is good for market competition but could also be a sign of reducing innovation and risk. Several smaller aircraft manufacturers are preparing to challenge this duopoly and are preparing their entrance into this jet airliner market. The aircraft development process is characterized, showing the long-term aspect of a new aircraft development program and the involved risks. Cost breakeven will normally not be achieved before 12–15 years. The industrial organization with the role of engine manufacturers, supply chain and the complex work breakdown structure is outlined. The cash flow principle is shown, which helps to understand major risk factors during the development of a new aircraft type. The importance of family and commonality aspects is introduced, leading to major cost savings for the operator.

The role of the aircraft manufacturer and the industrial supply chain is very important in the air transport system. The industry is responsible for the major innovations in the air transport sector, if it is the introduction of new aircraft designs (A380 and B787) or the introduction of new cockpit architectures or navigation systems, which are also interchanging with some other players like the airlines, the air navigation system or the airport infrastructure. New aircraft concepts, new aircraft cockpit architectures, new navigation systems, new engine concepts and new aircraft technologies can only be introduced via the industrial side.

In the past, nearly all innovative features have been pushed by the industry in order to improve their competitive situation with their customers, the airlines. This can be shown by two examples:

1. The introduction of the B747 in the beginning of the 1970s was leading to a major problem at the airports, as they were not properly and early enough informed that a new large aircraft vehicle had to be handled by the big airports. Major modifications at taxiways, gate positions, etc. were necessary to accommodate these new big aircraft types. A new aircraft class for airports had to be introduced (see Chap. 9 airports). To avoid a similar surprise when Airbus

D. Schmitt and V. Gollnick, *Air Transport System*,
DOI 10.1007/978-3-7091-1880-1_6

announced the introduction of a new big aircraft, the A380 [1], the airport association ACI had fixed a new category F with the famous 80m $*$ 80m $*$ 80 ft box, which should not be exceeded by any new aircraft design (details in Chap. 9).

2. Another example is the introduction of the standard glass cockpit and the "fly by wire"—concept for the A320. The real major benefit for all operators has been only realized later, when the big benefit on the cost side became visible (see Sect. 3.5), like reduction of crew cost, training cost, simulator cost, etc. It was not obvious by the view of the airlines that these new cockpit technologies will not have some additional risk and a lot of airline pilots were heavily opposing. The problematic and risk with a new aircraft program will be described in more detail in Sect. 6.3.

6.1 Role of Aircraft Manufacturer

The launch of a new aircraft program like the development of the A380 from Airbus or the B787 from Boeing is a major investment for the aircraft manufacturer. The cost estimation for the development of the A380 have been said in the press to be in the order of 12 billion $ without taking into account the interest for the investment [2]. With interest, Airbus is supposed to have spent for the A380 program in the order of 25 billion $. Even the upgrading of the A320 program with the introduction of new engines and some small airframe adaptations will cost around 4–5 B$!! No normal commercial company can take such a risk, as the return on their financial investment (ROI) will only start to happen after about 15 years. The Cash flow Sect. 6.5 will give some more details about the cost and risk involvement.
There are only two solutions to overcome such a commercially critical situation:

- The national government has to provide a financial guarantee in case of risk
- The risk has to be shared between a lot of other shareholders (engine manufacturers, supply chain, system suppliers, etc.)

In reality, both solutions are normally combined: the national government has to provide some financial guarantees and several risk sharing partners are integrated into the program.

In the airliner market (aircraft with more than 120 seats) there are today only two aircraft manufacturers active and successful, The Boeing Company from the USA and the Airbus SAS Company from Europe. They have about the same market share (50:50) with slight differences in the different market segments. The airlines and leasing companies as main customer are interested to keep the good competition between both manufacturers alive and are expecting from the fierce competition a continuous product improvement.

In the market of regional airliners and aircraft up to 130 passengers, there is starting a new and strong competition between seven companies:

- Bombardier/Canada: CRJ 200/700/900 and the new project CS100 [3]
- Embraer/Bresil: EJ 170 - 195 [4]
- Sukhoi/Russia: Superjet – 100 [5]
- UAC/Russia MS-21 [6]
- Mitsubishi/Japan MRJ [7]
- AVIC/China COMAC [8]
- Alenia + partners GRA [9, 10]

Several large and developing countries are showing a big interest to enter into this market. Aeronautical industry is seen as a strategic industry and it needs a lot of national support to bring the own industry into a position to compete successfully in this civil airliner market.

In this 100-seater market are too many new entrants and there will be a very hard and competitive situation. Some countries will see this design of an 80–100 seater only as the first step and entry card to develop the own industry further into the big airliner market. This is assumed to be the case for China and Russia. Only the future can tell, whether this approach was the right one and will be successful (see also Sect. 3.2). The World Trade Organization (WTO) [11] should take care, that all unfair subsidies in the civil aircraft market are not taking place. However there is a military market also which is not considered by WTO agreements, which puts all WTO decisions in question!

6.1.1 Industry Mergers

During the 1960s and 1970s started a wave of industrial regrouping and concentration in the civil market. The first steps were done in the US where Lockheed in 1981 left the civil aircraft market and McDonald Douglas was integrated into the Boeing Company in 1997, leaving only one company for large civil transport aircraft in the US.

Figure 6.1 shows the integration of the aeronautical industry in Europe within a period of 10 years, in order to achieve a reasonable company size. Most changes were done in two steps, first step to align the national strength by merging the national companies into a single unit (ex. In Germany is the creation of DASA in Germany). In second step, a European integration was started with the merger of Aerospatiale, DASA and CASA into the new EADS consortium. In 2014 the name EADS was changed into "Airbus group", to use the brand of airbus for all other aerospace activities.

It can be seen that there is in Europe a concentration of aerospace companies in France, which is part of the French national industrial strategy. The aerospace industry was always seen in some countries like USA and France as a strategic industry, where besides the military side also the civil side was seen as a strategic complementary part to the military autonomy. This explains that the Headquarters

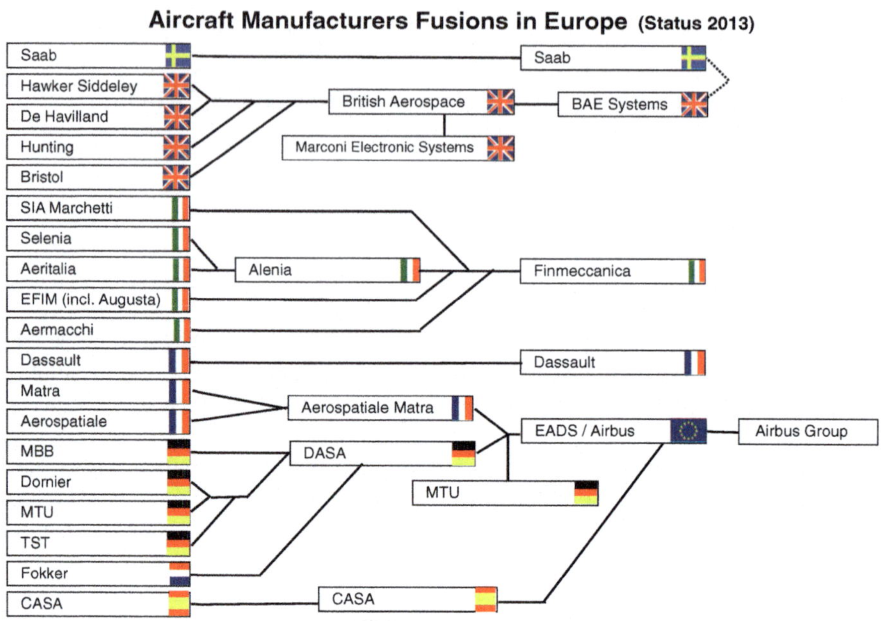

Fig. 6.1 History of aerospace mergers in Europe

for the civil aircraft industry are now in Seattle and in Toulouse. In Europe, there are also big competences and engineering skills in the UK, Germany, Italy and Spain. But Germany, UK and Spain have never seen the aerospace industry as their national strategic priority, but they have seen the need for a strong air transport operational system. Airlines, airports and related operational services are at top level in all bigger European countries.

6.1.2 Market Duopoly "Airbus Versus Boeing"

The civil aircraft market for aircraft bigger than 100 seats is today dominated by two manufacturers: Boeing and Airbus.

They have in total a fairly similar market share, in some sectors a 65–35 % share in others a 35–65 % share, but in total, both are at about equal level and airlines and leasing companies are interested to keep this head-on competition in the magnitude of 50–50 % and keep a strong competition of these two players.

The following Figs. 6.2 and 6.3 are showing the standardization of the aircraft program in Europe. In the year 2005 Airbus had 6 aircraft families in service, production and in development. In 2015, they have harmonized their production lines and are just producing only four aircraft families. But still two programs like A330 and A350 are covering a very similar market segment.

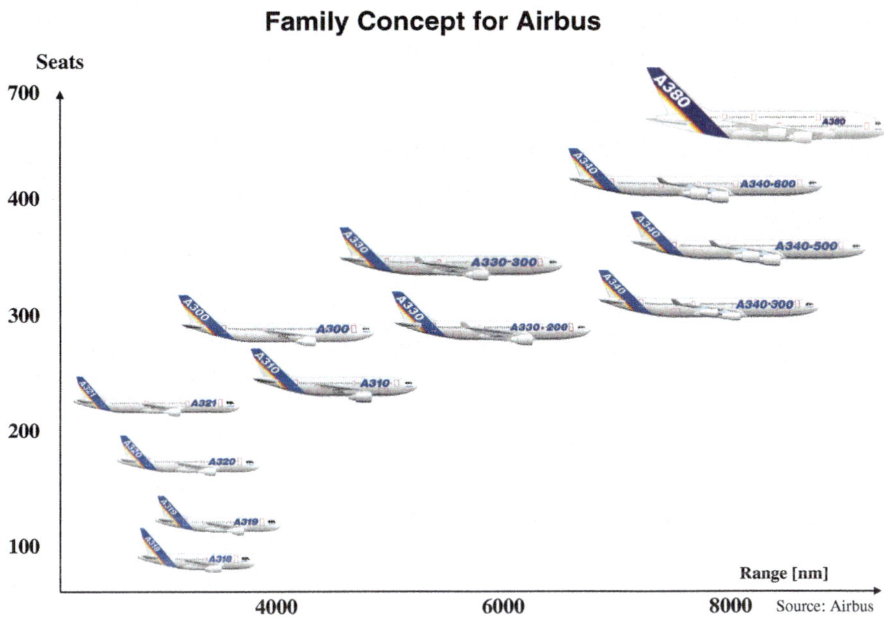

Fig. 6.2 Airbus aircraft family in 2000–2005

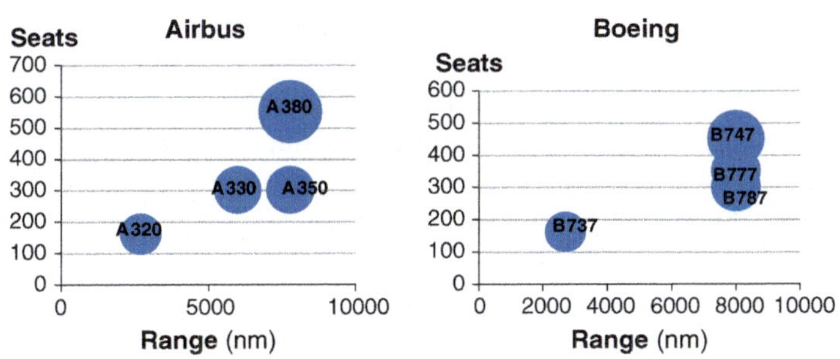

Fig. 6.3 Airbus and Boeing aircraft family in 2015

The aircraft families—as well from Airbus [12] as from Boeing [13]—are covering the whole range of aircraft sizes between 100 and 700 seats and a range of 2500 nm till more than 8000 nm (Fig. 6.3).
Both families show a clear tendency:

- small aircraft are only designed for short range missions
- the bigger aircraft have considerably more range.

This follows a clear engineering paradigm: Small aircraft with a certain technology need an optimum wing size, which allows naturally a fuel volume for normally 3000 nm. Taking the same technology standard the wing size in [m^2] increases

roughly with the number of passengers. The fuel, stored in the inner wing section, is increasing with the additional volume, i.e. with fuel volume $\sim$ (Wing size) $^{3/2}$. This is normally defined as "Square–Cube law", allowing larger aircraft to have more fuel volume and thus more range.
Figure 6.3 leaves two questions open (see also [14]):

a. Is there no market for optimized aircraft with short range, the so-called people mover??
b. Is there no market for small long range aircraft??

These two market segments are for the moment not covered by any optimum aircraft design.

6.1.2.1 The People Mover

In Japan for example, there are some routes (Tokyo–Osaka) where several daily B747 flights are scheduled! In Europe, there are some routes (Munich–London) or Madrid–Frankfurt or Paris Milano where more than 30 flights per day are scheduled, but no real big aircraft are used. Again the question: More daily frequencies against fewer flights with larger aircraft. The passenger clearly prefers more frequencies, but the constraints at the big hub airports are requesting larger aircraft to further reduce and meet the airport constraints.

Boeing is proposing a version of the B 787 with the designation—300, which is foreseen to operate best at 3000–4000 nm. A similar proposal is offered by Airbus with their A350-300.

However, the solution is not optimal, as both aircraft types will use the big wing from the long range version, which will not be optimal for the short range and also cause problems at the airports especially at the gate positions, where the big wing span will block some gate positions right and left of the aircraft (see Chap. 9).

6.1.2.2 The Small Long Range Aircraft

There are some specific flights, where airlines are using modified short range aircraft like B737 and A320 types, installing additional fuel tanks, but having therefore less seats—mainly business class seating—and offering transatlantic services with these "Business Jet type" of layout.

So even when the aircraft manufacturers are not offering specific optimized aircraft to address this market of small long range aircraft, there are possibilities to find solutions for airlines who are interested in this market segment and are exploring the possibilities. There are however also some constraints for airlines to offer these modified aircraft for long range routes. The speed is slightly different compared to the big long range aircraft and offering only business class seats may be a constraint in the airline policy.

In the past, airlines were buying the aircraft and there was always a direct contact between manufacturer and operator. With the establishment of big aircraft leasing companies like ILFC, GECAS, CIT, RBS, etc. [15], the relation between A/c manufacturer and airline is changing. The direct link between manufacturer and operator is vanishing and the big leasing companies are looking for standardized products, which can easily be transferred from one operator/airline to another, when the leasing contract has ended. This will/can change the market situation in the future. Then the leasing customers are looking for very standardized products to reduce the refurbishing cost when changing the customer/operator.

6.2 Industrial Organization

An aircraft manufacturer company is acting as overall system integrator. Very often this is also named as Overall Equipment Manufacturer (OEM). The system integrator is necessary at the top level of the aircraft industry and has the overall responsibility for the product. In this role the system integrator has at least to

- specify the aircraft
- market the aircraft
- final assemble the parts into a complete aircraft and test it
- integrate the different component and aircraft elements
- certify the aircraft and give guarantees for its performance
- act as a single interface to the customers (airlines)
- ensure a lifelong support to all flying aircraft.

In order to reduce the financial risk, the aircraft manufacturer is interested to find some strong financial partners, with whom he will/can share some technical, commercial and financial risks. Obvious risk sharing partners for the aircraft manufacturers are:

- engine manufacturers, where the engine is worth about 30 % of the total aircraft price.
- supply chain companies, be it from the structural side or the systems side.
- Financial investors from the airline side (Leasing companies, airlines, etc.).

The operational side is normally not a candidate for risk sharing. They prefer to stay independent from the different manufacturers. The obvious risk sharing partners are therefore mainly partners in the production chain.
Figure 6.4 is providing a rough scheme of the supply chain structure. Three main blocks are outlined:

- the aircraft structure,
- the systems and
- the propulsion units.

Each part is then further divided into components, elements, etc.

In order to keep the management of all the detailed parts at a reasonable level, there are different levels of suppliers.

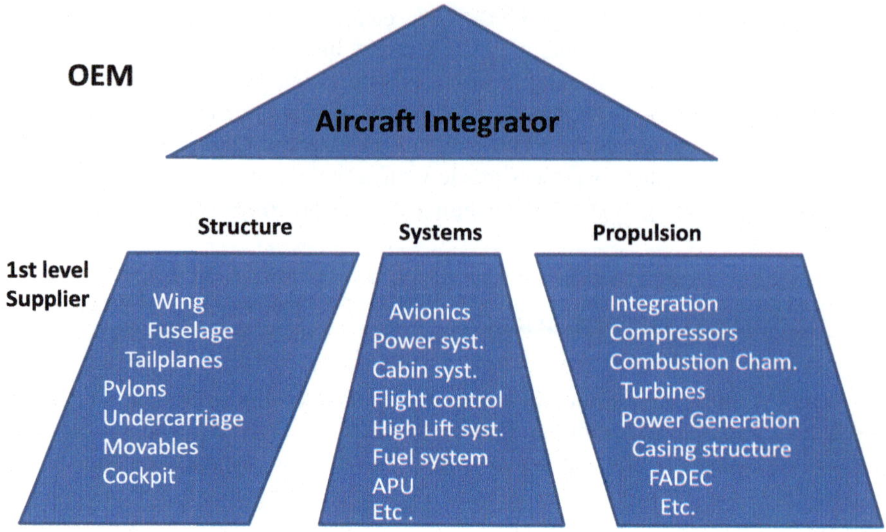

Fig. 6.4 The supply chain in the aerospace industry

The ***first level suppliers*** are strong and large companies who have strong competencies in

- Engine design
  - Combustion part, turbines, compressors, generators, systems
- structural component design
  - wing, fuselage, tailplanes, undercarriage, cabin interior (see also Fig. 6.10)
- system design.
  - Hydraulics, electrics, avionics, environmental system, flight controls, etc.

The level 1 supplier (sometimes also called "tear 1 supplier") can provide a complete section (wing, tailplane, etc.), a propulsion component or a complete subsystem design (hydraulic system, environmental system, undercarriage, etc.)

These first level suppliers are organizing themselves also in a way to have some component suppliers (level 2 suppliers) and those will have some lower level suppliers for detailed components, subsystem elements or specific services.

To stay efficient in such a supply chain, a lot of standards have to be fixed in this supply chain:

- A common language—English is the most common standard language
- A common metric, mass and design standard
- A common IT-management, design-, production-, industrial- standards
- A common quality assurance and selection of suppliers system

- A common change management system
- An agreed procedure for certification and approaching the certification authorities
- A common procedure who and when to approach the customer (airlines and operators)
- Etc.

Also some ***common management rules*** [16] have to be agreed like:

- What design features can be decided by the supply chain and what can only be decided by the upper level?
- How detailed should a specification be? (Very detailed and prefixed or open to allow innovative solutions from the lower supply chain partner?)
- Who will have to take care when deficiencies in quality are happening?
- Who has the right to contact the customer and to respond to which sort of questions?
- Interface management—Common teams to define, test and verify the interfaces?
- Use of patents, exclusivity rights or who has the rights for new patents, developed together?

6.3 Development Process (From Idea to Product)

The aircraft development process is fairly complex and it needs a lot of very experienced persons to keep the right balance between good standardized processes and flexible structures for further innovations.
Figure 6.5 specifies the four main aircraft process domains, which are:

- Research, technology development and Innovation
- Development of a new aircraft
- Production of aircraft
- Product support for the flying fleet.

Research is and should be a continuous process. Research has to cover the wide range of basic research up to project related research. Research can be grouped in different "Technology Readiness Levels", so-called TRL's. NASA has defined 9 TRL levels. However in the civil aircraft programs, there are often only six TRL levels defined [17].
The *aircraft development* process is normally divided into three main phases:

- The predevelopment phase, also called product definition or definition phase
- The real development phase, characterized by bringing the aircraft from the virtual definition into a real built and certified product
- The continuous improvement development process, leading to upgrades and recovery of definition deficiencies.

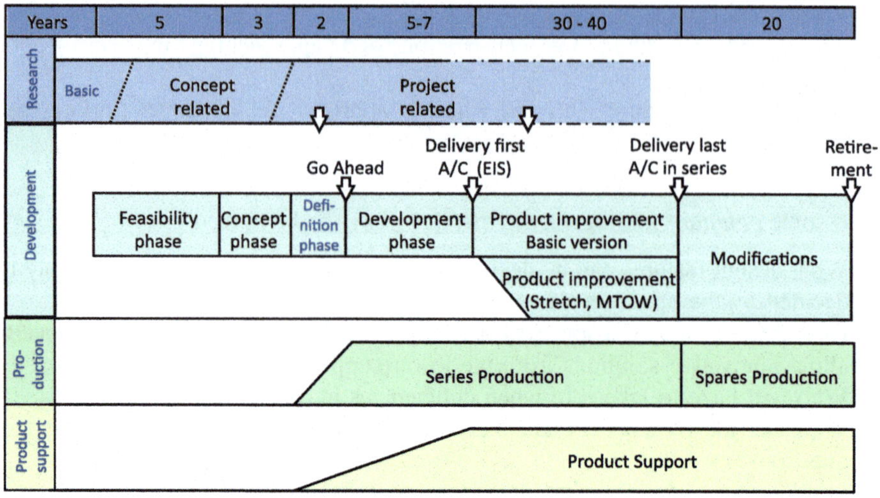

Fig. 6.5 Typical life cycle of a civil aircraft program

6.3.1 Product Definition

The predevelopment phase, often also named as product definition phase (Fig. 6.6), can again be subdivided into three phases:

- Feasibility phase
- Concept phase
- Definition phase.

Each sub-phase should start with a clear objective defining, which kind of aircraft should be specified and prepared, what resources and budget will be allocated and what level of detail is expected to be delivered at the end of each phase.

The Feasibility phase is the first part of a new aircraft development process, where still all possible aircraft concepts are open, i.e. engine location, wing sweep (forward or backwards), fuselage cross sections, tail design, under carriage, etc. At the end of this phase a detailed description of the aircraft concept, its geometry, its basic design features (material selection, system architecture, new technologies to be used, etc.) and the critical domains for further definition should be highlighted.

The Concept phase, which follows (see Fig. 6.5) will deepen the selected concept in such a way, that the all essential aircraft parameters will have to be specified precisely (cabin diameter, wing area, wing sweep, tail size, door concept, cockpit philosophy, system architectures, specifications for all subsystems available, etc.). At the end of this phase, the aircraft should be defined in such a detailed level with all performance calculations available to provide first proposals of this new aircraft to the airlines.

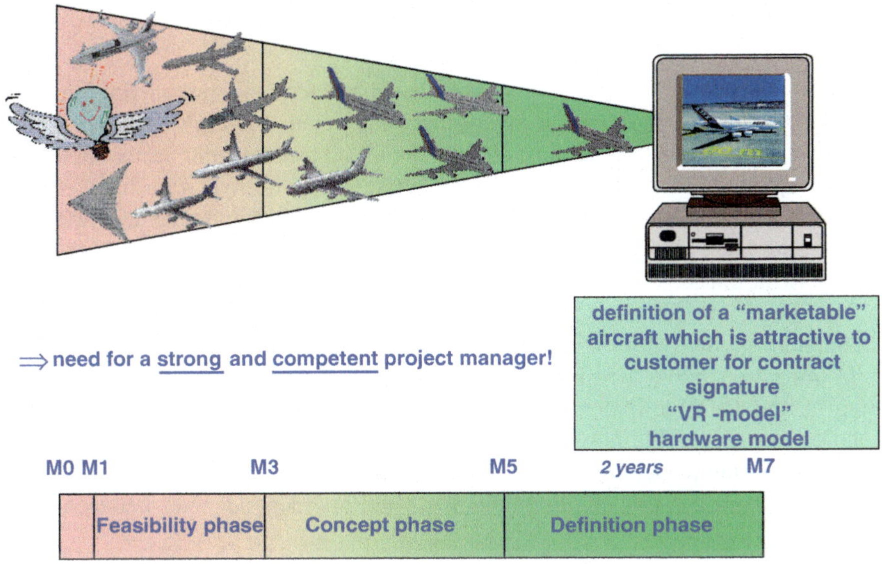

Fig. 6.6 Product definition phase—from the first idea to a valid aircraft definition

The Definition phase, which lasts about 2 years, will refine all definitions to such a detail, that the build process (production phase) can be launched immediately after. Before starting the definition phase, the company board has to authorize the sales department to propose the aircraft to the market and if the interest is strong enough, install an airline advisory board, where the airlines still have the chance to make some recommendations or proposals for improvement. Especially, the cabin and cabin systems architectures will be reviewed to be capable to accommodate a wide range of different cabin wishes within the proposed cabin volume and design.

The goal of the definition phase is:

- drawings have to be prepared to a level that all interfaces are defined,
- space for all systems is sufficiently large,
- main partners are identified,
- one engine MoU or better two engine proposals are available,
- the performances (range, mission fuel, weights,…) are sufficiently good to be attractive on the market and
- guarantees can be defined and met within the normal ($\sim$3 %) guarantee margin.

At the end of the definition phase, the aircraft manufacturer should have enough interest from the airline side to make a clear decision for starting the aircraft program, approve the "Go Ahead" milestone, commit to the development program and schedule and launch the aircraft program.

6.3.2 Aircraft Program Decision Point "Go Ahead"

The "**Go Ahead**" milestone is very critical within each new aircraft program.

As shown in Fig. 6.6, there is a very clear and detailed aircraft definition on the computer system, the "virtual aircraft", where all aircraft systems, external shapes, interior options, are defined to a "mm accuracy", but no hardware has yet been produced. Only after the official aircraft launch, the aircraft manufacturer will then start to produce all detailed drawings, select the main suppliers and start the development phase.

The most critical part in each new aircraft development process is the "Launching phase" with the milestone "Go Ahead".

In order to achieve the milestone "Go Ahead", the aircraft manufacturer has to bring in line the three major organizational parts of market, business/financial and technical aspects. (This is a simplified view, but helps to reduce complexity and improves the understanding of process elements!)
Figure 6.7 describes the simplified elements and interdependencies between

- Market/Sales;
- Business/Finance
- Engineering/Manufacturing.

It is obvious to start with the *market aspects* [18].

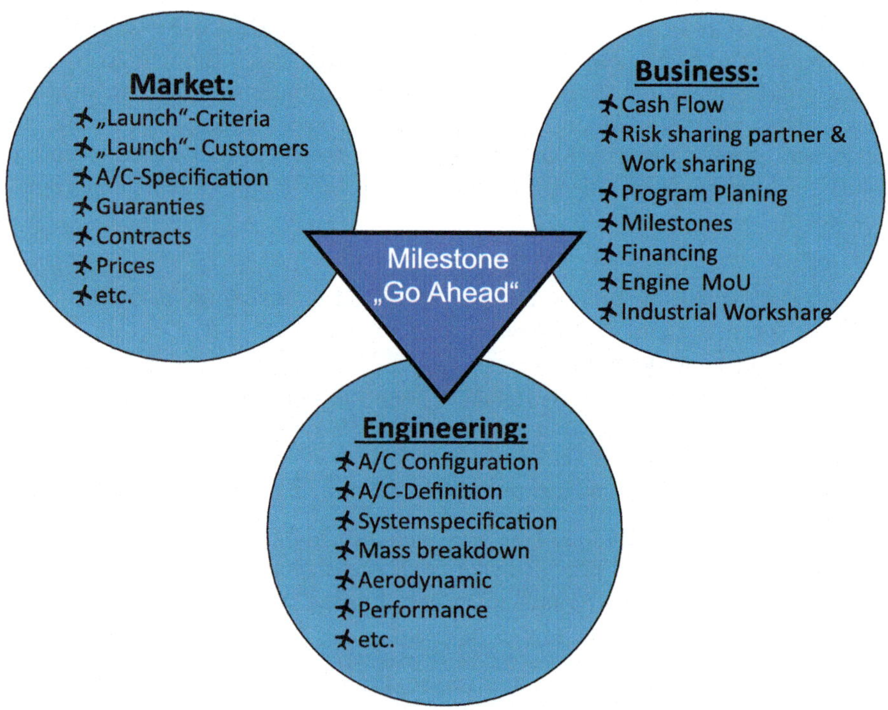

Fig. 6.7 The magic triangle for milestone "Go Ahead"

The aircraft manufacturer will only start his new aircraft program, if there are sufficient airlines and operators who are interested to buy this new aircraft and act as launching customers. The manufacturer has therefore to specify the launch criteria. Launch criteria are defining:

- how many airline orders are required?
- how many different customers are required?
- Which airline types (size, country, market image) should be amongst the launching customers.

When these launch conditions are fixed, the sales and marketing specialists of the aircraft manufacturers will approach the airlines and offer the new aircraft. They will have to provide the following informations:

- A detailed *aircraft specification*, defining the aircraft with all its major features
- A set of *performance guarantees*, that the aircraft can fly certain routes with adverse wind conditions, full payload and the maximum fuel required for a fixed mission. It is obvious, that the new aircraft should be better in fuel burn by at least 10 % compared to the existing fleet of the customer!
- *A price proposal* for the number of aircraft, envisaged by the customer.

Normally, the sales specialist will find several customers (airlines), who are ready to act as launch customers for this new aircraft and are ready to pay a certain percentage of the aircraft price in order to reserve a given date for the delivery of the new aircraft.

If these launch criteria are accomplished, the aircraft launch should be decided.

In parallel, the following *business* conditions have to be clarified:

- Which risk sharing partners are prepared to be onboard of the program and what will be there financial contribution?
- A Workshare breakdown, which specifies all aircraft parts, which can be given to outside partners.
- A list of possible partners
- An agreed milestone plan with fixed dates and a production plan
- A finance plan with all investments necessary for the production and assembly line
- A MoU with one or more engine manufacturers, who are prepared to provide engines, which match the aircraft specification with a guaranteed fuel flow for this aircraft program.
- A cash flow calculation, which shows that the aircraft program will be profitable after some x years, where x is in the order of 10–15 years (Details are given in the next chapter!).

Also the following *technical definitions* and documents have to be available for the aircraft launch like:

- Definition of aircraft configuration
- Specification of aircraft and all cabin configuration options

- Detailed systems specifications
- Detailed aircraft weight breakdown
- Aerodynamic configuration tested and validated
- Overall aircraft performances.

Taking these three major domains (market, business, engineering), it is clear that a new aircraft program will only be launched, if the fixed launching criteria are fulfilled. Launching a new aircraft is a huge undertaking and always will put the whole company at stake. It is a commitment from the aircraft manufacturer,

- to build an aircraft in a given time (4–6 years),
- give guarantees for the performance of this new vehicle 5 years before the first flight and
- the whole development will cost roughly 20 billion $!

Coming back to the milestone "Go Ahead", it is obvious, that the discussions with the airlines, to receive the commitment from them to become a launch customer will at least take 1 year. So all technical definitions (A/C specification, performance guarantees, etc.) have to be ready in such a detail and quality, that guarantees can be given. So the technical definitions have to be ready already at least 1 year before milestone "Go Ahead"! and all this without knowing, whether the aircraft will be launched at all. So the management has to invest a lot of money in order to prepare and define a new aircraft program and there is no certainty, that this money spent is not in vain! If the aircraft program will not find sufficient support from the market, all efforts are nearly lost [19].

Another aspect in the definition of the new aircraft is the development of an aircraft family. From the beginning, the family concept has to be part of the definition. This means, that the wing size, tail geometry, undercarriage have to be fixed that a later stretch or shortening of the fuselage can be done without redefining the major components later on (see also Sect. 3.5).

6.3.3 Product Development

The *development phase*—as shown in Figs. 6.5 and 6.8 in more detail—is normally lasting about 5 years; recent developments of the B787 and A350 have however led to much longer development times ($\sim$7 years).

In the development phase, all detailed drawings for the manufacturing process have to be prepared, the main suppliers have to be selected, big investment for all the new production facilities have to be done, the production concept and the final assembly concept and place, where all the big parts will be assembled, has to be fixed.

The numbering of the milestones as shown in Fig. 6.8 is arbitrary and was based on a former Airbus concept, but is today no longer applied in the internal Airbus processes. The authors have decided to keep the numbering, as it will help the readers and students to better understand and follow the logic of the process.

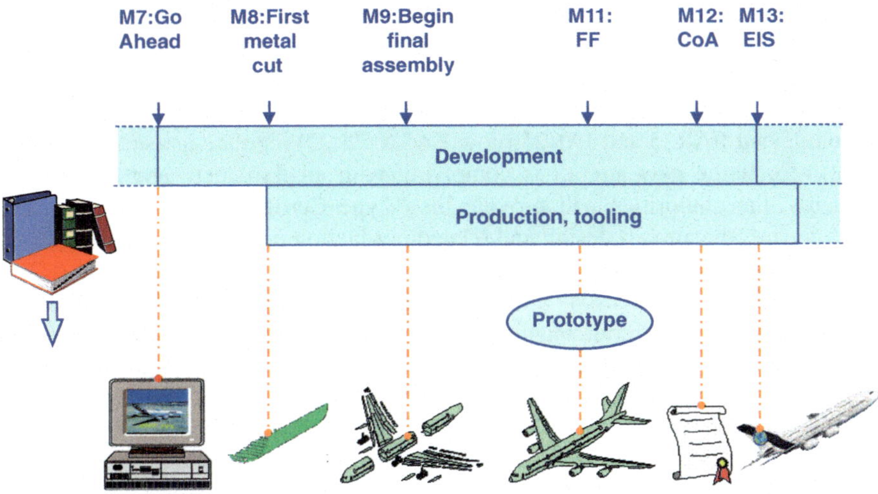

Fig. 6.8 The development phase—from aircraft launch to "Entry into service"

The following milestones have to be accomplished in order to bring the aircraft into airline service:

"*First metal cut*"—this term is still used, despite the big changes in the material from aluminium alloys to CFRP (black metal). This milestone defines, that first hardware parts are build and wait to be shipped to the final assembly line. In detail, there are several more internal milestones for the first wing, the first fuselage sections, the tail parts, the engine pylons and all the other large and long-lead aircraft parts.

Next milestone is the start of the "*final assembly line*". All aircraft parts will be at the Final Assembly line and the assembly of the first aircraft will start. In parallel to the first aircraft assembly there are several test benches and tests, where major aircraft parts or components are required and which have to be done to prepare the first flight of the aircraft like the static test, the fatigue test, and several system test benches. A cockpit-simulator combined with an "iron bird" will help to investigate and test all control laws for the aircraft handling and also will allow the pilots to familiarize with the new aircraft before their first flight. Structural tests have to be prepared and the maximum wing loading capability has to be tested and demonstrated to ensure that the engineering predictions and methods are in line with the actually produced aircraft hardware.

The "*First Flight*" is a major event in each new aircraft program. After several rolling tests on the runway, the aircraft will prepare for the first flight, will take off the first time and will be tested in real weather conditions. Major representatives from the launching customers, political persons and partners from the aircraft manufacturers will be invited to follow this event.

The flight testing of the aircraft will normally take about 1–2 years. The first part is dedicated to explore the Flight envelope (see Fig. 5.33), bring the aircraft to all critical flight conditions within and some specific points outside the normal flight envelope and fine-tune the flying characteristics of the aircraft. The second part of the flight test will involve the certification authorities (FAA, EASA, others) to show them the compliance of the aircraft with the certification requirements, as shown in the relevant FAR25 and JAR25 (now EASA CS 25) regulations (see Sect. 4.4.3, Table 4.5). If the new aircraft is complying with all the safety and certification standards, the authorities will provide the "Type Certification" for this aircraft, which means, the aircraft design and related production process fulfils all necessary safety features and the production process can start.

This certification process is a major element during the development phase. At the end of this, the aircraft manufacturer will have produced a huge amount of certification documents to demonstrate to the authorities, that all safety features have been respected and all operational instructions for pilots and crews are clear and easy to follow.

The next milestone will be the "Entry into Service" (EIS), where the first airline will provide the first flight and will have the first passengers onboard, either invited guests or already the first paying passengers. A certain time is needed between the CoA and the EIS milestone, about 3 months. The airline pilots have to be trained on the real aircraft, some "route proving" flights will be done to familiarize handling teams at the airports with the new features and procedures for this new aircraft type. Gate positions and gate access have to be studied and adapted for the new aircraft, emergency procedures have to be trained on the real aircraft, loading tests for trolleys and baggage will be accomplished and a lot more. With the EIS milestone, the development process stops normally. But between the ideal plan, shown above and the real life, there are always several features during the development phase which have to be improved (i.e. small structural reinforcements at critical areas and highly loaded junctions, systems updates for environmental control system etc., …) which will require some modifications at the aircraft. But these modifications cannot be introduced immediately at the 2nd or 3rd aircraft, as they have been already produced. These modifications will then have to be introduced at a later serial aircraft number and a specific aircraft modification system is needed, which documents all modifications, which have been applied to each individual aircraft.

In parallel to CoA and EIS milestones the aircraft production has already started and the next production aircraft are produced for the next customers. It is fairly obvious, that—once the new aircraft type has appeared on the market, the airlines will quickly ask to receive also their ordered models. A big production ramp up should be organized to satisfy the starting market demand and give some advantage to the launching airlines. There may be however the risk, that during the flight and system test phases some deficiencies have been discovered, which will require modifications. This will lead to major problems for the aircraft manufacturer, as he wants to deliver as produced, but the airline will go for major price reductions if their aircraft does not fully comply with the final aircraft standard and the specified aircraft performances, as guaranteed several years ago.

The two phases—product definition phase and development phase—are completely different in their process planning. The process aircraft development in

general is more clearly structured. It is a process to bring the "virtual aircraft", as defined in a complex Computer system, into a real aircraft. If the aircraft is well-defined upfront, then the process "develop the aircraft" can run with an experienced team and management structure straight forward with respect to cost and time.
The process "*product definition*" is completely different to all other product processes of the aircraft:

- The target is not clearly fixed!—engineering wise. When the aircraft manufacturer decides to replace an older aircraft version or define a completely new aircraft type, there is a possible target date, also some rough ideas about the aircraft definition like—define an aircraft configuration which is "marketable"!
- there is no clear "market specification"
- the payload-range capability is about fixed
- the technology level should be high but cost efficient for the user
- the competition will not wait for the final "product definition"
- the "product proposal" has to show a "significant" market benefit relative to existing products on the market
- the rough schedule to achieve "Go Ahead" is defined, but will depend on market situation

But the targets are not well expressed and the management is normally reluctant to spend the necessary money in advance.

6.3.4 Production Phase

As shown in Fig 6.5 the production process starts already parallel to the product definition and development phases. Development and production are linked today much closer than in the past. Design and production feasibility are closely linked to achieve the weight and cost targets for each aircraft component and all equipment items. As major aircraft components, and system equipment will be developed by partners a clear definition of the aircraft, the production process as well as all interfaces has to be developed. Competent and knowledgeable partners have to be identified and a "Make or Buy"—policy has to be established upfront.

6.4 Production Process and Work Share

The aircraft is a very complex product with several millions of different parts. In order to manage the complexity the total amount of work, the tasks for the engineering development, for the production of the components, for flight testing and certification etc. have to be defined. For the aircraft production a certain A/C decomposition in components, elements, parts and services, etc. has to be done.

Figure 6.9 is showing the main production work sharing of an aircraft (A321) and also the main contractors who will produce individual components of the aircraft.

Figure 6.9 looks like a nice colourful illustration of an aircraft, but the aircraft Breakdown structure needs much more elements than just the main production components and elements. Table 6.1 is showing one possible Breakdown structure for the production parts. Here all aircraft components are identified; each component is then broken down in the major level-1 parts and each level-1 part broken down further in the corresponding level-2 parts etc.

The bolt component is then further detailed [20–22]:

The General Cost Breakdown Structure includes besides the major hardware components additional cost elements (Fig. 6.10).

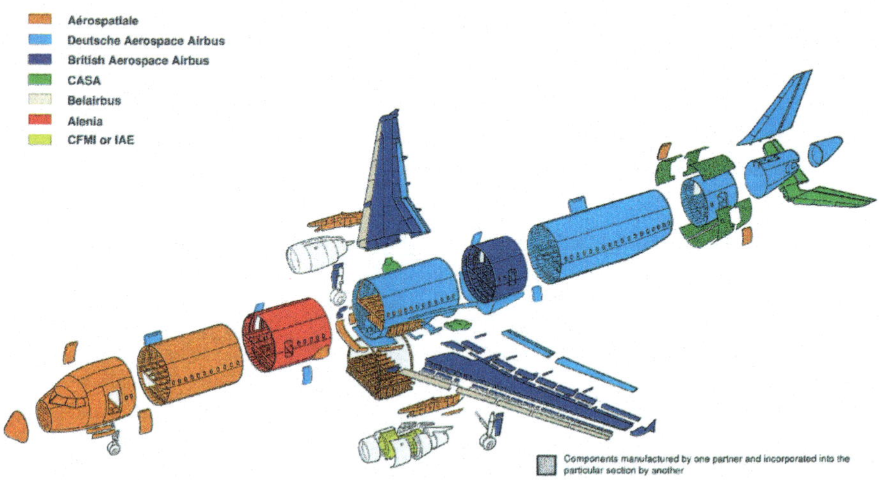

Fig. 6.9 Typical production work share for a new aircraft program (example Airbus A321)

Fig. 6.10 Typical Aircraft component breakdown structure

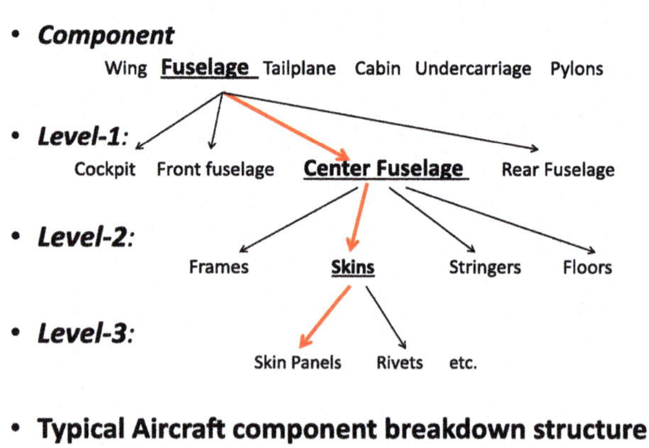

Table 6.1 Typical cost chapter breakdown for an aircraft program

Typical cost chapters for the aircraft development are	
1	Non-specific design
2	Specific design
3	Tests (aerodynamic [Windtunnel], structural, system tests)
4	Production
5	Jigs and tools
6	Modifications
7	Ground support equipment
8	Component development
9	System and equipment development
10	Documentation
11	Certification
12	Engine and nacelle development
13	etc.
50	Management

A typical Cost Breakdown structure is shown in Table 6.1:

Table 6.1 is not exhaustive but just highlighting the major big Cost chapter for development which has then to be defined in much more detail with all the necessary engineering tasks, the responsibilities in the organization, the necessary quality measures, the timeline and finally the expected or allowed/estimated cost for the task in a certain time frame.

An overview of the Engineering Cost Chapter is given in Table 6.2

For the production, a very similar breakdown per component will be needed with a component breakdown according to Tables 6.1 and 6.2.

These details of the aircraft development process are highlighting the complexity but also the necessary systematic definition of a well-specified Breakdown system in order to manage the complexity.

As usually in industry, it will need one or more successful aircraft developments till the aircraft manufacturer will have a database, which is realistic enough to develop the second or further aircraft in much better detail and with a more experienced and consolidated technical and financial database for design and production. This is one of the reasons, why it is said in the aeronautical industry, that the entry barriers for a new entrant are quite high, i.e. it will need about 20 years and one or two successful aircraft programs till the new company is well enough established in the market. Having seen all the long development cycles, this is not an easy and open market, where any new entrant can enter. It will need at the start a lot of national/governmental support to reach this acceptance level on the market side and a lot of engineering, production and management knowledge in the new company.

Table 6.2 Engineering cost chapters during aircraft development

Non specific design	Specific design
Configuration development	Includes all engineering tasks to develop the different aircraft parts like
Aerodynamics	
Performances	
Weight and balance	• Fuselage
Noise and accoustics	• Wings
Loads and aeroelastics	• Tails
Structural test programme and analysis	• Pylon and
Stability and control	• Nacelles
Design and certification philosophy systems	• Systems
• All ATA chapters like	• Section assembly
–Light	• Final assembly
–Environmental control system	• Furnishing installations
–Flight controls	• Ground tests
–Communication	• Customer-related flight aspects, etc.
–Water and waste	
–Navigation etc.	

6.5 Cash Flow and Manufacturing Cost

Main cost elements for the aircraft manufacturer are the development cost for an aircraft and the production cost per unit.

Development costs are also defined as "Non-Recurring cost" NRC. They are including all the cost elements from the predevelopment phase and the development phase, with all the investments for new hangars on different production sites, the transport infrastructure, etc. Table 6.1 is providing more details what sort of cost elements has to be included in the NRC.

The production costs are also called "Recurring cost" (RC). They include all cost linked with the production of one aircraft (material and labour mainly, see Sect. 3.6, Fig. 3.14). For the production of each additional aircraft the same amount of cost (labour and material) will be needed, therefore they are also called "Recurring Cost", recurring for each production unit!

In order to define the aircraft sales price, the RC have to be used, a certain percentage of the NRC has to be added and general overhead cost and a profit margin can be added. The percentage for the NRC part may change from programme to programme. For the big aircraft like A380, a number of 200–300 production units will be used. For the B787 and A350, where already more than 600–800 commands have been taken, the NRC may be divided over roughly 1000 units or even more.

For the large Short Range Aircraft programs like B737 and A320 and their future replacements, where already more than 5000 have been produced from each type, there could be even a higher number to split the NRC.

Some operators, mainly the military operators, are using the term Life Cycle Cost. Life cycle Cost are including the total cost, starting from developing and buying the aircraft, operating it over a certain time period and dismantling the product at the end of its life. This is normally a financial consideration, used by operators who are not flying the aircraft very often—in contrast to the commercial airlines—and therefore, the production and development cost are still significant and play a dominant part.

6.5.1 Cash Flow Calculation

For each new aircraft program, a so-called "Cash flow calculation" will be done to see, when and after how many years the profitability of the program will start and also to see, whenever it will become positive.

The Cash Flow is a purely income–outcome calculation. It specifies over time all the money, which is spent for development (NRC) and all the money spent for producing the aircraft (RC) on one side and all the income from the customers either as down payment when the contract is signed (x % of price) and the rest of the contractually fixed price when the aircraft is delivered.

Figure 6.11 is showing a cash flow calculation with the major elements like development cost (green triangles) and the final cash flow curve (blue line and red line with interest rates). It is obvious, that at the beginning of the program there are only expenses, expenses for development and later on for material, labour and parts to produce the aircraft and expenses to promote the program at all air shows and public events.

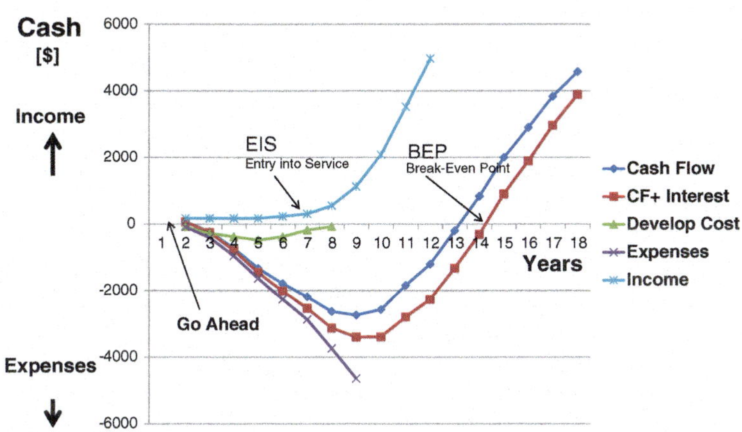

Fig. 6.11 Typical cash flow curve for an aircraft program

The money, received from the customers as "income" contains in our simplified calculation two elements:

- a 10 % of down payment of the contractual fixed price at the signature of the contract and
- and the rest of the price (90 %), when the aircraft is delivered to the customer.

At the start of the aircraft program—which corresponds normally with the milestone "Go Ahead" (see Fig. 6.6)—there will be some down payments from the launching customers! The real money income flow will only start with the delivery of the first aircraft to the customer. Before this EIS point there are only very small positive income cash figures, just for all the signed contracts.

From the delivery start of the first aircraft, the trend is slowly changing, but still a lot of material and labour cost for the production is visible. With the increase of aircraft deliveries the bottom of the cash flow curve will be seen and the cash flow is changing to the positive side after roughly 4 years after EIS (more income from the deliveries then expenses for material and labour!). It will take another 5 years at minimum, to see the cash flow curve entering from the negative part (more money spent than received as income) to the positive part. The time, when the neutral line between cost and profit will be reached, is called the Break-Even-Point. The Break-Even Point is an important number, as it indicates after how many years a new aircraft program will be beneficial. In the example of Fig. 6.11, this Break-Even-Point is reached after 14 years! Figure 6.11 is showing two cash flow lines, one as purely money flow with income and expenses, the lower and thicker curve as a consolidated curve including interest rates for all the money spent. This curve is the more important one, as it shows the realistic cost statement. The new aircraft program will only be beneficially when it will generate more cash income than the interest rates, which can be generated when putting the money to a classical bank or invest it differently.

The Break-Even-Point BEP occurs for good programs after 12–14 years, for normal programs after 15–17 years and for bad programs never!

Now just a short moment of reflection and comparison with development of other technical products! A product, which has a development cost of 5–20 billion $ and which needs at best 13 years to reach the Breakeven point is not a "normal business". No classical bank will be interested to invest in such a critical and uncertain program. So it is absolutely clear that a national support or backup is needed, in case there are major economical and environmental changes during this period between launch and Breakeven.

Once an aircraft program is running over 15 years and the production rate is still very high—like it is the case for the B737 and A320 programs—then this is a very good cash cow, which has to be kept on the market. In these cases are the manufacturers not very interested to disturb this situation, even when the market (i.e. the airlines) are asking for a better product. A better product (especially, in the Short range Narrow-body market) would be feasible today with all the technologies available today. But it needs an aircraft manufacturer, who is feeling, that the introduction of a new technological step in a new program will bring him some real advantage, compared to his competitor. The recent decision from Airbus to upgrade the A320 program and offer a new engine option for the A320, which is now named A320 neo (new engine option), is such a strategic move, which then will force the competitor (here Boeing) to propose either a similar solution (an upgrade of the existing B737 MAX) or really propose a complete new Short range aircraft with a different and improved fuselage cross section and also a new wing design in addition to the new engine option.

The preparation of a cash flow calculation needs besides the development cost, we have shown as an example in the former Sect. 6.3 some additional inputs which are not very easily to assess.

There must be an assumption, how many aircraft will be delivered over the next 20–30 years. Figure 6.12 is a possible scenario, which has to be elaborated already at the start of a new program in order to be capable of producing a cash flow assessment.

Another input for the cash flow calculation is the price assessment, which can be achieved on the market.

A rough statistical data collection is given in Fig. 6.13, which shows the a/c price per seat for different types of aircraft. These data may slightly change over time, but it gives a first good estimation and trend for any rough calculation (see also [23]).

For each new aircraft program, a cash flow calculation is mandatory from the management side, to allow a certain assessment about the validity of the new program. It should however be clear, that all these Cash flow assessment are not very precise and the history has shown—at least from the European Airbus side—that none of the cash flow calculations ever has been roughly correct. A good example is the program A320, launched in 1982 and with EIS in 1988. This program was calculated with development cost for 200 aircraft and a monthly production rate of maximum eight aircraft. After more than 20 years of production and a complete family in production, the production rate today is more than 40 aircraft per month and this program is the cash cow for the company, allowing the company to develop new aircraft like the A380 and the A350.

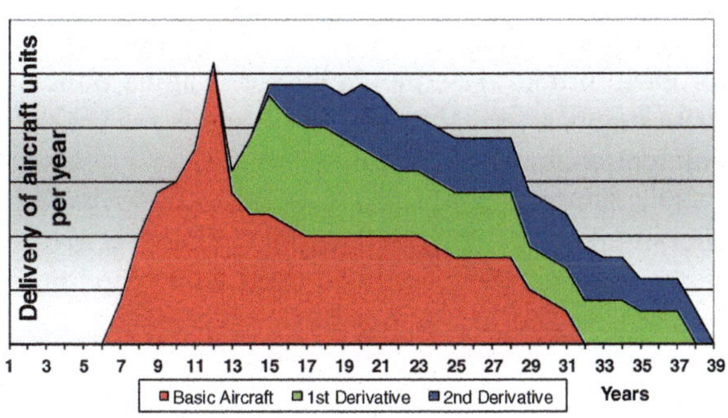

Fig. 6.12 Aircraft deliveries over the next 35 years

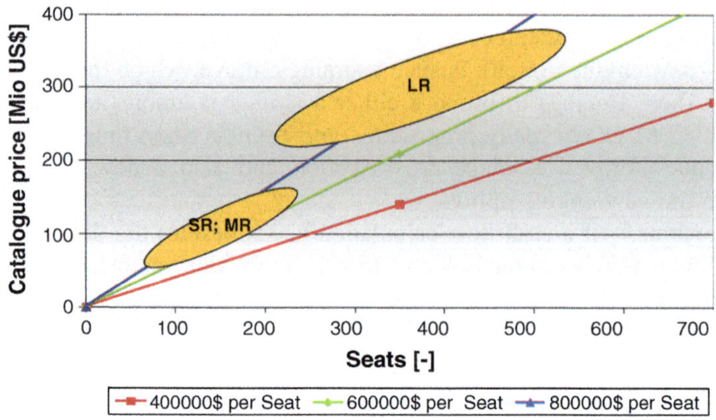

Fig. 6.13 Statistics for aircraft price per seat

In short: Cash flow calculations are an important element to get a feeling about the validity of a program, but it should never been taken as guaranteed, as the long period of an aircraft program can never be properly assessed at the start of a new program. The best decision: prepare a very high technology product and improve it constantly over the time [24], which will give you the best value for money. As can only be shown in a simplified manner, all the data needed for a proper cash flow assessment are very critical and the dynamic of the market will be difficult to anticipate correctly. Scenario technics may help, to better identify the risks and the chances.

6.6 Engine Manufacturer

The aircraft engine manufacturers are playing a very important and critical role in the aircraft development process. The engine is representing roughly 1/3 of the aircraft price. There are at least 2 engines on each aircraft and there is also a need for each airline to define their spare engine policy, i.e. have a certain number of spare engines available. But there is also a very strong competition between the different engine manufacturers [25–27].

The aircraft manufacturers prefer to have two competing engines for their new aircraft programs. This will help them later on to develop the aircraft further in size and range and they will also need some modifications or improvements on the engine side and in a competitive environment, this can be achieved much easier.

Some past examples show the problematic of having only one engine available: The BAe RJ85 had only one engine available, they needed four to have sufficient thrust, but a real good competitive aircraft was never achieved, despite a fairly good chance in this market segment of 80–100 places and with a fairly weak direct competition at this time.

The ranking of the engine manufacturers has changed considerably over the last 30 years. PW was the market leader in the 1980s, and GE and RR had a fairly weak market penetration. In the meantime GE has become the market leader via a strong liaison with Boeing and a good strategic alliance with SNECMA in the market of the 25–35 klbs thrust class by forming the CFM consortium. The competitor engine consortium is the V2500, where PW and RR are the leaders with three other partners involved to become the V (latin figure for five partners).

Strategic decisions are always more difficult, when five partners have to agree and those five are in strong competition in some other thrust classes and a transfer of technology will not be so easily exchanged.

The ranking today in terms of engines sold has changed and the market leader is GE; followed by RR and PW. The different engines available for the different aircraft types are listed in Fig. 6.14.

New technologies, which are studied at the moment, will be of interest for the future success. RR and PW are relying on the GTF concept, while GE is relying to further develop the classical engine concept by further increasing the BPR.

There is also a strong improvement seen by the so-called Open rotor concept, which at the end is a modern propeller, but with its new technologies, will be capable of flying at fairly high Mach numbers (Ma = 0.7–0.75), with a further improvement potential of 10 % in sfc compared to the classical jet engines with BPR of 10. Figure 5.25 is showing the conflict of the development. The OR concept has a clear advantage in sfc, however, it has a higher noise level and the balance between noise and emissions will be one difficulty and it will need a very careful assessment to do the right choices. The new models, following the A320 Neo and B737 Max will have to make the difficult choice of better SFC against better noise level.

Fig. 6.14 Aero-engine families and their thrust classes

	GE	P&W	RR
turnover			
Single aisle A 320/ B 737	CFM 56	V 2500	V 2500
A 330	CF6-80 C/E	PW 4164/70	Trent700
B 787	GEnX		Trent 1000
A 350			Trent XWB
B777	GE 90	PW 4164/70	Trent
B 747	CF6 –80C/E		RB211-524
A 380	GP 7200	GP7200	Trent 900

6.7 Supply Chain

The aircraft and engine manufacturers have specialized during the last years and are changing and adapting their business models continuously. At the moment, there is a clear trend, to concentrate on the aircraft resp. engine integration task. This means, they need strong partners to support them as integrators or so-called OEMs. The selected suppliers should take a major share of the risk and cost in the development and production and partly also in the responsibility for the overall aircraft program, i.e. finance their own development and even participate in the cost of certification. but will be participating by the sales of each aircraft, according to their negotiation of their work share in the overall aircraft program (Sect. 6.5).

Referring to the overall aircraft Work breakdown structure (Sect. 6.4), the integrator (engine or aircraft manufacturer) has to clearly identify, which parts of the aircraft resp. engine he will produce under his own responsibility and which are open to be offered to the market and ask for proposals.

This is part of the "Make or Buy"—policy, which has to be established.

This "Make or Buy"—policy needs a set of criteria, which will help to establish the company policy.

Criteria for Make or Buy:

- Is my own company structure capable to produce the part in an cost efficient way?
- Is the necessary Know how, to produce this part/component so specific that it should not be shared with the supply chain (Be aware: all the technology you are expecting from your supply chain will also be available for your competitor!).
- Look at your competitor and his philosophy and define your own technology strategy
- Quality of supplier
- Is the Supplier financially strong enough to support the program over the long A/c development process??

For most of the system components, this "Make or Buy"—policy is already fairly clear: The OEM has to define the overall system architecture, send it to his partners and waits for proposals. But there are very different levels of specifications possible:

- The OEM is specifying all expected technical solutions in all details and is just waiting for the most cost-effective proposal
- The OEM is just defining the performance parameters the physical interfaces and the certification requirements and the modifications processes for necessary customer adaptations as well as weight, size, cost and leaves several technological options open for the supplier.

In reality OEM and suppliers have already good relations and a certain know-how, trust and respect between the partners is established and the spec will depend on the OEMs philosophy in this area.

The world of the supplier industry is also changing dramatically and a certain concentration to very strong and big first level suppliers can be seen. Honeywell Aerospace, UTC-systems, Rockwell Collins in the USA and Thales and SAFRAN in France are well-established strong supplier companies [28–33]. Several old names have been disappeared from the market like Sundstrand, Garrett, Smiths, Litton, Messier-Dowty, have been bought or being integrated by bigger companies and have disappeared from the market. Figure 6.15 is giving a rough overview about the main systems and the important companies for this domain.

For the Structural parts and components, this could be done in principle in a similar way; however, recent problems during the development of B787 have identified, that it requires a much more detailed look in the technology of manufacturing, the manufacturing tolerances, the capabilities of the supplier to produce sufficient units at constant quality.
The supply chain will only provide the expected benefit if a very sophisticated "Supply Chain Quality Control"—process will be defined and properly managed. As shown in [33] the supply chain controlling has to integrate at least

	1st level Supplier	
Avionics	Thales, Honeywell; UTC –S; Safran; Rockwell Collins	
Com – Nav - Surv	Honeywell, UTC-S, Rockwell Collins,Thales; Safran	
Electrical System	Honeywell, Thales, Hamilton-Sundstrand; Rockwell/Collins	Safran,
Hydraulic System	Parker, UTC -S	
Flight Control system	Honeywell; Rockwell-Collins; Parker, Liebherr, Moog, etc.	Nabtesco,
Fuel system	Parker; Moog	
Landing Gear	Safran, Liebherr, UTC-S, Goodrich	SPP,
Wheels & Break	Goodrich, Safran, Honeywell	Messier-Dowty
Environmental Control system	Liebherr, UTC-Systems, Honeywell	
In-Flight Entert. IFE	Honeywell, Thales, Rockwell-Collins	

Fig. 6.15 Some major suppliers for system components

- the development management,
- the material purchase management,
- the production management,
- the transport—and distribution management,
- the product support management and
- the specification modification management.

At the end, the Integrator has to make a financial calculation of all cost aspects involved, the own quality control cost included to decide whether it is cost efficient to outsource the system/component to the market and make a careful assessment of all offers received. Price should not be the main criteria for the selection of the supplier, it is only one amongst others.

One parameter amongst all the system features needs to be highlighted here. It is the "In-flight entertainment system", which is in so far important, as it is a parameter, where each airline is trying to differentiate from the competitors and where a lot of innovation seems still possible in the future, following the intensive market development of Smart phones, mobile phones, tablet PC's, mobile audio and video systems, etc.

Some overview can be gained from [34, 35].

6.8 Offset Agreements

An *offset agreement* is an agreement between two parties whereby a supplier agrees to buy products from the party to whom it is selling, in order to win the buyer as a customer and offset the buyer's outlay [34]. Generally the seller is a foreign company and the buyer is a government that stipulates that the seller must then agree to buy products from companies within their country. The aim of this "Offset agreement-process" is to improve and harmonize a country's balance of trade. This is frequently an integral part of international defense contracts.

The key question is: to what extent is the offset proposal a factor in the consideration of defense contractor's tender during the evaluation and the decision procedures? "Transparency International" clearly summarizes the risks of corruption of offsets as marketing tools, that makes offsets "the ideal playground for corruption" [36]:

Direct versus Indirect Obligations

Every country and obligor has their own definitions of these terms. Strictly defined "Direct" offset obligations consist of the local foreign supplier producing or servicing the actual products being sold into the country. Some countries support a broader definition that covers aerospace-and defense-related production opportunities. This might also include technology transfer and training that supports the country's military requirements.

Generally speaking, one may define "Indirect" offsets as everything else that a governmental offset authority may decide to credit. Some specific examples are manufacturing opportunities offered by the obligor from other seller's business units, suppliers, and global partners. These exports need to be high technology oriented and show potential for job expansion and export growth.

References

1. Airbus A380; http://www.airbus.com/aircraftfamilies/passengeraircraft/a380family/. Accessed 29 Nov 2014
2. Sperl, A.: A380 Financial Update; Global Investor Forum, 19_10_2006
3. Bombardier C-series. http://www.bombardier.com/en/aerospace/commercial-aircraft.html. Accessed 29 Nov 2014
4. Embraer, Bresil. http://www.embraercommercialaviation.com/Pages/Ejets-195-E2.aspx. Accessed 29 Nov 2014
5. Sukhoi Superjet. http://www.scac.ru/en/products/sukhoi-superjet100/. Accessed 29 Nov 2014
6. UAC Russia. http://uacrussia.ru/en/models/civil/ms-21/. Accessed 29 Nov 2014
7. Mitsubishi/Japan. http://www.mrj-japan.com/mrjfamily.html. Accessed 29 Nov 2014
8. Chinese COMAC Aircraft: http://english.comac.cc/products/ca/. Accessed 28 Nov 2014
9. Finmeccanica; Alenia; http://www.finmeccanica.com/en/-/cleanskyjti. Accessed 29 Nov 2014
10. CleanSky—Green Regional Aircraft; http://www.cleansky.eu/content/page/gra-green-regional-aircraft. Accessed 29 Nov 2014
11. World Trade Organization: http://www.wto.org/english/tratop_e/civair_e/civair_e.htm. Accessed 29 Nov 2014
12. Airbus. http://www.airbus.com/aircraftfamilies/. Accessed 15 Oct 2014
13. Boeing: http://www.boeing.com/boeing/product_list.page. Accessed 28 Nov 2014
14. Schmitt, D.: Lecture on "air transport system"; Institute for Transport, TU Munich as part of MSc Course on "transport", TU Munich (2011)
15. Aircraft leasing companies: http://www.flightglobal.com/news/articles/top-50-leasing-companies-2008-221028/. Accessed 20 Jun 2012
16. Schäffler, J.: Lecture on "International Management in Aeronautics"; Lehrstuhl für Luftfahrttechnik at TU Munich (2000–2004)
17. Schmitt, D.: ECATA lecture "Business Reengineering and Change Management" during ECATA-ABI course in Munich at TUM, 10.3.1997; in Stockholm at KTH, 26.3.1998
18. Schmitt, D.: The Importance of an Integrated Product Definition in Civil Aircraft Programs; 1st CME Symposium in Bremen; 17.6.1998
19. Buttazzo, G., Frediani, A.: Variational analysis and aerospace engineering, pp. 379–394, Springer, ISBN 978-1-4614-2434-5 (2010)
20. Work Breakdown Structure; http://dictionary.sensagent.com/Work%20breakdown%20structure/en-en/ . Accessed 20 Oct 2014
21. Pritchard, C.L.: Nuts and Bolts Series 1: How to Build a Work Breakdown Structure. ISBN 1-890367-12-5
22. MIL-STD-881C, Work Breakdown Structures for Defense Materiel Items, 3 Oct 2011
23. Airbus Aircraft Price List: http://www.airbus.com/presscentre/pressreleases/press-release-detail/detail/new-airbus-aircraft-list-prices-for-2014/. Accessed 20 Nov 2014
24. Steiner, J.: How Decisions Are Made—Major Considerations for Aircraft Programs, ICAS 1984
25. Arndt, N.: Environmentally friendly aero-engines for the 21st century, CEAS Conference Berlin (Sept 2007)
26. Daly, M.: Jane's Aero Engines; HIS Janes (2011)
27. Gunston, B.: Encyclopaedia of Aero Engines, Cambridge, England. ISBN 1 – 85260 – 163-9 (1989)

28. Supply Chain companies; http://www.airframer.com/sector_page.html?cat2=96. Accessed 28 Nov 2014
29. Honeywell: http://aerospace.honeywell.com/products. Accessed 20 Nov 2014
30. SAFRAN: http://www.safran-group.com/site-safran/aerospatial/equipements-aeronautiques/. Accessed 28 Nov 2014
31. Aerospace Suppliers: http://aerospacesuppliers.com/. Accessed 28 Nov 2014
32. Thales: https://www.thalesgroup.com/en/worldwide/aerospace. Accessed 20 Nov 2014
33. Stocker, S., Radtke, P.: Supply chain quality. In: Kamiske (Hrsg.) Pocket Power. Carl Hanser Verlag (2000)
34. In-flight Entertainment; http://en.wikipedia.org/wiki/In-flight_entertainment. Accessed 28 Nov 2014
35. Inflight Entertainment: http://www.air-valid.co.uk/in-flight-entertainment.html. Accessed 20 Nov 2014
36. Transparency Barometer 2009, http://www.transparency.org/whatwedo/pub/global_corruption_barometer_20091. Accessed 29 Nov 2014

Chapter 7
Airlines

Abstract This chapter describes the role of airlines in the air transport system. It starts with a description of the various airline types and the associated network structures. The development of local and global operation strategies including the different concepts of low cost carrier (LCC) and flag carrier are discussed. A major part is also dedicated to flight planning and ticket pricing as core elements of the airline business models. Also, different aspects are discussed that drive the setup of airline fleets and the selection of aircraft. For operational issues, the organizational setup of an airline and its related stations are discussed. The chapter ends with a description of aircraft maintenance as a major driver for aircraft availability. Here the major activities and strategies for improvement are introduced.

7.1 Overview

Worldwide Airlines carry about 3.3 billion paying passengers per year along more than 50.000 routes, [1]. To deliver these services airlines employ about 4 million staff and operate approximately 25 thousand aircraft. The yearly revenue of all IATA airlines accounts for 500 billion US$, [1]. In Germany about 750 thousand jobs are directly associated to air transportation.

Air Transportation has been a growing market, which will go on at least for a while. Starting at annual growth rates during the seventies of around 9 %, the increase was damped to 6 % during the eighties and 4–5 % actually. This development resulted in a growing number of journeys from 1.5 to 2.3 billion between 1999 and 2010 for example.

Referring to the main product or service an airline offers the individual air travel allows people to move between two points. Here various options exist for the traveller, e.g. to take one direct flight if available, or fly from A (origin) to B (destination) stopping over C and eventually changing from flight 1–2 to reach B.

D. Schmitt and V. Gollnick, *Air Transport System*,
DOI 10.1007/978-3-7091-1880-1_7

The link between Origin and Destination is called "OD" or "OD pair". For example Deutsche Lufthansa provides air transportation services for more than 25.000 ODs.

Various options exist to select the takeoff airport as well as the transport mode to reach it, as shown in Fig. 7.1. Main reasons for the choice are schedule quality regarding time of day and day of week, ticket price, airline image, loyalty programs or even recommendation from the travel agent.

Medium to long range flights are often offered by different airlines. The passenger has the choice between different options (airlines and routings) to reach his destination. Figure 7.2 gives a typical example for various options to travel from

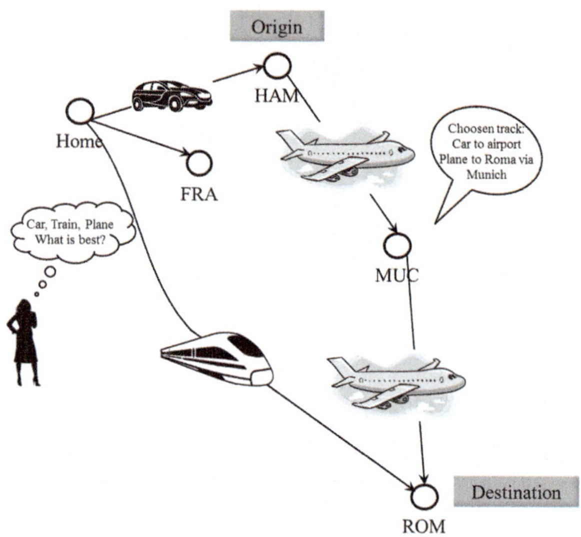

Fig. 7.1 Customers routing choice between modes of transportation

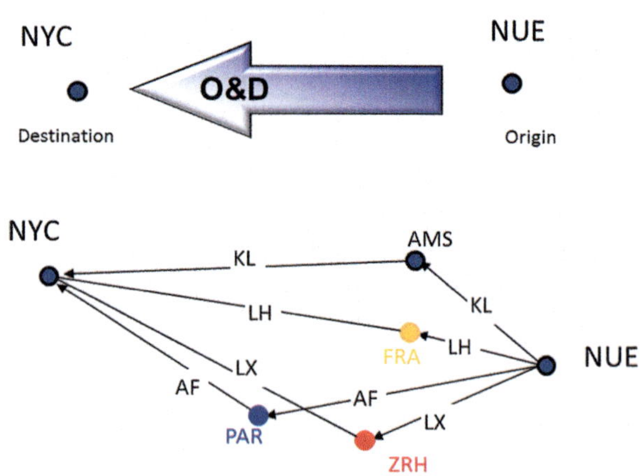

Fig. 7.2 Customers routing choice between different airline networks

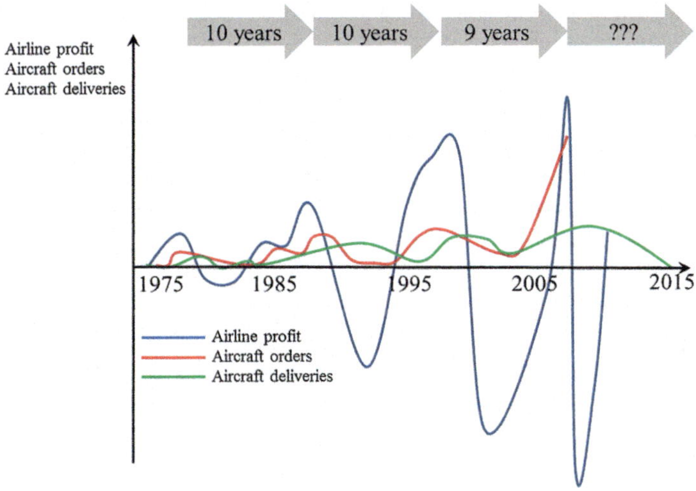

Fig. 7.3 Cyclic behavior of worldwide airline profit and aircraft orders

Nuremberg (Germany) to New York. Different airlines offer different flights, which also incorporate several transfer airports. In addition, one can see that airlines are in strong competition on a route (here: NUE—NYC) and try to offer interesting options.

Airline business is very sensitive and cyclic. The last decades showed a pattern, which repeats every 10 years roughly, Fig. 7.3. The increasing amplitudes of profit and loss are remarkable in this figure. For instance, the all-time record in profit of 13 Billion US$ achieved by all IATA airlines in 2006 was followed by an even higher loss of 16 US$ in the subsequent economic crisis in 2009, [1].

From the seventies up today airline profit was increasing at its maxima but also the sensitivity to economic disturbances became more intensive. The development of aircraft orders follows the profit with a delay of about 3–5 years, while approximately 1200 aircraft are delivered per year since 2007.

Looking at the airline size in terms passengers carried, the world biggest airlines are still located in the USA as shown in Table 7.1.

The American airlines are also those who actually placed the overall largest aircraft orders of about 2500 short range and 500 long range aircraft for the next years.

7.2 Airline Types

In the beginning of commercial aviation air transport was considered as a strategic national task to provide trading opportunities for the national society. Therefore nearly all airlines, which came up were owned and operated by a nation. Further,

Table 7.1 Worldwide top ten airlines based on scheduled passenger carried, [2]

Rank	Airline	2012	2011	2010	2009
1	Delta Air Lines[1]	164,571,529	161,379,000	162,614,714	161,049,000
2	United Airlines[2]	140,441,000	141,799,000	145,550,000	81,421,000
3	Southwest Airlines[3]	133,978,100	135,274,464	130,948,747	101,339,000
4	American Airlines	107,806,578	106,013,737	105,163,576	85,719,000
5	Lufthansa Group	103,050,000	100,602,000	90,173,000	76,543,000
6	China Southern Airlines	86,483,380	80,674,800	76,500,000	66,280,000
7	Ryanair[4]	79,600,000	76,400,000	80,553,580	66,503,999
8	Air France-KLM	77,450,000	75,780,000	70,750,000	71,394,000
9	China Eastern Airlines	-	68,724,960	64,877,800	44,042,990
10	US Airways	62,524,816	60,854,368	59,809,367	58,921,521

Notes

- Based on International Air Transport Association and individual companies' published figures
- ^1 Numbers on and after 2009 are for the merged Delta/Northwest
- ^2 Numbers on and after 2010 are for the merged United/Continental
- ^3 Numbers on and after 2010 are for the merged Southwest/AirTran
- ^4 Ryanair number based on actual tickets sold

the military aspect to provide transport capacity in case of crisis influenced the national airlines as well as national image and power. As already described in Chaps. 3 and 4, the national authorities are providing licenses for possible operators and therefore have an interest to use the national airline as a flag carrier.

The core business of an airline is carrying people and/or cargo over long distances to enable business and trade. From the beginning commercial aviation always targeted to long distances, like airmail services in the 1920s and 1930s of the last century.

Only in the last decades, when global air transport liberalization was further developing, business and market elements became much more relevant. Nevertheless the aforementioned aspects are still influencing airline development [3]. Today four main market segments in commercial air transport are considered:

- National or Flag Carrier (FC)
- Charter Carrier (CC)
- Low Cost Carrier
- Air Cargo Provider (ACP)

Figure 7.4 gives some typical representatives for national, charter and LCC.

Fig. 7.4 Overview about principle airline market segments

In the following these principle market segments and associated airline business models will be introduced.

7.2.1 National or Flag Carrier

The core business of civil aviation is air transportation of passengers and cargo between countries and continents, which is organized by FC. Due to their operational setup they are also recognized as network carrier. They operate complex global network systems with very different flight legs in terms of capacity, length and frequency. Central airports called hubs, where the long-haul flights start off, characterize the network structure of a Flag Carrier. These hubs are fed by a lot of short and medium-haul flights from so-called spoke airports. Further specialized terminals in hub airports as nodes for connecting the flight legs are operated to ensure a more or less seamless transfer from these feeding short haul flights to the long-haul flights.

A typical intercontinental flight of Deutsche Lufthansa carries 65 % transfer passengers, Fig. 7.5. In the given example about 43 different flights feed the flight from Frankfurt to New York. These transfer passengers are collected by short haul aircraft at airports all over Europe, change aircraft at Frankfurt hub and fly with a long-haul aircraft to New York. 10 % of the passengers change the aircraft a second time to travel to their final destination.

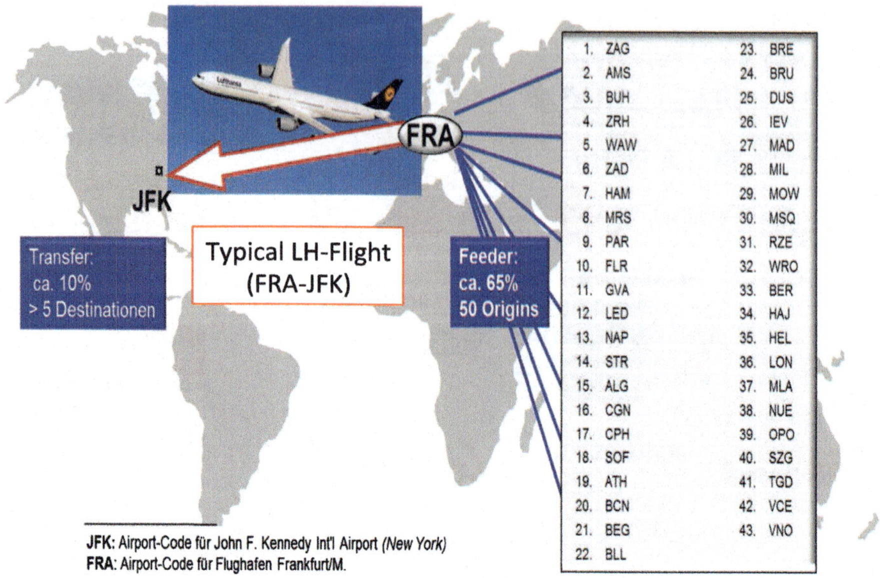

Fig. 7.5 Passenger composition in a typical Lufthansa long-haul flight [4]

The concentration on long-haul flights is the key market for Flag Carrier, where they get maximum yield. The short haul market is mainly to feed this profitable long-haul flights. Therefore the business is set up by the following attributes: Customer segment

- International and intercontinental passenger and cargo
- Concentration on time sensitive business travellers and intercontinental flights

Product and service

- Flights from main airports with good accessibility and airline lounges
- Connecting flights at hubs
- Seat reservation
- At least two cabin classes, three on long-haul flights
- Highly differentiated on board services

Production

- Use of main airports and hubs, therefore high airports fees and risk of delays
- Highly differentiated fleet to cope for different flight legs
- Aircraft utilization of feeder flights limited by connectivity requirements at hubs
- More and better paid crew to provide adequate service
- Ticket sales through computer reservation systems and own website
- Utilization: varying (75 % (Feeder) – >90 % (long haul))

In order to realize this business a typical network carrier fleet consists of short and long-haul aircraft.

7.2.2 Charter Carrier

Airlines delivering air transport services for passengers and goods on occasion for a specific demand are denoted as charter carriers. In international air law the notion is "non-scheduled traffic", because these carriers do not provide a regular public scheduled transport service. Charter operation is very common in holiday traffic. Travel agencies buy seats in flights from a CC at own risk in order to combine them with hotel stays and transfer transport services to a vacation package, which is sold to travellers for a lump sum.

Typical attributes of a CC include

Customer segment

- International and intercontinental passengers
- Concentration on holiday and leisure travellers

Product and service

- Flights offered on a seasonal basis mainly
- Flights from airports with good accessibility and seasonal capacities
- Direct flights mainly
- Seating considerably more dense in comparison to scheduled flights
- Minimum or on request board services only

Production

- Aircraft capacity blockwise sold to travel agencies
- Seats are guaranteed for whole or part of aircraft capacity
- Ticketing mainly done by travel agencies
- Utilization: Very High (load factor of 80–90 %)

Recently the boundary between charter and scheduled carriers diminishes as charter carriers progressively position themselves as scheduled airlines by selling seats directly via their own websites like LCCs.

7.2.3 Low Cost Carrier

Low Cost Carrier were only able to develop due to the liberalization and strong economic orientation of civil aviation in late seventies. Therefore special importance has gained the differentiation of low cost and network carriers. While LCCs concentrate on providing continental air transport passenger services to selected destinations, network carriers organize the intercontinental transportation of passenger and cargo.

The pioneer of the low cost segment, Southwest Airlines, started operations in June 1971. Due to missing traffic rights the company was restricted to Texas at first, Fig. 7.6. Nevertheless, this was the beginning of a massive structural change in

Fig. 7.6 The beginning of low cost air transport—Southwest Airlines

commercial air transportation. After the liberalization of air traffic in the USA by the Airline Deregulation Act in 1978 Southwest and other low cost airlines quickly expanded.

In Europe liberalization of air transport began in 1987 and took 10 years until reaching a certain level of dissemination. Therefore European LCCs developed in the 1990s. In 1996, a few LCCs connected 20 airports with 800 flights per week, Fig. 7.7. Today LCCs are well established as an important part of the air

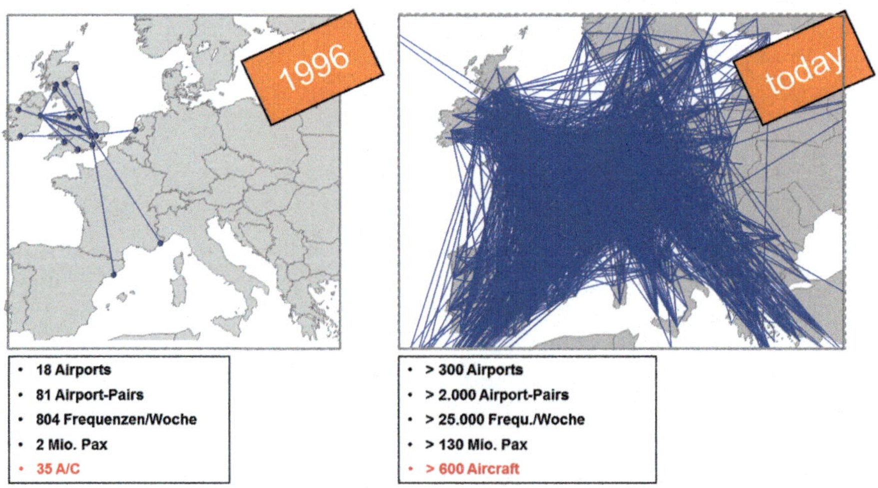

Fig. 7.7 Low cost carrier network development in Europe

transportation system. In Europe alone in this market segment 600 aircraft provide 25,000 flights per week.

The reason for the compelling success of LCCs is the consistent business model. They address mainly price sensitive customers, provide only transportation as core service and reduce complexity and cost on the production side to a minimum. Typical attributes of a LCC, where the numbers in brackets indicate cost savings compared to network carrier, include:
Customer Segment

- Only passengers (no belly cargo)
- Concentration on price sensitive travellers (best price strategy)

Product and Service

- Only one cabin class at high seat density (16 %)
- No seat reservation (3 %)
- Catering and other services on board only against additional pay (6 %)

Production

- Use of smaller airports in the vicinity of metropolis for lower airport fees and less delays (Lübeck (Hamburg), Frankfurt-Hahn, Gerona (Barcelona) Paris Beauvais-Tillé),
- Fast aircraft turnaround (6 %)
- Used airport often less accessible, but provision of airport shuttle from city center
- Standardized fleet with only very few aircraft types, often only 1 (2 %)
- No connecting traffic
- Minimum crew (3 %)
- Distribution of tickets on own internet websites (8 %)
- High utilization of aircraft >80 % (3 %)
- Simple price structure

Both, reducing the scope of the product and delivering the remaining core product more efficient, result in only 40 % of the cost per passenger kilometer compared to a network carrier.

Looking at the typical leg length LCC flights concentrate very much on short and partly medium ranges of 600–5000 km mainly.

There were some approaches in the past to establish the low cost flight business model also on long range legs like AirAsia, AirBerlin, [5]. These attempts failed because comfort becomes more and more relevant the longer the distance is.

For a given average stage length LCCs operate higher numbers of flights and therefore achieve higher aircraft utilization, Fig. 7.8.

As shown in Fig. 7.8 LCC are able to perform 1–2 additional flights per day compared to a flag carrier like Lufthansa, which increases utilization and productivity of the airline. However it must be noticed at this point, that short range flights of a flag carrier are essential to serve the long-haul flights with passengers. So the strategy and business of short range flights in this case is quite different.

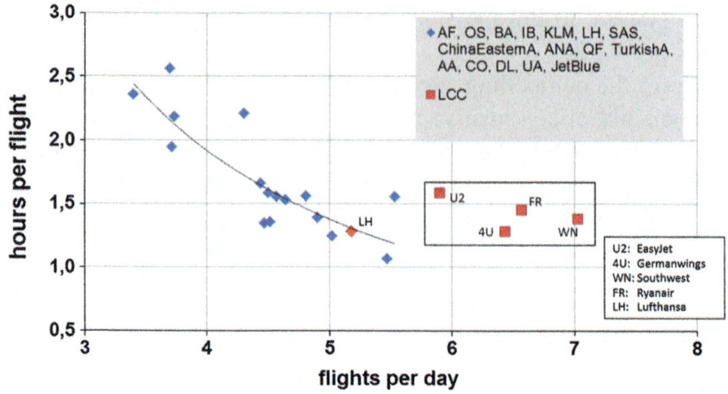

Fig. 7.8 Productivity of low cost carrier and flag carrier

7.2.4 *Alliances*

Due to the freedoms of the air it was not easy or possible in the past for a national airline to operate in another country. One way to extend its own network and to enter new markets is the creation of alliances. Alliances like StarAlliance, OneWorld or SkyTeam were created in the past on this basis and allowed the member airlines to extend their flight product portfolio significantly. As an example Lufthansa as one founder of StarAlliance is now able to extend their own destination network from approximately 400 to about 1200 destinations provided by the other 24 partners, Fig. 7.9.

It is a common strategy of all alliances to cover all globally interesting market segments by selecting partners with appropriate networks. Based on common quality

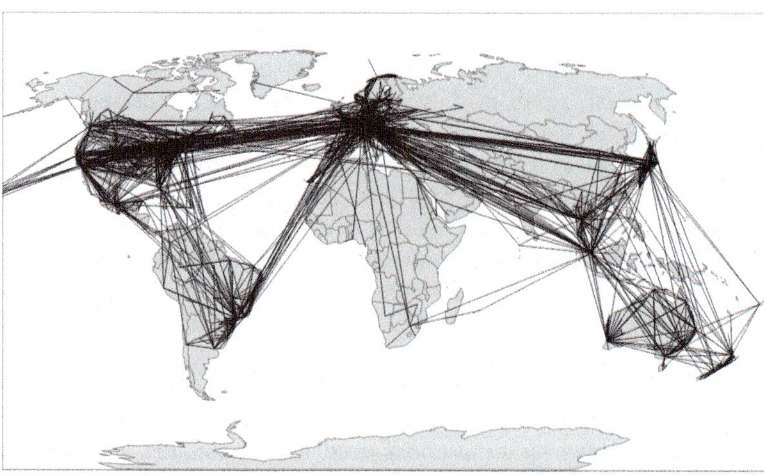

Fig. 7.9 Star Alliance global network

standards member airlines adapt their flight plans to each other in order to make operations more efficient. Such cooperation is also useful to manage slot limitations at airports and provide access to destinations without own flights. Fleet sizes of individual airlines can be reduced and individual flight load factors are increased. Also market presence of an individual airline is increasing through the brand of the alliance. Table 7.2 provides a comparison of the main characteristics of these alliances.

Table 7.2 Airline alliances and market share

	One World	Star Alliance	Sky Team
Members	28	13	19
Destinations	883	1328	1024
Countries	195	151	178
Daily flights	21,900	10,117	15,207
Overall fleet size	4701		2853
Passengerp.a.	727 Mio. Pax	353.5 Mio. Pax	569 Mio. Pax
Members	Aeroflot	Adria Airways	Air Berlin
	Aeromexico	Aegean Airlines	American Airlines
	Air Europa	Air Canada	British Airways
	Air France/KLM	Air China	Cathay Pacific
	(Northwest)	Air New Zealand	Finnair
	Alitalia	ANA	Iberia
	China Airlines	Asiana Airlines	Japan Airlines
	China Eastern	Austrian Airlines	LAN
	China Southern	Blue 1	Qantas Airways
	Czech Airlines	Brussels Airlines	Royal Jordanian
	Delta Air Lines	Croatia Airlines	S7 Airlines
	Kenya Airways	EgyptAir	
	Korean Air	Ethiopian Airlines	
	Tarom	LOT Polish Airlines	
	Vietnam Airlines	Lufthansa	
		Scandinavian Airlines	
		Singapore Airlines	
		South African Airways	
		SWISS	
		TAM Airlines	
		TAP Portugal	
		THAI	
		Turkish Airlines	
		United Airlines	
		US Airways	

It is vital for the success of airline alliances that a common culture of work is established, which is really challenging and critical because of the partly different national cultures and attitudes of the various airlines. This affects for example:

- Vision and strategy
- Quality and standards
- Market positioning and orientation
- Establishing a win-win-situation
- Through cooperation trust in place of dominance

From passenger perspective alliances are valuable because a passenger can book a flight from A to B in one step although several airlines are involved. So the entire alliance network is available to the passenger and the entire air travel becomes more seamless, especially, if transfers between flights and airlines are involved. Mileage and bonus programs including all services are valid across all partner airlines and also the use of all lounges is possible.

Based on this two perspectives the air traffic market of today shows strong global growth in passenger movements and tightening competition. The mobility of customer increases and companies act cumulative globally. In this environment awareness for price and service strongly increases.

Similar trends are visible in the air cargo market, where integrated logistic demands press for air cargo alliances. But cargo is partly more complex, depending on the time aspects and the additional competition between freight forwarders and cargo airlines, Sect. 3.5.

So the originally national-oriented airline industry moves more and more into globalization. Also the growth of the low cost market drives the network carrier more into extended cooperation. On the other hand the upcoming LCC and the massive decrease in ticket price created a strong increase in passengers. Deregulation intensifies the competitive situation in all four main airline segments mentioned before.

In the future airlines are turning their focus in their development much more on customer orientation. While in the beginning of air transport the general task to move people on airways was the prime strategic task, today and tomorrow services become much more relevant and key buying factors. Therefore marketing and distribution concepts will be modernized. Further strategic factors for success will be:

- More orientation towards customer expectations and customer binding
- Partners become more important for market penetration
- Revenue strategies become vital

Due to the increasing competition a consolidation of the airline market will be seen in the next years, as it happened in the past in the United States, Fig. 7.10:

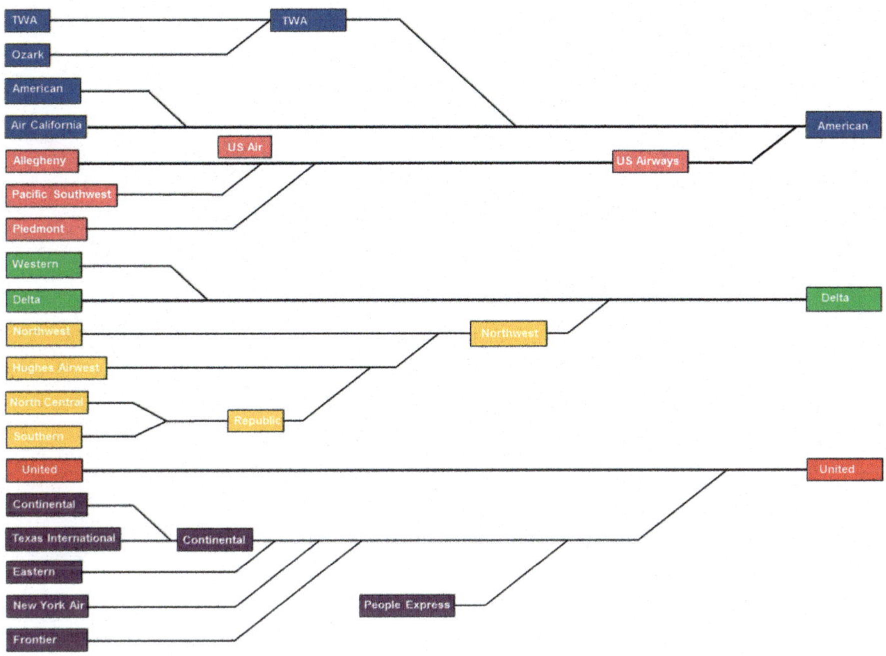

Fig. 7.10 Consolidation of the US airline market

Looking at the US market actually three major airlines exist despite the LCC. Also in Europe first mergers are observed like Air France/KLM and Lufthansa with Swiss, Austrian Airlines and SN Brussels Airlines Other flag carrier like Alitalia will follow.

7.2.5 Air Cargo Provider

Air transportation of cargo is a relevant business in civil aviation. It is done via designated cargo aircraft or in combination with passenger transport, when the lower deck cargo compartment (belly cargo) is used. As mentioned in Sect. 3.5 pure cargo aircraft are mostly refitted from former passenger aircraft. Only a small share of worldwide cargo aircraft are newly built aircraft. Remarkably air cargo covers only 1 % of the worldwide amount of goods being carried. But this very small share represents about 40 % of the worldwide cargo value. For certain types of goods the value of time is crucial and therefore air transportation pays off. Figure 7.11 shows the main air cargo flows worldwide, which are linking Asia with Europe and the United States mainly.

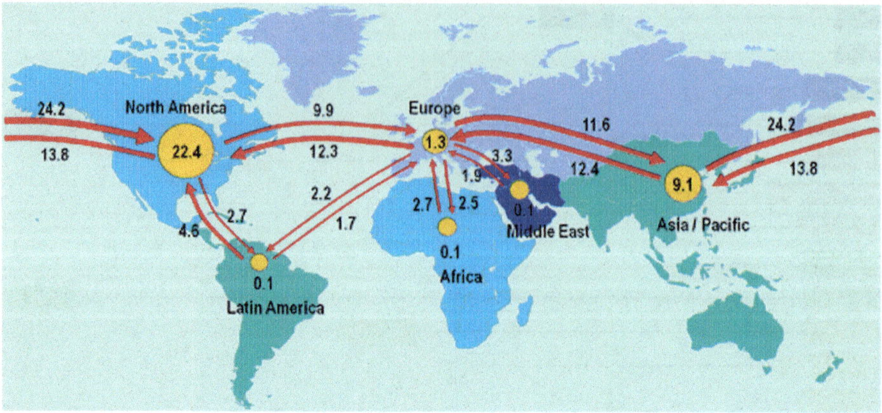

Fig. 7.11 Worldwide air cargo flow in billion tons, [6]

Typical goods for air transport are those, which lose their customer benefit over time quickly like:

- physically perishable goods would not sustain a transport process for weeks.
- news papers, mail, movies.
- seasonal clothes, and products with a high frequency of new models due to rapid technology development.
- goods of high value per weight and volume in general.

Air cargo greatly reduces the capital lockup cost and the risk of thievery for these goods while being transported. Some examples are:

- Electronic consumer articles like smart phones or HiFi equipment
- Aircraft spare parts
- Medical goods like medicine and medical technical equipment
- Chemical products
- Soft goods
- Flowers
- Animals
- Fruits

Also the business of mail-order companies (integrator) creates a significant part of air cargo, Sect. 3.5.

Most of the air cargo is carried as belly freight with passenger aircraft, while pure cargo aircraft transports approximately 40 %. Looking at the air cargo market it is very sensitive to the global economic development. The development of air cargo is an early indicator the economic evolution. Actually global air cargo is growing about 5–6 % per year, which is slightly higher than the passenger transport growth. Air transportation of cargo is more costly and has a bigger environmental impact than transportation on ground by truck, rail and ship. Nevertheless, the competitive advantages of air cargo versus ground transportation are:

- *Speed*—lowest runtime in comparison to ground transport
- *Reliability*—outmost in time transportation due to fixed flight schedules
- *Security*—low possibility of unauthorized intrusion and lower exposition to environment and vibrations.

At company level it can be differentiated between cargo airlines, providing only air transportation and so-called integrators, Table 7.3. In the first case, a shipper is managing the entire transport process and also orders the air cargo service, while in the other business model the air cargo company is managing the entire transport from door to door. While the shipper handles all kind of freight the air cargo integrator mainly offer services for freight up to 50 kg.

Air cargo is mainly competing with ship, land and rail transport. While on short ranges up to about 1500 km trucks and trains dominate the market on long haul tracks air cargo is highly competitive due to the timely advantage. For example transport of electronic equipment between Hamburg and Shanghai lasts about 30–35 days if it is carried by ship. Rail will take about 17 days, while the aircraft can make it within 3–4 days [7]. On the other hand transport cost are inversely related.

The achievable yield of air cargo is significantly decreasing since a couple of years since there are strong overcapacities. In 2008 about $4 yield per kg freight was achieved, while today it decreased to 0.1 $/kg approximately.

Compared to passenger transport air cargo typically is a one way track. From logistical perspective this means, that devices like pallets and containers might be

Table 7.3 Comparison of air cargo shipper and integrator process chain

Classical door-to-door air cargo transport	Integrated door-to-door air cargo transport
Product/service: all types of freight	Product/service: max 50 kg packages
Sender	Sender
Carrier	Integrator
Haulage company	Duty
Handling agent carrier	Integrator/airline
Duty	Duty
Handling agent airline	Integrator
Airline	*Recipient*
Handling agent airline	
Duty	
Handling agent carrier	
Haulage company	
Carrier	
Recipient	
Overall duration: 5–6 days	Overall duration: 2–3 days

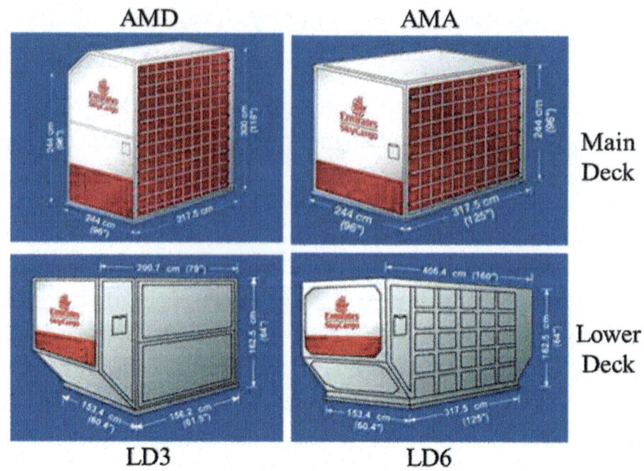

Fig. 7.12 Examples of air cargo container

carried back empty, which creates cost. Some typical types of air cargo container are shown in Fig. 7.12.

Despite other aspects this leads to a concentration of the air cargo network which is mainly based on 15 big air cargo hubs, where Frankfurt, Paris, Memphis, Anchorage, Hong Kong, Tokyo, Shanghai and Singapore are some major representatives.

7.3 Network Management

7.3.1 Traffic Flows and Networks

The way the airline is offering its service depends directly on the business model. Since network or flag carrier had its historic origin in the national wish to link foreign countries over long distances, they typically offer long range flights across and between countries and continents. Those flights are today performed with large high capacity aircraft like the A380 or B747-8. In order to fill them as much as possible they are typically operated from a limited set of very big airports. As those airports never have a catchment area, which provides sufficient passenger, the passengers have to be carried to and redistributed from these very big airports—the hub airport, Fig. 7.13. This logic leads to the so-called "hub and spoke" airline network concept, which is typical for flag carriers.

The principle of such a network is to feed central hub airports with passengers by using short range aircraft. Those short range aircraft pick up people in a certain area, which may have different destination targets and bring them to the hub. Here these passengers can transfer to the long range flights, which bring them to the final destination.

A good hub features a design with many connections and short transfer time. The incoming and outgoing flights are batched in two waves following each other

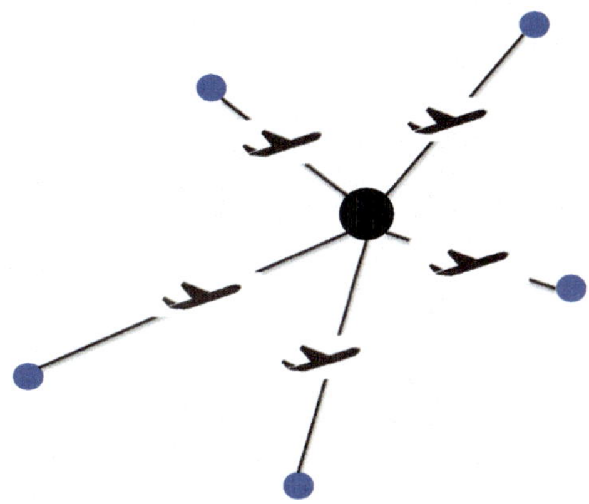

Fig. 7.13 Principle structure of a hub and spoke airline network

with an adequate time lag to ensure a balance of a short transfer time and the safeguarding of the passengers connection. Following this strategy short range feeder and long range flights are connected. This pattern of in- and outgoing waves repeats at different times of a day, especially early in the morning, afternoon and in the late afternoon.

From a quality of service (QSI) point of view, it could be best to schedule all incoming flights and shortly after all outgoing flights at the same time, respectively. But apparently this is not possible due to restricted airport capacity. The concentration of flights in waves and the necessary connectivity of passengers leave a hub airport vulnerable to delays. With each wave a lot of aircraft—from a 500 seat A380 to a 70 seat short haul aircraft—want to arrive and depart at the hub simultaneously to provide all the connections for passengers. Only a small irregularity can produce many subsequent delays. For that reason some hubs have shifted to the rolling hub concept which spread a higher number of smaller waves over the day using airport capacity more evenly, reduces delay and provide passengers with more alternative connections but also slightly longer transfer times.

The business requires the airline to offer a lot of long distance flights (continental and intercontinental) from the hub and further it always implies at least a two legs air travel (A to B over D). Hence, the hub enables the bundling of traffic flows between different ODs in one flight. This again allows the airline to increase aircraft load factor, increase frequency of service or to use bigger and mostly more efficient aircraft. So hubs and spokes are vitally linked. In real business such network airlines operate one or more hubs, which are linked to same or different spokes.

Deutsche Lufthansa operated Frankfurt as the only hub over many years until Munich has been assigned the second hub within the flight network. Since the foundation of StarAlliance (Table 7.1), Lufthansa has synchronized further hubs from alliance partners with its own network. The advantages are obvious—it is now possible to offer travellers manifold routings in an integrated system due to coordination of schedules worldwide.

Fig. 7.14 Three hub models—global overview

There is no unique hub type, but rather three different models as shown I Fig. 7.14 for reasons related as much to the markets as to geography [4]:

- United States
- Europe
- Asia Pacific

The American model is essentially connecting medium-haul flights with other medium-haul flights, primarily domestic. This enabled national carriers to grow over many years, but it seems no longer appropriate today. In a market where very large volumes of traffic between major cities exist, direct flights are economically viable.

Taking over these routes and bypassing the major hubs, let low cost carrier have successfully penetrating the U.S. market. This is a major factor behind recent difficulties encountered by traditional U.S. carriers, as 80 % of their turnover is generated by domestic traffic.

A second family of hubs is found in Europe, connecting medium-haul flights with long-haul routes. The European model links medium-haul flights with long-haul flights, thus enabling operators to channel small traffic flows which alone would not justify the opening of new services. This model still has a promising future ahead of it, as direct long-haul services out of Europe continue to be economically viable. By 2013, only London, Paris, Moscow and Frankfurt will have the potential for carrying more than 100,000 point-to-point passengers a year on more than five intercontinental routes.

Third, carriers in Middle East and Southeast Asia, linking up three continents, Europe, Asia and Africa, develop the third type of hub. Connecting long-haul flights with other long-haul flights was developed first by carriers in Southeast Asia (particularly Singapore Airlines), due to their specific geographical situation. Today

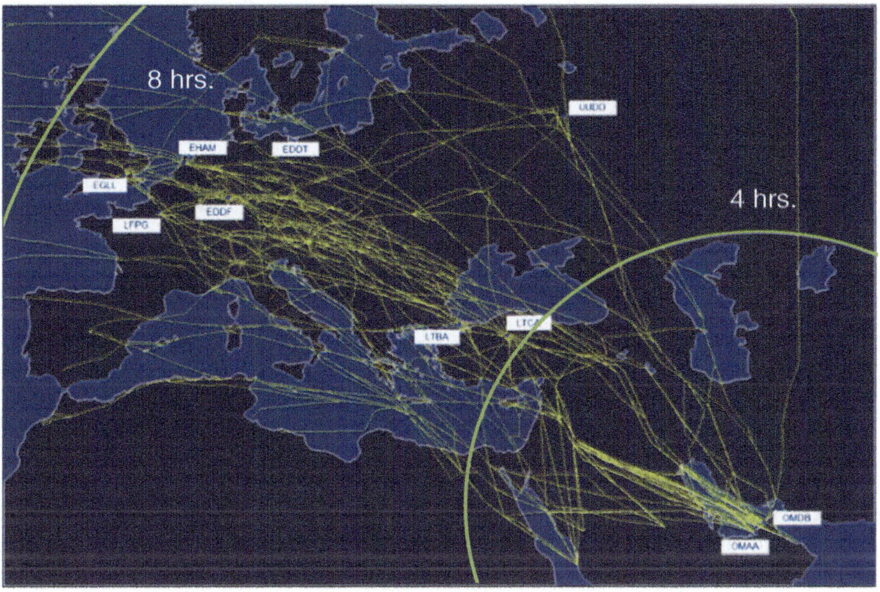

Fig. 7.15 Global destination catchment area of Middle East airports

also Emirates and Qatar Airways are using this model, which concentrates on linking several continents. Due to time shifts of several hours on most of the long-haul flights, operational restrictions like airport curfew hours and passenger preferences determine the possible and valuable scheduling of flights.

It must be mentioned at this point that the upcoming Middle East region will change the world air traffic flows.

As shown in Fig. 7.15 within 8 h of flight Middle East airports and related airlines cover 2/3 of the world population, which is a clear geographic strategic advantage of this area. Today and much more in the future especially European and American airlines will be in strong competition to these airlines, offering a lot of attractive flights between Europe and Australia/East Asia.

A further major advantage of Middle East airlines (Emirates, Etihad, Qatar et alii) is a very low fuel price, as their home countries are holding major crude oil reserves. Since airlines, airport and oil resources are in one hand, the operational business model of those airlines can have some significant advantages. Alliances of the other global airlines may help them to survive in that market. The amount of transfers within an OD will decide of the success in terms of seamlessness and comfort, but also price.

A different network business concept is the so-called "point-to-point" concept, as shown in principle in Fig. 7.16. It is very typical for continental operator and especially LCC. In this case, no long range flights are fed but various airports are connected to each other. In general the point-to-point model is the preferred option for all passengers. The passenger prefers to have a direct flight from O to D. But for

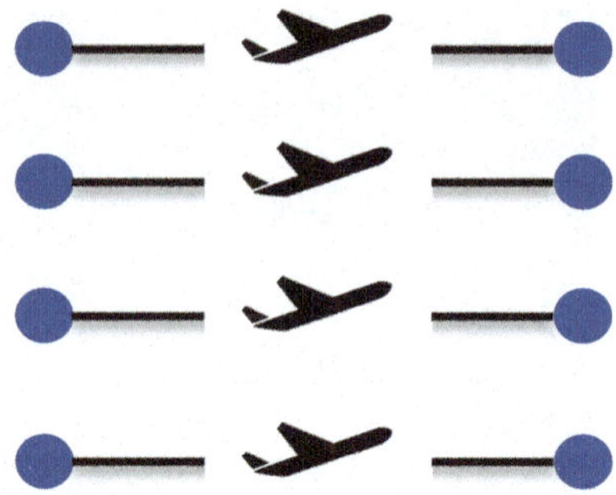

Fig. 7.16 Principle structure of point-to-point airline network

direct long range connections in most cases there is no sufficient demand on such a single leg to create a viable business with an adequate load factor, justifying a daily connection with a long range aircraft, [5].

The rotation of aircraft in such a network is different to that for a hub and spoke network. In a point-to-point network aircraft are mainly flying between the two connected locations, while in a hub-and-spoke system aircraft rotate across different airports. Low Cost Carriers are exclusively using point-to-point networks. For economic reasons, they are using smaller airports in the vicinity of big cities and call them London Gatwick, Barcelone Gerona or Frankfurt-Hahn, where in fact the real airport is sometimes up to 100 km remote from the advertised real city!

7.3.2 Flight Planning

7.3.2.1 Time-Based Steps in Flight Planning

Flight planning can be divided into four time scales with increasing accuracy, which are described in Fig. 7.17. The first scale focuses at long-term market considerations 2–10 years ahead and offers a high degree of freedom to elaborate the flight schedule. Traffic flow information is retrieved from mass raw data by extensive data processing. Based on this market and competitor information, targets on market shares are defined for the strategic traffic flow bundles.

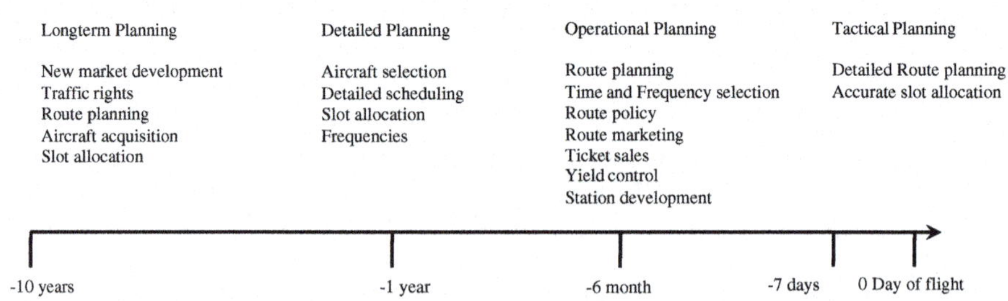

Fig. 7.17 Time horizons of flight planning

One year to six month before operation the strategic flight planning creates real global production scenarios (scenario flight schedules). Balanced airport and airspace capacities, slot allocation, traffic rights and curfews between airlines and airports are the major objectives of this planning phase. The ability to change the individual flight being offered decreases as the publication of the schedule approaches. One year prior departure of respective flights ticket sales begins according to the published schedule. Now the offered flights are fixed, an aircraft type has to be selected for each flight and the aircraft turnaround plan has to be established. The focus of the airline switches from "setting capacity" to "control the demand". A new station might be developed, marketing for the new route has to be started, demand has to be controlled, the development plan for this new route has to be established and ticket sales have to be carefully monitored and incentives have to be defined.

The third time scale aims at the pre-tactical planning some days before the day of operation. Detailed slot allocation is performed and ticket prices have to be adjusted in order to maximize seat utilization of aircraft (aircraft load-factor).

The fourth time scale is the accurate planning of the flight at the day of operation. Here the actual weather situation, airport capacity, aircraft status and ATC capacity are considered to operate the flight as close as possible to the scheduled flight plan.

7.3.2.2 Influencing Factors in Flight Planning

Different factors have an impact on flight planning and have to be coordinated [4]. These factors can be related to demand, operations and restrictions as described in Fig. 7.18. Demand and market factors determine the revenue an airline can achieve with a given flight plan.

The consideration of these factors increases the chance to choose attractive markets and serve them with good flight products. To compare the economic attractiveness of different ODs airline network planers use the yield as a metric. It is defined as revenue divided by passenger and kilometers. In order to maximize the revenue a network planer will strive to offer transportation services for ODs having a high number of potential passengers. These passengers should be willing to pay a reasonable price for this direct route. The passenger segmentation (business, leisure) and respective preferences for the ODs need to be analysed as basis for decisions on the mix of cabin classes and the schedule. Finally, the potential demand for air cargo needs to be taken into account to assess its revenue contribution. All these analysis have to consider also the competitive situation on the single route, because the amount of airlines offering flights as well as their individual market strategy on that route drive the ticket price.

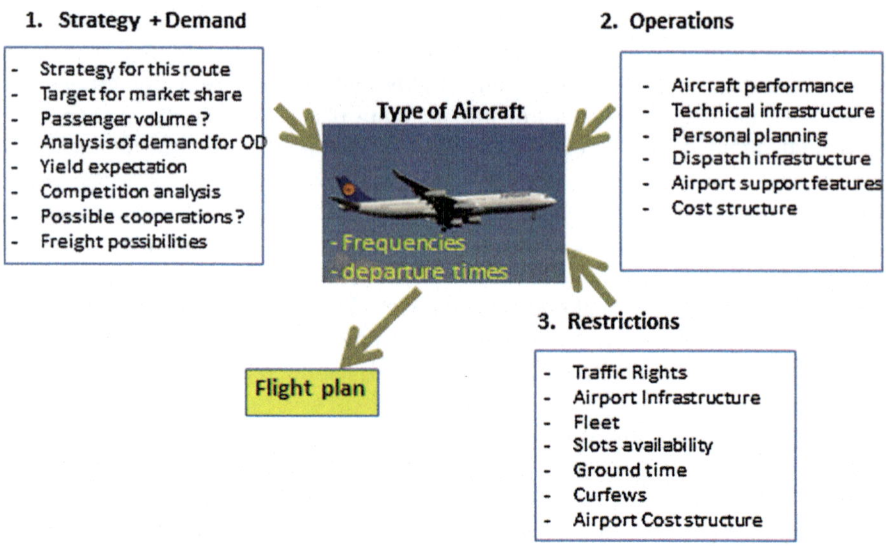

Fig. 7.18 Influencing factors for flight plan development

Airline supply factors determine which flight plan can be realized at which cost? Operation planers have to make sure that a flight plan can be flown with available aircraft. This comprises the number of aircraft with the needed payload (passengers and cargo), range and speed capabilities. Operation planning has to make sure, that aircraft are available at the correct airport when needed, sufficient ground time for maintenance and overhaul is retained and reserves are contained to ensure punctuality. The result is an aircraft rotation plan for every aircraft for every single day, Fig. 7.19.

Aircraft and ground stations also have to be crewed with pilots, cabin and service staff, taking into account working time restrictions given by law and labour agreements. Thus, from every marketing-optimized schedule a "flyable" and resource efficient rotation plan has to be developed.

7.3.3 Flight Plan Utilization and Ticket Pricing

Airlines strive to maximize revenues by offering customized transportation products and controlling demand and price with sophisticated IT-driven systems. "Product" means in this context the flight itself with a certain level of comfort and the associated services.

It is characteristically for a travel product, that this product is produced in time of travel.

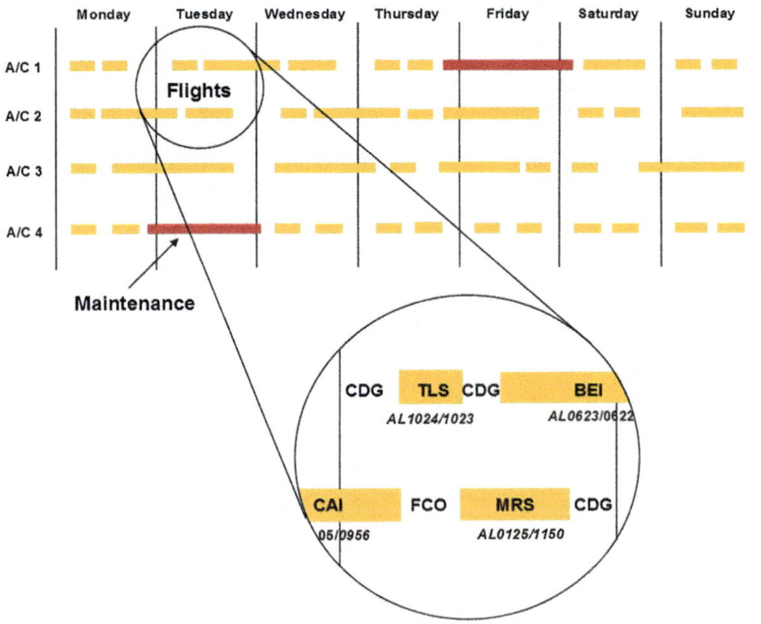

Fig. 7.19 Aircraft rotation and maintenance planning

7.3.3.1 Control of Demand and Price

From the time the schedule is published an airline is committed to its planning. Although some aircraft assignment may still be swapped, the capacity on most legs and therefore ODs is fixed.

A main characteristic driving the airline industry, but also other transportation sectors and, e.g. hotels, is the fact that it produces non-storable goods.

When an aircraft departs every seat is a "produced" available seat kilometer (ASK).

If the seat has been sold it turns into a revenue passenger kilometer (RPK). If not, the flown empty seat kilometers create economic losses, since they affect fuel burn by their weight, but cause no revenue. Moreover, the marginal cost of taking on board an additional passenger on an otherwise empty seat is very low. This is the main reason an airline partly sells its tickets for very low ticket prices, far below average production cost, to ensure a sufficient aircraft load factor. Inevitably, the airline has to sell another part of the tickets—the major part—at normal and much higher prices to achieve average revenue for the whole aircraft, which exceeds the costs of the flight. Hence, airlines sell seats of a flight at different prices, even in the same cabin class.

During the flight planning process airlines try to forecast the demand of passengers for each flight and cabin class with complex demand modelling (QSI, logit). After the schedule is published ticket sales are monitored continuously and

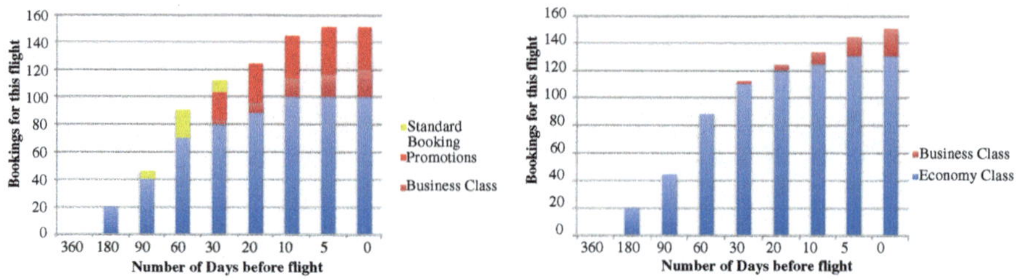

Fig. 7.20 Booking pattern for economy class

compared to historical booking patterns, Fig. 7.20 left. At the right side of Fig. 7.20 the reactions to underbooking situations are shown, where especially promotions drive the booking rate to a higher load factor.

If the current booking level is below the historical value for the respective day before departure, the offered ticket price is lowered and some specific promotion campaigns will be launched and vice versa. Given, a sufficient price elasticity of demand the airline can achieve an acceptable load factor by this simple mechanism. Airlines even practice a slight overbooking of flights, because they frequently experience passengers who have booked, but do not arrive for boarding (no-shows). The use of statistical models ensures that additional revenues from overbooking are not overcompensated by the potential cost for compensation and accommodation of passengers with denied boarding [4].

The airlines revenue management systems are in reality much more sophisticated with hundreds of "nested" booking classes (baskets). Normally, a number of seats are reserved for certain baskets. Nested means, that if a high price basket is sold out, it can access seats from low price ones (an upgrade of passengers for a specific flight), but not the other way round. The nesting concept prevents to refuse high value passengers and avoids blocking high price baskets by low price bookings.

For network carrier revenue management is even more complex, as each flight is commonly part of different ODs. As intercontinental flights are the cash cow of network carriers, revenue management systems must retain sufficient seats in the various feeder flights. This could even mean to operate feeder flights, which do not earn their direct costs themselves, but carry passengers for a high yield intercontinental flight.

7.3.3.2 Customer Segmentation and Product Development

In order to maximize airline revenues a price should be charged the passenger is barely accepting to take the flight. This economic concept is called full price discrimination, which seeks to capture the differing maximum willingness to pay from every single customer. The price-demand curve in Fig. 7.21 can be interpreted

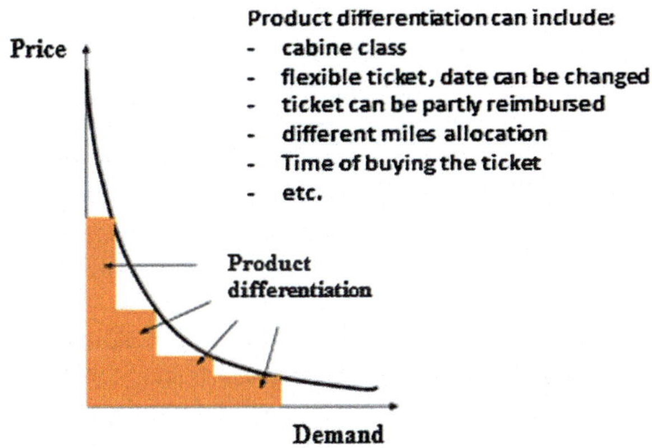

Fig. 7.21 Product differentiation and price allocation

as an infinite number of points which each represent a customer, willing to pay a certain price for the offered product [4].

As it is difficult to sell identical units at various prices, airlines use a wide mixture of product differentiation. The most obvious are cabin classes with different service levels (first, business, economy plus, economy).

Within a cabin class there are rates, which differ in additional services like baggage limit, on-board meal or priority boarding. The less visible, but still very important measures of product differentiation are those, which are related to:

- flexibility of ticket (to be partly reimbursed)
- departure time of the flight
- flight date can be changed with a certain amount of cost
- different bonus miles allocation for the flight
- etc.

From this perspective, seats of a flight in the same cabin class with the same sub-rate, but booked at different days are also defined as "different products".

Based on these opportunities airlines attempt to define transportation products, which are attractive to customer segments with different willingness to pay and therefore sell theses customized products at different ticket prices (steps in Fig. 7.21). The most obvious distinction is between leisure and business travellers, where the first is very much price oriented, while business traveller are more in favour of time and comfort.

7.4 Fleet Strategy and Aircraft Selection

The airline operates an aircraft fleet on a certain route network to produce its travel services.

Depending on the leg length share, the different destinations, and the expected demand of passengers and cargo different types of aircraft are needed for economic profitable operation. The aircraft selection process is therefore a very important and difficult task for each airline.

While LCCs are typically operating fleets of one or a very few types of aircraft (Ryanair uses only B737 and Easyjet only A319), network carriers operate very heterogeneous fleets with many different aircraft. For example in 2012 Lufthansa operated 12 different types or variants of aircraft.

Depending on the overall route network, several types of aircraft are needed, regional and short range aircraft for the continental routes and medium and long range aircraft for the intercontinental routes.

Also the size of aircraft will differ, depending on the demand at different routes. The shortage of slots and busy airports may force airlines to use bigger aircraft with lower frequencies at their preferred "best routes".
The philosophy of each airline, how to choose the different types of aircraft can be quite different.

- Some airlines prefer to have only one aircraft manufacturer, which may lead to special purchase deals with this preferred manufacturer.
- Some strategies go to use aircraft from several manufacturers, but use as much as possible always engines from the same engine manufacturer.
- Some airlines try to be open and just choose the best product of the market as defined by their specific network requirements.

There is no general strategy, as a lot of parameters, sometimes very soft and difficult to quantify, are involved like national, political, traditional aspects.

If different capacities are needed a more rational approach is possible. Depending on the route a certain aircraft size is used. If now demand will increase further, there will be a time where a bigger aircraft may be reasonable in order to fulfil the increasing market demand. This graduation in capacity is economically best if the next aircraft offers 20–25 % more capacity. As shown in Fig. 7.22 each

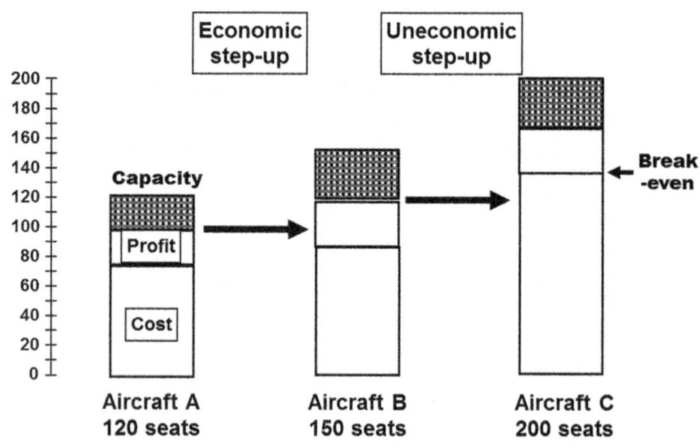

Fig. 7.22 Economic graduation in aircraft capacity

aircraft needs a minimum load factor to cover the cost of the flight. More seats additionally sold are leading to a profit on this route.

If the load factor will increase so strongly, that a lot of demand can no longer be fulfilled, a bigger aircraft has to be used. If the aircraft is too big, the basic cost will not be covered and the bigger aircraft will create no profit. An optimal step up in capacity can be achieved when the bigger aircraft will still be capable to earn money with the max. load factor of the smaller aircraft. This staggering in aircraft size is normally in the order of about 25 %, which can also be seen by the family concept of Boeing B737 family as well as Airbus A320 family concept, Fig. 3.16.

7.5 Flight Operations

The flight operation of an airline is closely linked to the airports and their capabilities, which are used in the airline network. The airline has their home base, which is the hub, but can also have several hubs in their network.

7.5.1 Stations

At each airport an airline is operating at various services must be fulfilled [7]. This is done either by the airline itself with own staff or, if by cooperating airlines or service agencies. The size of the airline station at an airport depends on the number of flights, departing from this airport. The following services need to be provided at a station (Fig. 7.23):

- Passenger services, sales and special services
- A/C handling
- Cargo handling

The airline station must be clearly visible to the passenger, when he is arriving at the airport by whatever transport mean. There must be a check-in counter to provide

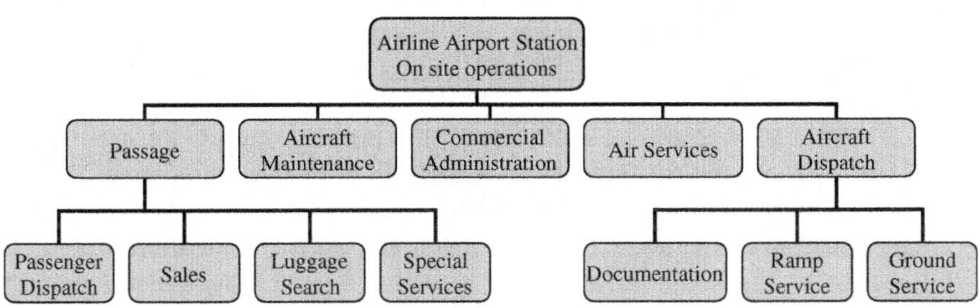

Fig. 7.23 Principle setup of an airline airport station

the boarding pass and baggage drop off. A ticket counter for all rebooking, baggage overweight, children support or similar passenger requests is also needed.

All ticket and check-in counter infrastructures are normally part of the airport, but these facilities will be rented to the airlines upon their needs.

If the airport is an airline hub, the airline will dominate the overall operation at this airport. Sometimes complete terminals will be owned and operated completely by the hub-airline.

The headquarter and airline administration will be located here as well as other major functions like crew training, maintenance center, training simulators, etc.

In case the airline requires only a small operation activity at the airport the question of "make or buy" occurs. The airline has to decide, whether own staff will be needed at this airport or the airport or other services might provide all required functions.

If the airline is part of a global alliance, they may negotiate with another partner airline to take care or join the efforts for their operational services at this airport.

One advantage of a global airline strategy is the provision of common services at outside stations and also the standardization of service quality.

7.5.2 Passenger Services, Sales and Special Services

The whole package of workload can be listed as follows:

- Passenger handling (check-in)

  In Europe most of the airlines have installed machines where the boarding pass will be issued automatically. The passenger has to type in either his reservation code, his E-ticket number or his passport/identity Card, and the machine will react and ask for a seat selection and provide the boarding pass. The baggage still has to be registered at the baggage drop off counter.

- Ticket sales and Customer support

  A counter for ticket sales is always required, to help and support the passenger in case of flight confirmations and rearrangements, if overweight or outsized baggage hast to be transported, if unaccompanied children (UMs) are travelling, if handicapped persons need a special support etc.

- VIP-service, lounges

  For the VIP persons, special lounges are normally offered. Even differentiations of lounges for frequent travellers and VIPs (Very Important Persons and Honorary members) will be offered at the airport. This feature is often provided in cooperation with other members of the global alliance.

- Baggage handling "Lost and Found"

  A specific service at the baggage arrival hall is required, where all passengers can address their problem, if a baggage did not arrive at the final destination. This service will then check and inquire, where this specific baggage has stranded and take care to send the lost baggage when finally arriving at the airport to the hotel or the address of the owner. This service is a unique service at the airport and normally handled for all airlines. However, this service is part of the airline operation and has to be paid by the airlines.

7.5.3 Aircraft Handling—Turnaround

Besides the passenger related services there are quite a lot of other functions and services, which an airline need at an airport station like:

- Timely preparation of all necessary handling documents for the crews (e.g. weight and balance, e.g. load plan of the aircraft with loading space for baggage and freight)
- Apron services: coordination of fuel, cleaning, water, control of loading and catering as prescribed (often done by the airport or other service providers)
- Aircraft handling has to be carried out during the scheduled ground time

Figure 7.24 shows a summary of activities to be performed during aircraft ground handling.

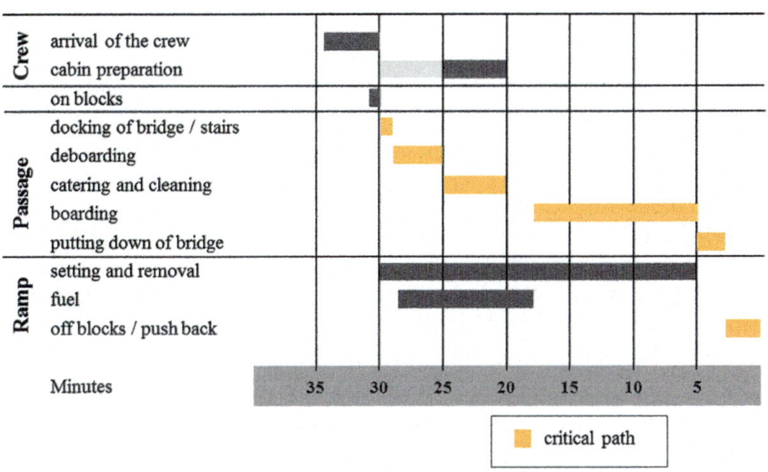

Fig. 7.24 Minimal ground time for a short range aircraft

A flight does not start just when the passengers are entering the aircraft. Before each flight several preparations are necessary, amongst others also planning and calculation of the best route for the flight. The flight route can change depending on wind and weather conditions and on traffic situations. Target of the route planning is to arrive at the destination on time, but also economic considerations have to be taken into account. Besides the fuel consumption on the different routes, there can also be quite different fees for en-route charges, which will be considered. In areas with high traffic volume, there may be less freedom for detailed optimizations. But on long-haul routes there are more operational options and the optimization process may bring some benefits to the airline. The shortest route may not always be the most efficient route to fly.

The dispatcher is supporting the pilots in this topic. As the dispatcher is doing this flight planning job for various aircraft of an airline the whole day he may be better informed about a lot of weather and traffic situation. He is normally also less paid and therefore beneficial for the airline operation.
Another responsibility of the airline on the airport is the decision process, when to close a flight. In case an incoming aircraft is delayed and there are several transit passengers, who may lose their connecting flight, there must be a well-balanced decision like:

- would it make sense to delay the start of the connecting flight for some minutes to take the delayed passengers still on board
- or would any delay of the aircraft cause following delays later on, so that delays should be avoided in any cases.

An interesting case is a connecting flight in the evening—the last leg of this aircraft for the day—where a delay of the aircraft would have no further consequences for the following flights. But if then a group of several persons is arriving with some delay at the transit airport, it would make sense to let the outgoing aircraft wait, so that the passengers will still receive their connection and come home at night before paying them a hotel at the transit city and get negative reactions from the passengers.
The process for dispatching a flight is rather complex, as there are several partners involved:

- the airport with their personal like bus drivers, drivers of aircraft push-back vehicles, gate bridge personal
- the ATM personal for flight clearance processes who are optimizing the slot distribution for takeoff and landing
- the airline who are trying to minimize the down-time at the airport.
- the "no-show" of passengers, who have registered their baggage but are not showing up at the flight gate are providing additional trouble for the airlines.

There is no generalized unique process for this dispatch process. It depends on the airport and its gate-apron structure, on the actual traffic situation on the airport, weather conditions, etc.

7.5.4 Cargo and Baggage Handling

Air cargo transport is a very specific area, which is part of aircraft ground handling. For completeness it is listed here. However, air cargo and baggage is described in Sect. 3.5 (air cargo market), in Sect. 7.2.5 (ACP) in Sect. 8.3.5 (airport baggage handling), in Sect. 8.3.6 (freight handling) and in Sect. 8.4.4 (planning of baggage and cargo handling). The important aspects of cargo handling are described in these chapters.

7.6 Aircraft Maintenance

Keep aircraft safe and operational is a fundamental pre-requisite of aviation. Therefore, legal baselines laid down in certification requirements like EASA CS25 1529 for continued airworthiness and the related appendices request for a detailed description of all relevant maintenance actions [7].

Accordingly EASA part 145 "Maintenance Organization Approval" sets standards for the company, which is performing maintenance tasks.

Often aircraft maintenance is a division of an airline but today also independent maintenance companies are acting especially for smaller airlines. Ensuring safe and seamless operation of the entire fleet is also a fundamental economic interest of an airline.

7.6.1 Maintenance, Repair, Overhaul

Aircraft maintenance in general covers all activities to keep the aircraft safe and operational. This includes inspections, services, repair but also modifications of components. All these activities are summarized also as Maintenance, Repair, Overhaul (MRO).

More in detail *inspections* cover pre and aft flight checks of aircraft and systems status and functions. These activities are performed at aircraft during turnaround. Mainly the cockpit and ground crews are in charge of this task. But also during hangar checks components are investigated concerning its operational health state.

Aircraft *services* comprise refill of lubricants, cleaning, and intensified functional checks of the aircraft and its systems. Also exchange of defect parts or components are part of maintenance services. The airline ground station maintenance crew in a hangar performs these tasks typically over night on a regular time base.

Repair of defects is a further level of maintenance where defect parts or components are either replaced or overhauled. Depending on the affected component repair tasks are performed on or off aircraft by the maintenance organization.

Modifications and improvements cover small adaptions of components and parts but also software, which are initiated by the aircraft design organization due to safety reasons, life time, functionality or performance improvements. Those modifications are finally requested or recommended by the certification authority.

Overhaul of aircraft and engine systems, components and parts summarizes all activities mentioned above to recover the safe operational state. It also includes exchange of material and equipment.

7.6.2 Maintenance Management and Organization

Maintaining an aircraft to keep it air worthy and operational is a time and resources consuming task. As mentioned before there very different activities, which have to be performed either on aircraft or off aircraft, Fig. 7.25. Off aircraft maintenance tasks refer to components, parts and systems, which are removed from the aircraft and maintained in dedicated shops.

Further these activities are distributed over long time schedules and also summarized to various maintenance blocks.

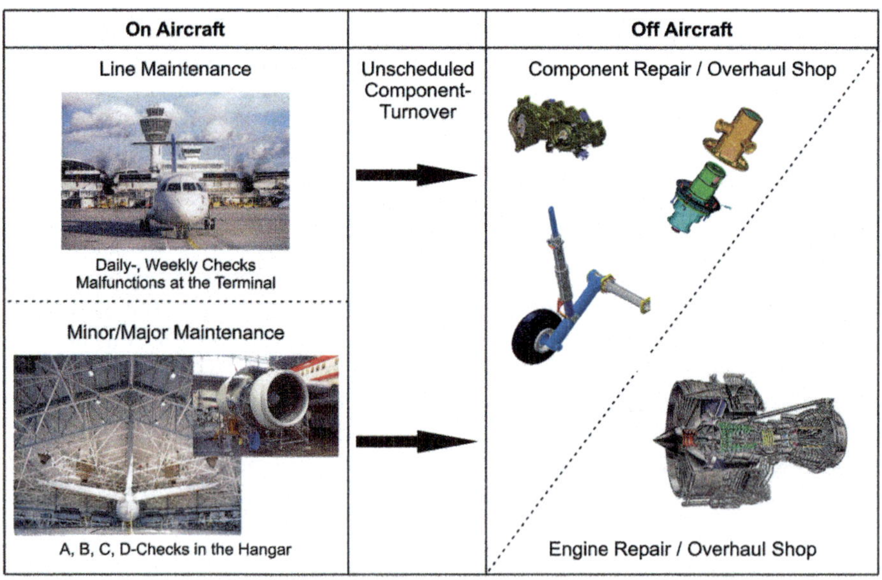

Fig. 7.25 Overview of On and Off aircraft maintenance activities

7.6.2.1 Line Maintenance and Technical Handling

Line Maintenance covers mainly services to keep the aircraft operational.
Mostly daily maintenance is connected with aircraft rotation planning (ground times, turn around)

- Visual checks and analysis of the documented measures values of the integrated systems during the flight
- Fueling, oil supply
- Cabin cleaning
- Electricity supply
- Removal of small claims
- Removal of so-called No-Go-claims.

7.6.2.2 Light Maintenance (A- or C-Check)

For light maintenance activities the aircraft will be taken out of operation. The following tasks are performed:

- All work orders, which have to be done in intervals of 50–1000 h of operation
- Controlling of the essential components and removal of findings
- Limited change of spares
- Development is only possible on special stations and home bases.

7.6.2.3 Heavy Maintenance and Aircraft Overhaul (D-Check)

The D-check is the most comprehensive maintenance block being performed every 8 years with

- All work orders, which have to be done in intervals of 1000 operation hours
- Detailed control of components which are difficult to access (structure parts, cells) partly with special equipment and test methods
- Layover days lasting several days at the dock

Table 7.4 provides an overview of different maintenance blocks and the related effort.

There are different strategies to organize the required maintenance tasks in the most efficient way.

7.6.2.4 Scheduled Maintenance

Traditionally maintenance tasks are performed as defined by the design organization according to section 1529 of the design standards on a scheduled basis. That

Table 7.4 Overview of different maintenance blocks

Event	Interval	Volume	Ground time	Effort (h)
Trip-check	Before each flight	• Walkabout the aircraft • Cabin and cockpit checks • Control of lubricants • Cabin cleaning	35 min	0.5
Service-check	Weekly	• Refill of lubricants • Thorough cabin cleaning	4 h	20
A-check	Every 230 Fh	• Service-check • Additional cabin and systems check	6 h	40
C-check	Every 13 month	• A-check • Detailled structural overhaul and system tests • Removal of fairings	30 h	700
R-check	Every 15 month	• Cabin overhaul	in parallel to other checks	10
IL-check (Intermediate Layover)	Every 4 years	• Thorough structural and cabin overhaul • Repair and polish of painting	2 weeks	12,000
D-check	Every 8 years	• Fucelage overhaul incl. all systems • Large parts exchange • New painting • Intensive cabin overhaul • All tasks of the previous checks	4 weeks	30,000

means after fixed flight hours inspections, services, replacements, etc. are defined. This approach is used to monitor the component state development and life time consumption. Following this philosophy all tasks are planned fixed and there is no real flexibility in aircraft rotation planning. An advantage of this traditional maintenance philosophy is the reliable planning of resources. However, a lot of resources are required which are varying over the different maintenance blocks. Also the workload of the staff is varying much because the required effort of various checks is very different.

7.6.2.5 On Condition Maintenance

Over time life time consumption of various components is reduced but also sometimes increased and during the overall aircraft life the scheduling of components is adapted. Very often maintenance of components can be shifted to a later block and component life time can be used more efficiently leading to more operation hours and less maintenance cost. As a consequence the related maintenance resources in terms of material, shops and staff must be organized more flexible.

Upcoming maintenance activities must be grouped on the time schedule so that the aircraft will have only minimum ground time. If it would be possible to forecast the component's state a flexible on condition maintenance strategy can be realized. This will lead to a maximum use of life time and minimum ground times and effort.

7.6.2.6 Unscheduled Maintenance

Although for all relevant components life times, mean times between failures (MTBF) and main time between overhaul (MTBO) are defined, components can fail before due to over load and over stress or simple material failure. In such cases unscheduled repair, replacement or overhaul has to be performed. Unscheduled maintenance causes often additional ground times and cost.

7.7 Airline Organization

The operation of an airline has to be authorized, which is in Europe under EASA part OPS "Air Operation Requirements" defined, [8]. The main objective is to ensure safe and stable operation of the airline. Therefore the following topics are addressed:

1. General pre-requisites
2. Financial pre-requisites
3. Flight operations requirements
4. Technical requirements
5. Disclaim of licence
6. Further requirements

A general pre-requisite for European airlines is the main business place being in a European member state. Further the main business of the company has to be air transportation. A member state or a citizen of a member state must be the main owner of the airline, so that they can control it every time. At last the operated aircraft must be registered in a member state.

From financial perspective the company must guarantee, that at least for the first 24 months the economic obligations can be fulfilled also without any revenue in the first three month. Therefore an economic plan for the first three years has to be provided.

Regarding flight operations the company has to establish an appropriate organization and management (Fig. 7.26) [4].

Also main responsible staff for operations, crew training and ground operations has to be nominated. A quality management system as well as a flight safety program must be provided. At last all organizational and flight operations procedures have to be documented.

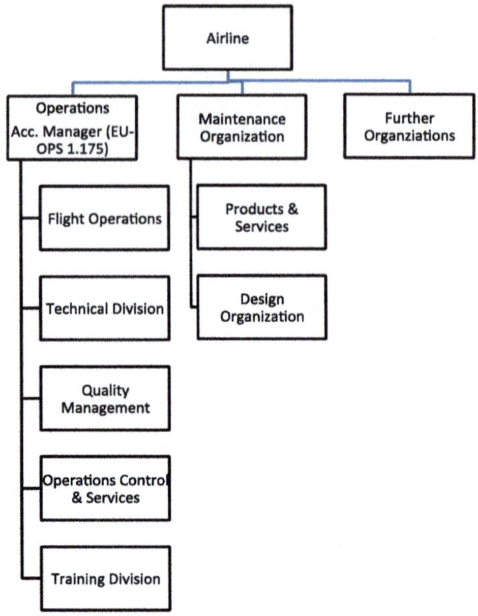

Fig. 7.26 Typical and principle airline organization

The technical requirements address mainly qualification for continuous airworthiness (Continuing Airworthiness Management Organisation, CAMO).

This contains a maintenance program for each aircraft being operated. Further also repair and overhaul procedures and a system for documentation (tech-log system).

Finally, an airline has to provide an insurance to cover losses of passenger, luggage, cargo and accidents.

References

1. IATA: Annual review 2014. International Air Transport Association, Montreal, Canada. http://www.iata.org/2014-review/reader.html?r=29/569 (2015). Accessed 01 Feb 2015
2. Wikipedia: World's largest airlines. http://en.wikipedia.org/wiki/World's_largest_airlines (2015). Accessed 01 Feb 2015
3. Belobaba, P., Odoni, A., Barnhart, C.: Global Airline Industry. AIAA, Alexander Bell Drive, Suite 500, Reston VA 20191-4344, USA, 1st edn. (2009). ISBN:978-1-60086-702-6
4. Echtermeyer, K.: Airline operations. Lecture at the Technical University Hamburg-Harburg, Institute for Air Transportation Systems, Hamburg, (Summer 2014)
5. Gülden, K.F.: Low-cost-airlines on long-haul routes—identification of impact factors for sustainable economic success. Diploma thesis, DLR Institute for Air Transportation Systems, LK—DA 15/2009, University Witten-Herdecke, Hamburg (2009)
6. Clancy, B., Hoppin D.: The MergeGlobal 2004–2008 world air freight forecast, MergeGlobal, Inc (2004)

7. Gollnick, V.: Air transportation systems. Lecture at the Technical University Hamburg-Harburg, Institute for Air Transportation Systems, Hamburg, (Summer 2014)
8. EASA: European regulations about aviation. European Aviation Safety Agency, Cologne, Germany. http://easa.europa.eu/document-library/regulations (2014). Accessed 7 Dec 2014
9. ACARE: European aeronautics: vision for 2020. www.acare4europe.org/docs/Vision%202020.pdf (2011). Accessed 27 Feb 2011
10. Ashford, N., Stanton, H.P.M., Moore, C.A.: Airport Operations, 2nd edn. Wiley, New York (1997)
11. Gollnick, V.: Stagnation in aviation industry—consequences and opportunities for SME, 3. In: Symposium "Air Transport of the Future—Less Growth, More Innovation", Institute for Air Transportation Systems, pp. 27–28, TU Hamburg-Harburg, Hamburg. 8 Aug 2014, German (2014)
12. Hirst, M.: The air transport system. AIAA, Alexander Bell Drive, Suite 500, Reston VA 20191-4344, USA, 1st edn. (2009). ISBN:978-1-56347-964-9

Chapter 8
Airport and Infrastructure

Abstract The airport is an essential element in the air transport system for all payload—passenger as well as cargo payload—to get access to the aircraft for transport from origin to destination. The airport can have a very simple structure, with a small runway for the aircraft for take-off and landing and a type of hangar to prepare the passenger boarding, baggage treatment, formalities such as customs or passenger checks and the infrastructure to allow for the preparation of the flight with meteorological information, route planning and aircraft loading.

On the other hand, there are the big airports which handle several hundred thousand passengers per day, have up to six parallel runways and can handle some thousand aircraft per day, with a very sophisticated infrastructure, hotels, conference centres and business areas as integral part of the airport.

All airports are organized in the Association of Airports (ACI), which is located in Montreal/Canada and has five regional offices in nearly all continents. Similar to IATA, ACI is a non-profit organization, but trying to support the air transport worldwide and representing the interests of the airports [1, 2].

8.1 Role of Airport

Analogous to other transport modes, the airport has a similar role in the air transport system like the harbour in the maritime transport system or the railway station in the railway system or the bus terminal in the road transport system.

Some major differences from the railway system exist. Most of the national railway stations are owned and operated by the national railway companies. There is normally no independent railway station or railway station operator. This may sometimes cause problems, when the local community or city would like to further develop their own connection point—their gateway to the national railway system! But all the development plans for railway stations are organized and decided by the (national) railway system. In some European countries, there are now some new railway operators. This may help to further develop the railway system and the

D. Schmitt and V. Gollnick, *Air Transport System*,
DOI 10.1007/978-3-7091-1880-1_8

railway stations, contribute to better efficiency of the railway system and improve their competitive situation with other modes like road and air.

The air transport system has a clear differentiation. The airlines—mainly international airlines—have no direct financial or organizational link with the airports they are operating. They have a clear commercial link with the airport by negotiating the time slots at the airport for their flights to and out of the airport, the services the airport will provide, and the airport fees for landing, take-off and station cost at the gates or apron positions.

The airport plays an important role for a city or a region. It is the window to the outside world. It allows and facilitates easy access to other cities, regions, countries or continents. Therefore, the city and the region have a very clear interest to develop their airport and most of them have direct shares in their airport or are the owners of the airport. A typical airport view is given in Fig. 8.1.

The airport is also a source of annoyance for a lot of persons living in its vicinity. Persons living close to the airport suffer from the take-off and landing noise, have a less positive view about the airport and are very skeptical with further air transport development and air traffic increase. This classical conflict—the users of the airport live away from the airport while the criticizers of the airport live close to the airport —is a constant challenge leading to a lot of conflicts, especially with the still expanding air traffic and causing major problems if an airport wants to further increase the capacity and develop its infrastructure (see also [3, 4]).

Fig. 8.1 Munich airport and its connection to road and rail

8.1.1 Location of the Airport

The best location of an airport is close to the city centre, with direct access to the railway station and the underground transportation system. But, as the airport needs a certain development potential, the city centres will not accept such a location where the noise impact will be too severe.
In one word, the best location does not exist. A compromise solution has to be defined, depending on the following constraints:

- The airport should be as close to the city as possible.
- The airport needs a development potential for the next 20 years, allowing at least a duplication of air traffic.
- A fast public transport system (Underground, metro, Maglev, etc.) is needed to connect airport and city centre with a travel time of less than 20 min!
- No obstacles for aerial development (Chap. 9)

As most airports already exist and are located close to the big cities, it is the main task of the airport and its shareholders to ensure the growth, development potential of the airport and the optimal connection to the city centre. But also the link to the motorways and the direct connection to a high-speed train system should be envisaged (More details are given in Sect. 8.5).

8.1.2 Intermodality Aspects

Intermodality is a very important aspect in transportation. It is always announced as major research topic for the further development of transport systems in the world. But first of all, we have to define the word intermodality precisely:

- *Intermodality* is defined as mode change between air, rail, road and water transport systems. Intermodality means a seamless transfer/connection from one transportation mode to another mode with a minimum of time delay and a maximum of comfort for passenger and/or payload.

With this definition of intermodality, the airport is the only unique element in the air transport system responsible for and ensuring the intermodal transfer of passenger and cargo. For other transport modes, the railway station, the bus terminal and the harbour are the interchanging areas for a mode change!

It is very rare that a passenger is really living close to the airport, so that he can walk to the airport or use his bicycle, like he can do to board on the metro system, tramways or buses in the cities. The passenger either comes in his private car to the airport or he may use public transport means like metro, underground, bus or taxi. So the airport is the place where the mode shift occurs (see also Fig. 1.3).

It is mandatory for an airport to have access to the road system (motorways, highways, autoroute, autobahn, whatever you name it!) and to a railway system (metro, tram, railway, high-speed railway). Harbour and airport are in normal operation not connected. Maritime transport and air transport are normally in direct competition only for freight transport, especially on medium and long range routes. In the freight market, air and maritime transport are alternatives which have to be carefully selected by the shipper, depending on cost and time criteria. Some more details about this will be given in Sect. 8.5.4. For passenger transport, the selection of the best transport mode between maritime and air is very restricted, mainly for people living close to the sea or on islands, where these alternatives are existing. Maritime transport offers more comfort, normally less cost, but takes more time. The strength of the air transport system is the speed and therefore quick access to nearly all destinations around the world, however at higher cost and normally reduced comfort, due to its volume and weight restrictions.

8.1.3 Classification of Airports

There are different systems to classify airports: It can be done either by the size of the airport, number of runways, number of passengers, number of intercontinental flights, number of international airlines, users (military, civil, etc.) or the number of employees in the airport.
ICAO, FAA and EASA are specifying the airports in the following way:

- International airports
- National airports
- Military airfields
- Heliports

Each airport has a Code number, which is used as "location identifier".

At the beginning, the ICAO airport Code consisted of a three digits code. But in the meantime, the number of airports have increased drastically and a 4 digit code wass required. In this 4-digit code the first digit defines the region of the globe; the second defines the country and the third and fourth digits are chosen by each country individually. The US however has a problem of not having sufficient free codes and they use their own codes, the FAA codes.

But for the daily operation of airlines, there is still the three-digit IATA code which is used. For details, the website [5] will provide all major civil airports with the country code, airport code and specific information about each airport.

Table 8.1 gives a summary of the most important airports worldwide.

Table 8.1 Ranking of major airports worldwide (status 2012)

Ranking	City	Airport name	ICAO code	Runways	Number of passengers	Number of movements	Altitude
					(in Mill.—2011)	(in thousands)	(m)
1	Atlanta	International	KATL		89.3	970	
2	Beijing	International	ZBAA		73.9	489	
3	Chicago	O'Hare	KORD		66.7	828	
4	London	Heathrow	EGLL		65.9	466	
5	Tokyo	International	RJTT		64		
6	Los Angeles	International	KLAX		58.9	634	
7	Paris	Charles de Gaulle	LFPG	4	58.2	525	
8	Dallas	International	KDFW	6	56.9	639	
9	Frankfurt	Rhein Main Int.	EDDF	4	53	463	
10	Denver	International	KDEN		52.2	607	
11	Hongkong	International	VHHH		50.4		
12	Madrid	Barajas	LEMD	4	49.8	435	610
13	Dubai	International	OMDB		47.2		
14	New York	John F. Kennedy	KJFK		46.5		
15	Singapore	Changi	WSSS	3	46.5	301	
	Munich	Franz Josef Strauss	EDDM	2	37.8	405	7
	Sydney	Kingsford Smith	YSSY	3	35.6	289	1694

8.1.4 *Important Airport Elements and Characteristics*

The most important characteristics of an airport are the number of runways, the apron space, the number of passengers and the number of aircraft movements handled per year. In the following, Munich Airport is taken as example to describe the principal airport characteristics (Fig. 8.2).

In addition to Table 8.1 the main data for a selected airport like Munich are given in Table 8.2 [6].

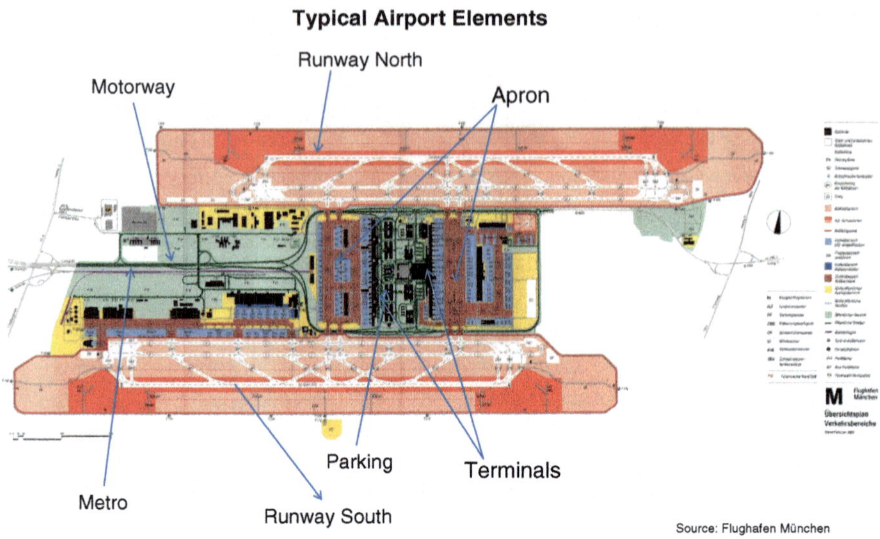

Fig. 8.2 Schematic view of airport Munich in Germany

Table 8.2 Statistical data for Munich Airport

Munich Airport—Facts and figures			
Owners		*Statistical Data* (2011):	
• State of Bavaria	51 %	• Passengers	37.8 Mio
• Germany	26 %	• Aircraft Movements	405.700
• City of Munich	23 %	• Air Cargo	395.000 t
		• Employees	29.600
Take-off and Landing Runways:			
Two parallel 4000 m long and 60 m wide runways with a distance of 2300 m;			
Staggering 1500 m 08R/26L and 08L/26R surface: concrete; PCN 90			
Infrastructure at Apron/Terminal:			
19 + 24 Boarding Bridges at Terminal 1 + 2			
14 + 47 Aircraft positions on the apron West + East			
Airport area:	1618 ha		
Airport Altitude:	448 m		

8.1.5 Airport as Economy Driver

The airport is a strong economic driver for his region due to:

1. Generating new jobs directly at the airport and indirectly in the region.
2. Offering good international connectivity, a prerequisite in a region to attract and settle new enterprises, develop tourism and becoming a strategic element to add economic value.

Statistical data show that the development and growth of an airport will also lead to an increase in jobs at the airport. Statistical data from several airports show the following tendency. For an increase of additional one millon passengers at an airport, the airport roughly needs an additional 1000 jobs to handle the additional volume of transport.

Most of the tasks at an airport, such as positioning the aircraft at a parking position, offloading passengers and baggage, refueling the aircraft, providing all sorts of services on the aircraft (water filling, waste deployment, galley loading, etc.) are all work to be done by trained personal. There is little chance for additional automation for these timely constrained processes at the airport turnaround (see Sect. 8.4.3).

If the airport is also increasing its capability to attract more long range flights, there will be additional jobs for the operating airlines at the airport. The airport Munich has issued figures that the operation of a new long haul route, operated by an Airbus A340-600 for example, will generate about 200 new jobs at the airline. This is similar to the creating an SME (small and medium enterprise) in this region.

Each job at the airport will create in the region about one–two additional jobs. This is depending on the wealth of a region and particularly the airport situation.

In addition to the created jobs at an airport, the airport plays an important element in the selection of a site for a new company. In the worldwide trade environment, fast connections are an important factor for each industrial company. So a well-functioning airport is a prerequisite for a company's installation in a region.

A lot of reports can be found which have analyzed the situation of specific airports and its economic impact on the region. The summary of most of these reports is lead to the conclusion that the airport plays a crucial role in the development of a region and country. The following references are a short summary to prove this fact [7–9].

However, the real answer for a specific airport and its development is often more complex, as it can be seen from the ESPON report.

A study was initiated by the European Commission in 2011 under the theme "Airports as drivers of economic success in peripheral regions" (ADES) and the results are published under [10].

The report states that the answer is not a clear yes or no, but depends on the regional factors. The main conclusions indicate:

- accessibility in general is an important location factor;
- for some remote regions, airports with enough scheduled flights are crucial for economic development;
- the bottleneck usually is not lacking infrastructure, but lacking scheduled flights to relevant destinations;
- the limiting factor (or bottleneck) for economic prosperity is often not accessibility but rather the availability of qualified manpower;
- it is better to use a larger airport in a neighboring region than to develop an airport of its own (if accessible within some three hours);
- not all existing airports are needed—some of them can be closed and the territory can be used for something more efficient;

- the airports can often be improved (to make them more attractive);
- good airport policy and strategy can make a huge difference.

This subject can be treated in this more technical-oriented book only in a rough summary. But the ADES study [10] gives a good general summary.

8.2 Regulatory Issues, Safety and Security

Air transport mainly happens on an international basis. So each country has its national rules of safety and security, which has to be followed. There is an international standard, defined by ICAO, but a lot of additional national standards still exist which have to be respected by each country.

8.3 Regulatory Issues

The airport council (ACI)—an international organization, representing all airports in the world—has set up a lot of committees and regulatory rules for their customers —the airports—to follow the international standards. Safety seminars are run and offered to their customers to inform, update and train the personal of their customers.
In a similar way, the airlines take responsibility for their safe operation and also provide safety seminars. The following two websites provide additional details about the airport and airline approaches [1, 2, 11].
The national law is on one hand following the national historical legal development and on the other side including the international rule-making procedures. Specific airport regulations deal with

- Safety management (see Sect. 4.4)
- Security (see Sect. 4.6)
- Local airport emissions (see Sect. 10.2)
- Airport noise emission (see Sect. 10.3)

8.3.1 Airport Safety and Security

Airport safety figures from ACI Europe for the year 2010 show that there have been 62 emergency landings, four major accidents and more than 40 bird strikes during take-off and landing.

The analysis of several flight accident reports clearly indicates a lack of safety discipline and culture. This can be improved by applying an active safety management system.

Fig. 8.3 Incident reporting as prevention for accident

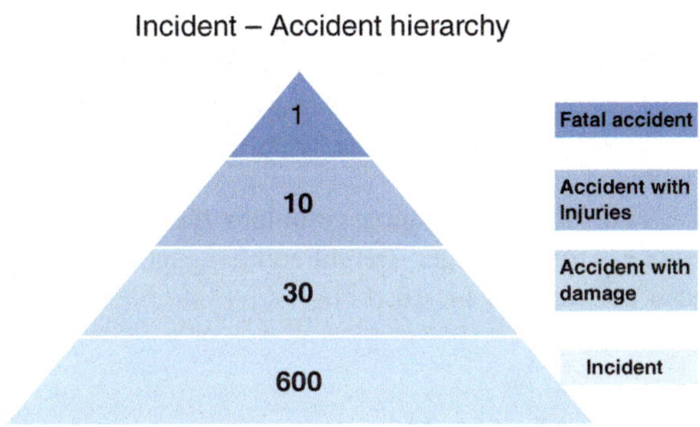

The airport has an obligation to invest in a safety structure, which brings the awareness of risk involved in all sort of activities in an airport to all personnel working at the airport.

As shown in Fig. 8.3 there are a lot of small incidents that are happening daily and which are then accumulating to accidents and at the end to fatal accidents. This can be improved considerably if a culture of incident reporting is applied and actively pursued. If all incidents are openly reported in a system (at an airport or at an airline) an active safety management system can be installed, to analyse these incidents and to draw conclusions for further operational improvements. This incident reporting system has to be set up in an open way, encouraging people to report and not to give them the blame that they have done some strange things. Normally, this safety management system has to be set up parallel to the operational system with good links to the working staff, providing an atmosphere of openness and developing and encouraging training courses for all safety critical domains.

More details can be found in different publications from the ACI [2, 11, 12]. General security aspects are handled in Chap. 4, specific passenger security checking will be handled in the next chapter.

8.4 Airport Operation and Services

The airport operation has to be done in a way that the services that are provided will be at least balanced by the fees the passengers and airlines are paying for these services. Each airport is forced by their shareholders to operate efficiently and provide reasonable benefit. This pushes the airports to look for other income sources and most airports are changing today their character from pure passenger and cargo handling to modern service providers. Several modern and big airports prepare specific events—cinema areas, fair and trade events, show rooms for luxury cars and beauty articles, etc.—and attract a lot of public, who are not travelling but only using the entertainment possibilities of modern airports [13].

The ACI systematically provides some data from all their members about the economics of airports. These reports can be ordered via [14, 15].

The focus of this book is more dealing with technical and procedural aspects of air travel and therefore, less reflecting the economic challenges and chances of this sector.

The following chapter combines the necessary functions of an airport and the process for passenger, freight and baggage handling, as defined today. More details can be found in [4, 16, 17].

8.4.1 Aircraft Handling Process at the Airport

Basically, the passenger embarking and disembarking process can be divided between the airside and the landside part. As shown in Fig. 8.6, the airside cycle starts with the landing of the aircraft, taxiing, docking at the gate or at the apron. Once the aircraft is parked safely, the landside part starts. It begins with the gate or staircase positioning. After opening of the doors the passengers disembark either through the passenger bridge directly to the terminal or via a staircase and a bus in front of the aircraft, which will bring the passenger to the terminal. When the passenger arrives at the terminal there are two options:

- The transit passengers will be directed either to the transfer desks or directly to the departure zone, where they will look for the gate for their next flight. This process is normally handled within the security zone and no additional security check for the passengers is required. There are however some countries and depending on where the passengers are coming from, an additional security check is required.
- Those passengers who have arrived at their final destination will be guided to the baggage claim zone, pass through customs and immigration controls and will then arrive with his baggage at the open airport side. Here he will find information desks, which will guide him to car parks, the bus- or metro-station, the high-speed railway or the taxi stand.

The embarking process is reversibly happening. The passenger is arriving at the airport by bus, taxi, metro, private car etc. and will go to the registration zone. The airlines have here their check-in counters, which will have either machines to distribute the boarding pass and/or have counters, where the "baggage drop" will happen and baggage will be checked in for the flight, but also controlling the allowable weight and outsize baggage. With his hand luggage, the passenger will then pass through the security check, which today is still a fairly time-consuming step and can take up to 30 min at big airports during peak times. Behind the security check, the passenger will then have to go to his gate and wait for the start of the aircraft boarding. In this area of the airport, a lot of small bistros, shops and service providers are located. As the normal passenger will take quite some reserve time as he is not so familiar with all the time delays for security check and finding the right gate for boarding, he normally has quite some time behind the security control zone to walk slowly to his departure gate, have a look at some shops, take a snack and is ready to spend some money! This is quite an important element for the airport to install sufficient space in this area for all sorts of shops bistros etc.

Figure 8.4 is showing the airport processes and the air and landside part of it. This definition of airside and landside process helps to differentiate also between the passenger movements and the aircraft movements during take-off and landing.

Once the passenger is seated on his place in the aircraft, he is waiting for all the other actors to do their duty and bring him safely to the destination he wants to go.

When the passenger is boarding the aircraft, the difficult process of interaction of pilots and crews from the airline side and the airport services, and the air traffic control starts. So three different organizations are involved and have to coordinate the next steps.

When all passengers are on board, the pilot will request from the ATC (air traffic control, see Chap. 9) the necessary instructions for his take-off procedure. The air traffic control will check his destination and will clear the aircraft for take-off at a certain time slot. The aircraft will either need a push-back vehicle to leave his gate position or will need a person to check and release the parking brakes before the wheels and give the pilot the green light to start the engines. Then the pilot will

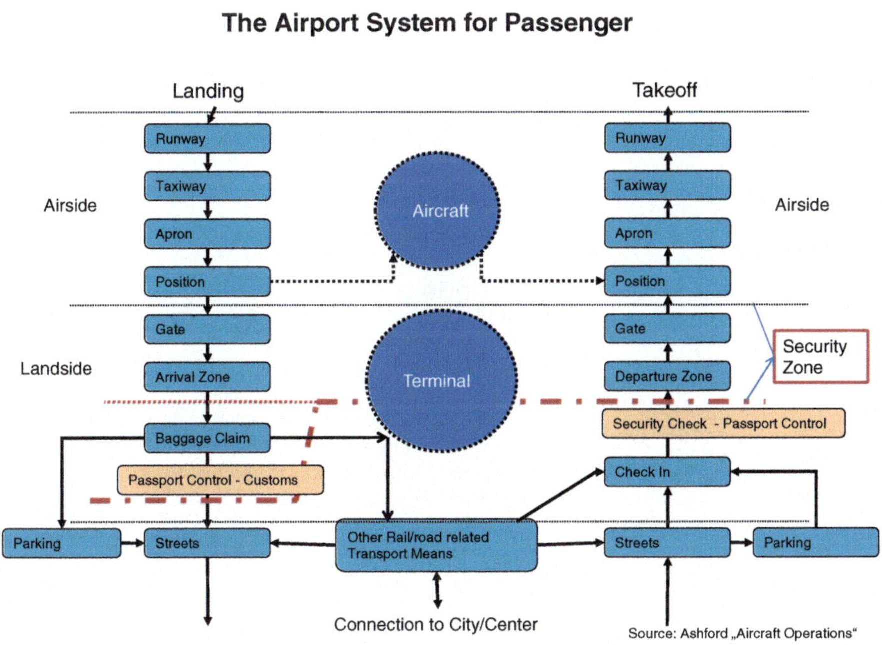

Fig. 8.4 Passenger/aircraft process during take-off and landing cycle

follow the instructions from the ATC till the aircraft reaches the indicated runway and his take-off position. He will then switch to the air traffic control and follow the instructions for take-off and the instructed flight path after take-off. When the aircraft is at the runway and cleared for take-off, the pilot will then apply full take-off thrust, accelerate the aircraft till it has reached the rotation speed, rotate the aircraft and follow the predefined route out of the terminal area to the upper air-space. The detailed take-off procedure is described in Chap. 9 and the aircraft mission profile is shown in Fig. 5.35. When the aircraft has reached its climb phase the pilot will already follow his flight path to his final destination, while negotiating during his cruise phase with the air traffic services the optimum speeds and altitudes for the best cruise performance for his flight.

8.4.2 Definition of Major Airport Elements and Services

The smooth operation of an airport during day and night during all seasons of the year will require specific equipment and services. Table 8.3 is defining all the elements which might be necessary for this continuous operation across all seasons and meteorological conditions.

Table 8.3 List of all elements, mandatory for a smooth operation

• Take off/Landing-Runway(s)
• Taxiways • Terminal and aprons for passenger-, cargo- and General Aviation transport • Tower for Air Trafic Control • Navigation means and installations • Illumination of all airport areas and buildings • Kerosin Reservoirs • Aircraft maintenance hangars • Airport maintenance and winter service • Shortterm-, Longterm-parking • Catering-Services • Motorway-, Railway-, Metro-connection • Energy systems (electricity, air, heating, etc.) • Water and waster system • Safety fences and doors, surveillance cameras • Medical Care system • Fire brigade • Services like hotels, restaurants, meeting rooms, etc. • Etc.

As can be seen in Fig. 8.2 the most important elements of an airport are the

- *runway*, which is needed for take-off and landing
- *taxiway* which is allowing the aircraft to access the apron and the terminal gates
- *apron*, which is the area for parking the aircraft during their downtime
- *terminal* which is providing the access of passengers to the aircraft
- *parking area*, allowing the private parking at the airport
- *access of metro/public transport, trains*, integrated in the terminal area
- *road connection* (car, bus taxis) to the motorway and fast city link

A more complete list of all the airport elements is shown in Table 8.3.

Figure 8.2 is showing a newly developed airport (Munich airport, MUC in Germany), where the basic layout is done in a way that there is still sufficient space for further extension of the airport. There is place to install two additional runways north and south of the existing two runways. The northern 3rd runway is just in the planning phase, however leading also to a lot of political discussion within the local community. The access to the terminal area by road and public transport is clearly structured from east and west. Main access to the airport by road and metro is coming from West, but plans for a better connection also from the eastern side are available and in negotiation. Parallel to the motorway connection is a metro connection installed, which should also leave enough space for an extension for a train/high-speed-train line directly to the terminal.

The *runways* are normally directed in such a way to be in line with the dominant wind directions in this region. In Western Europe and North America, most runways have an east west direction, in line with the typical westerly winds on the northern hemisphere. There should also be no obstacles in the direction of the runway, at least within the next 15–20 km to allow a 1 degree minimum take-off for possible emergency procedure during take-off. More details are given in Sect. 8.5.3.

The *taxiways* are connecting the runways with the terminal area, the apron. There are two parallel taxiways for each runway, allowing a one-way operation for each taxiway, improving thus the throughput of the airport. Several exits from the runway to the taxiway can be seen allowing the smaller aircraft, approaching at a lower speed, to leave the runway at the earliest possibility after landing.

The *Apron area* is the place in front of the passenger terminals where the aircraft can be parked or via gate bridges can be directly connected to the passenger terminal. Depending on the overall airport architecture, there could be more than one apron area.

The example of Munich airport (Fig. 8.2) is showing two apron areas for passenger embarking on two terminals plus an additional apron area for cargo handling.

Each airport tries to define as much direct gate positions at the terminal as possible. There is however, a compromise to be found between the numbers of gate positions at the airport, the maximum passenger movements in the terminal while changing the aircraft. More details will be given in the Sect. 8.5 (airport planning). Besides the direct gate positions there are a lot of additional parking positions at the apron, where passengers can (de)board the aircraft and will be carried to the terminal via a bus, a procedure which normally foreseen for smaller and short range aircraft at the airport.

The *terminal* is a big building or a complex of several buildings, needed to direct the passenger flows for all the passengers landing at the airport from the aircraft gates via the baggage hall to the exit ,i.e. car park, metro station train station, taxi station, buses etc.

For the departure of passengers, the terminal will direct the passengers from the landside areas (Bus, metro, car park, etc.) to the Check-in counters, through the security check area to the departure gates and finally to board the aircraft.

This terminal building should protect the passenger from any bad weather situations when changing from the land to the airside and vice versa. The terminal area has therefore, to integrate all the necessary facilities like access to the parking areas, access to the metro, train and bus stations, the taxi positions and also some hotel and conference facilities, which are a part of nearly all modern airports. It also has to integrate all passport checks and customs facilities for all international flights, all security checking positions for passengers entering the protected and secured terminal part.

There is a tendency to separate the arrival passenger flow and the departing passenger flow within the terminal area. More details are given in Sect. 8.4.2.

The areas for the departure will also have a lot of restaurants, bistros and shopping boutiques, while keeping the passenger waiting before boarding the aircraft. The arrival area in contrary has little shopping areas, as the passengers have the normal tendency to leave the airport as fast as possible. But the access to taxis, buses, metros, car parks and also the rental car stations should be located close to the arrival exit areas and should be indicated clearly in the national language and in English!.

Parking areas are fundamental for all airports. Most of the big hub airports are amongst the biggest park house owners in their region, with partly direct access to the terminal area for short term travels and specific long term parking areas remotely located, but with bus connections to the terminal. The pricing policy is intended to steer the parking flows. Access to the public road system is mandatory for all big airports. The parking areas are connected to a multi-lane road system, leading to the normal motorway system of the region.

Parking fees are specific issues and are normally, a function of parking time, with some fee-reduction at specific long-term parking areas.

Access to public transport is mandatory for each international airport. Normally a fast metro/underground/fast train-line will connect the airport with the city. Some of the big European airports have also direct access to the national high-speed rail system like Paris CDG, Frankfurt-Rhein-Main, London-Heathrow, Madrid, Amsterdam and Brussels and are preparing this connection. This is a real progress for a better transport mode change and a big step for "intermodality in passenger transport". For big airports like Frankfurt, this could really bring big benefits as some short haul flights can be cancelled like Frankfurt-Cologne and Frankfurt-Stuttgart, where the high-speed train will take a bit more than an hour and bring the passengers directly to the city centre. This option seems to be well accepted by a lot of passengers and Lufthansa has now cancelled all flights between these cities. Nevertheless, Lufthansa is offering flight tickets, but the flight is done on ground in a High-speed train (ICE) in some specific Lufthansa wagons, where the specific Lufthansa service quality can be provided. Some specific arrangements are needed as how the baggage problem is handled. But mostly, the passenger will receive his baggage at the airport baggage claim and then has to take care of it like a usual train passenger.

Navigation Installations are needed at all airports. Dominant signal at each airport is always the tower, where the air traffic control services are installed, having a good visibility to all the operations at the apron area, the taxiways and runways. All runways are normally equipped with automatic landing systems, helping the pilots and aircraft to allow weather operations. These will be described in more detail in Sect. 9.4.

General Aviation will also have their separate area at most airports. Only the very big hubs have banned "General Aviation" (General Aviation is the description for all private aircraft—company business jets, private jets and propeller aircraft—not used for regular air transport).

Cargo terminal: At big airports a specific cargo terminal area is installed, where all sorts of goods will be prepared, packed into containers or stored in pallets. As most of air cargo will be transported by normal passenger aircraft in the lower cargo bay, these specific cargo containers will have to be transported between the aircraft and the cargo centre. Specific cargo aircraft will be stored at the cargo terminal for loading and unloading (see also Sect. 8.5.4).

Safety and security means have become more important after the 9/11 event in the US. Security checks have been mandatory in Europe long before the 9/11 event, but procedures have been standardized afterwards and commonly agreed at a worldwide level. Security checks are mandatory for all passengers, when they depart from an airport. But security checks are also obligatory for all personal at the airport, who are working at the aircraft (baggage handling at the aircraft,) or in the secured areas. Also the airfield is surrounded by a big fence and specific means of surveillance (video cameras, detectors etc.) are installed to protect the airfield from visiters of unauthorized person. (see also Chap. 4).

Medical care systems are mandatory at each airport. Persons during the boarding and deboarding process may need medical help or the aircraft have to land in an emergency situation, where a passenger got a medical problem during flight.

Fire brigade is also mandatory at each airport. In specific emergency cases an aircraft may have to be evacuated outside the apron, then there is a need for the fire brigade to help from the groundside at this emergency evacuation. The fire brigade will have to be trained frequently in specific exercises to be effective, in case they will be needed. More details are given in Sect. 8.5.5.2.

Winter operation: Those airports, based on the northern areas of the hemisphere, which are facing during winter time due to snow and ice conditions, will have to take precautions to maintain a 24 h operation. Amongst those are a series of snow trucks, which will be capable to free the runway within 5–15 min and also keep the taxiway and apron area in operable conditions. Other mandatory installations for winter-operation are specific deicing trucks or fixed installations close to the runway, which will allow to de-ice the exposed wings and surfaces of the aircraft before take-off. Specific fluids, sprayed on the upper side of the wings and tail planes of an aircraft, will prevent the falling snow to fix itself on those upper surfaces and might change the contour of the wing profile and change, therefore the aerodynamic flow conditions around the wing during take-off. Those specific fluids are ensuring, that at all critical temperatures in the airport including cold rain and falling snow, will not freeze on the cold wing and tailplane upper surfaces for at least the next 30–45 min. During the take-off run for the aircraft, this fluid layer will then be washed away by the wind and will ensure the right airfoil shape at the wing and tailplane for the critical flight phases during take-off and initial climb (see also Sect. 8.5.5.3).

Fuel and Power systems: (see Sect. 8.4.7)
Airport Services: (see Sect. 8.4.8)

More details about all these elements, installations and services can be found in [3, 4, 16, 17].

8.4.3 Turnaround Process

A very important feature for the operator—the airline—is the "turn around"- process for each aircraft at an individual airport. For the airline the turn-around starts when the aircraft has left the runway and has arrived at the parking position which could either be a gate position or an outside parking position, where staircases have to be provided to disembark the passengers.

When the aircraft has been parked at its position and the engine is shut off, the real turn-around process starts, which is mainly related and limited by the aircraft, its doors, and can be used for disembarking the cabin arrangement.

The aircraft manufacturer is already providing for each aircraft a certain master plan for the turn-around process (Fig. 8.5).

This turnaround process starts when the engines are shut off and the ground power is connected to the aircraft. Then the bridges or staircases will be brought close to the aircraft door(s), the door will be opened and when the safety check is done, the passengers will be allowed to disembark the aircraft. There are a lot of statistical data available for all different aircraft types to account the mean time for this disembarking process for a fully loaded aircraft. The aircraft deboarding process can be also be simulated and this simulation capability can also be used to calculate the disembarking time for newly developed aircraft types [18].

Parallel to the deboarding of the passengers, the cargo doors will be opened and all containers will be taken off, similar to the baggage located in the bulk room. In parallel the waste recovery can start, followed by the water refilling.

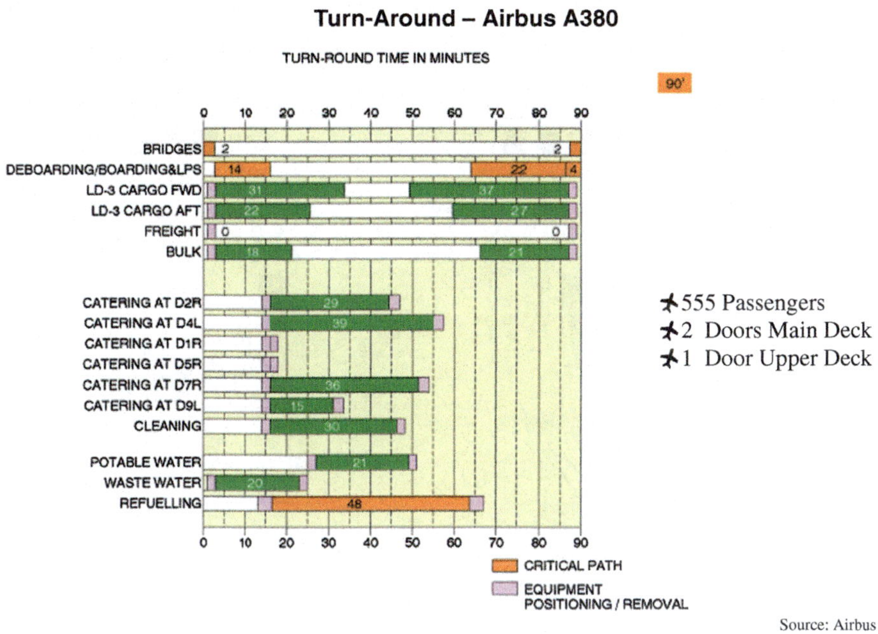

Fig. 8.5 Typical Turn Around process with the time critical path

When the passengers have left the aircraft, the cabin cleaning process can be started and all the catering trucks will be put in position to offload the used trolleys and will provide the new trolleys for the next flight. Also the refueling process will start. The refueling process is often on the critical path, but it can be only started, when all passengers have left the aircraft. It is a safe feature and is defined in the ICAO operational rules [19]!

After the aircraft is cleaned and refueled, the passengers can board the aircraft again for the next flight. Depending on the size of the aircraft, the time for boarding can be calculated and is fixed for each type of aircraft. However, all these time for boarding, deboarding are very much depending on the experience of the passengers, their knowledge about their seat location, the hand baggage to be stowed in the overhead bins, their cooperation to free the aisle as quickly as possible to allow others to pass etc. So all given times in Fig. 8.5 are estimated mean times which can be improved or further expanded, depending on a lot of factors, mentioned above.

The aircraft is making money only when it is flying in the air. The turnaround time is therefore a loss of time for the airline to make money. There is a strong interest for each airline to keep the boarding and deboarding phase as short as possible, which highlights the importance of an optimized turn-around process. Especially the big aircraft like the B747 and A380 need a careful analysis of the

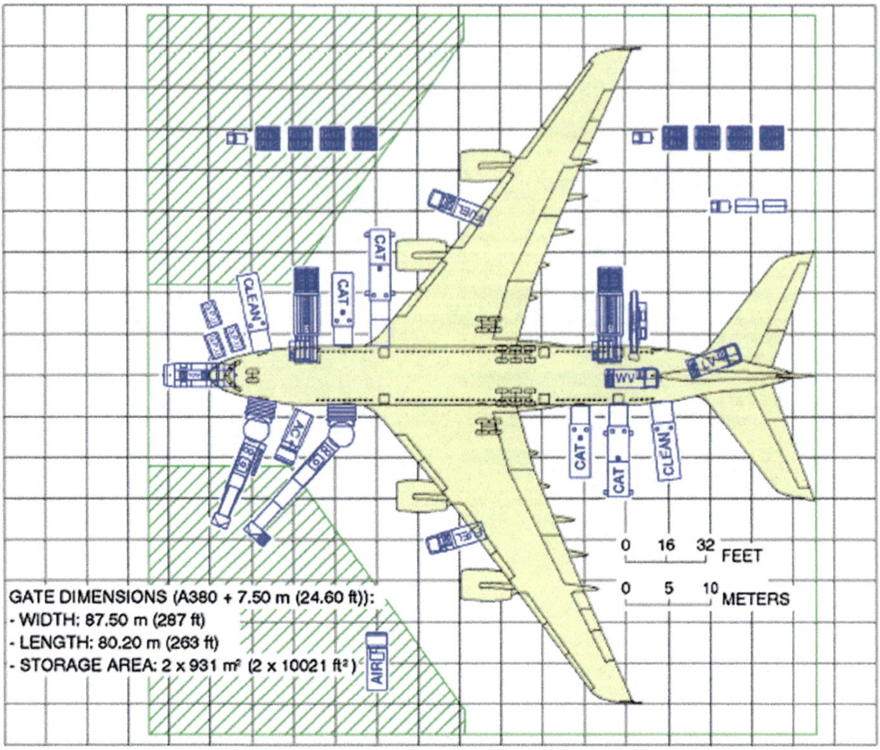

Fig. 8.6 Positioning of ground vehicles around an aircraft (A380 as example)

turnaround. Figure 8.5 shows the schematic turnaround process for the A380 aircraft, as defined by the aircraft manufacturer. Typical turn-around times (TAT) are:

~20–30 min for SR aircraft, operated by LCC at small airports
~30–60 min for SR aircraft, operated by line carriers
~60–120 min for LR aircraft

In addition to the turnaround process, there has to be a careful process during the aircraft development for the positioning of all different doors and service points at the aircraft. Figure 8.6 shows the very dense positioning of vehicles around the A380. There are some clear standards in the aircraft handling. The left doors in flight direction are used for passenger boarding and the right doors are used for catering vehicles. But for the double deck A380, the rear left hand side is used for catering vehicles.

The services provided for each aircraft during the turnaround process at the airport (i.e. cargo loading, bus services, aircraft dispatching, etc.) can either be provided by the airport team itself or by the airlines themselves or by some service providers, who are authorized by the airport. The third option is recommended by IATA, with a clear recommendation to have at least two or more service providers accredited at the airport to allow some competition between the service teams.

Role of dispatcher

At the airport, airlines often task specific dispatcher to ensure and organize the difficult task between the

- boarding-desk (are all persons on board, are some transit passengers expected and is it worth to wait for them ?),
- the pilot (contact the air traffic control to ask for the optimal take-off slot!),
- the airport services (the push-back vehicle to be in place!)
- the airport control to allow the push back as quick as possible

and finally to decide when the boarding counter should be closed to allow a quick start of the aircraft (more details are given in Sect. 7.5).

8.4.4 Airport Check-in

Check-in is usually the first process for a passenger when arriving at an airport, as airline regulations require passenger to check-in by certain times prior to the departure of a flight. This duration differs from 15 min to 3 h depending on the airport, the destination and the airline. The check-in is normally handled by the airline itself. At outside stations the check-in can also be subcontracted to a handling agent working on behalf of the airline.

Passengers are normally giving their travel documents to the airline, showing their passport or identity card to receive the boarding pass.

The check-in for the airline is the point, where bigger baggage items—the passenger do not wish or the airline do not allow to carry on to the aircraft's cabin—are separated from the passenger and are transported separately to the aircraft and stowed in the lower baggage compartment.

During this check-in process, the passenger has the ability to ask for special accommodations such as seating preferences, inquire about flight or destination information, make changes to reservations, accumulate *frequent flyer program* miles, or pay for upgrades.

Check-in is often possible or even required to be done at specific machines, which are issuing the boarding pass and then only the baggage has to be given to specific *baggage drop-off* counters. Even automatic luggage check-in counters are in use, supporting the airlines to reduce further their personal.

Check-in options and procedures vary per airline, with some airlines allowing certain restrictions, other airlines have in place, and occasionally the same airline at two separate airports may have different check-in procedures. Such differences are usually not noted by the average passenger and occasionally lead to service interruptions when one carrier refuses to abide by the procedure that another carrier normally would be willing to do.

But the automatic check-in can only be done for normal passengers. There should always be a need to provide a specific service for children, elder persons, disabled persons, which has to be done by qualified personal.

8.4.5 *Baggage Handling at the Airport*

The word baggage is used in this book simultaneously with the word 'luggage'.

In the air transport system, there is a clear principle, that at the airport and specifically the check-in counter, the baggage and the passenger have to be separated. The passenger is only allowed to take one piece of hand luggage with him into the cabin. The big baggage will be separated and stowed in the lower cargo compartment of the aircraft. This is partly due to the aircraft design, where the cabin is used to the maximum for passenger seating and the lower cargo compartments will be used to store all baggage, first of all the baggage from the passengers and if there is empty space, also some containers with additional cargo items.

This separation of passenger and baggage at the airport terminal requires a fairly sophisticated system, which will ensure that all the baggage items from the check-in counters will be transported to specific places, where all baggage for one specific aircraft will be collected and then transported by small baggage wagons to the dedicated aircraft. At big airports, this baggage handling system (BHS) is fairly complex and sophisticated to guarantee a more than 99 % correct delivery to the right aircraft. In addition, the baggage handling and transportation system (BHS) will also be used for the arriving aircraft (see [4, 17]). All baggage is then taken from the aircraft—partly containers, partly individual suit cases from the bulk area—and will then be brought to the BHS again, which will distribute all baggage from one flight to one specific belt or baggage distribution system in the arrival area. Specific companies like Webb, Herbert and Beumler have specialized amongst others to develop complete baggage handling systems. [20–23].

Of importance is also the technology, which is used to give each luggage a specific code and then track the luggage during its way throughout the system to the final point for the dedicated aircraft. Several systems and technologies can be used like RFID technology, as shown in [22].

8.4.6 *Freight Handling*

Most of the large airports are also handling specific cargo aircraft. The airport has normally a specific area dedicated for all cargo preparation, loading and off-loading the cargo aircraft and handling the incoming and outgoing freight.

Some basic market aspects of freight are already covered in Sect. 3.5 like the difference of Integrator and Cargo Carrier and the process between consignee and shipper. Figure 3.12 is showing that quite a lot of partners are involved in the cargo handling process. Most freight is collected by the freight forwarder, who is specialized to negotiate on one hand with the shipper to transport the goods and on the other hand to negotiate with the cargo airline to get best price conditions.

The airport has often own service providers or they are renting specific areas of the cargo area to airlines or Ground handling Agents (GHA).
The following items have to be offered from the GHA:

- Handling of import, export and transfer cargo, including all documentation
- Complete handling of special freight such as hazardous goods, express and courier shipments, perishable and refrigerated goods, animals, valuables and airmail
- Picking and deconsolidation services ("fast lane" accelerated handling)
- Interim storage and "ready-to-go" preparation of freight consignments
- Provision of trucks
- Last-minute services for urgent freight

The air cargo supply chain is a bit more complex, compared to the passenger handling at the airport. Cargo has no own intelligence and has to be managed in all details. Due to the different partners involved, IATA has issued an initiative, called "Cargo 2000" or abbreviated C2 K [24]. This C2 K initiative provides a quality management system for the worldwide air cargo industry to standardize and optimize the transportation process within the air cargo supply chain from shipper to consignee with the overall objective to increase service performance and thus satisfy customer expectations. Members of the C2 K are carriers (airlines), freight forwarders, Ground handling agents, airports, trucking companies and IT-providers, who committed themselves to implement agreed standard processes. As all the members in the air cargo supply chain are operating with different IT systems, the C2 K process achieves transparency and visibility of the actual freight movement for the customer by applying C2 K measurement of milestones and alert setting procedures during transportation. Main benefits are:

- Improving processes towards paperless shipping management
- Reducing claims through improved visibility, control and quality
- Ensuring reliable and timely delivery of freight through harmonized processes and standards of airlines and forwarders
- Training of operational staff on identical standardized processes

Freight was originally carried loosely in the cargo hold of the aircraft. But with the introduction of bigger aircraft and specific all freight aircraft, a more standardized transport device was needed, the so called "Unit Load Devices (ULD)" or standard container. Different container types are defined (LD1—LD11).

In addition, pallets were introduced which have some advantage, when there is mainly a one way direction of goods transport. Also pallets are standardized (96 inch, half sizes etc.). Details can be found in [25].

8.4.7 Fuel and Energy Needs

Fuel supply is one of the major requests from airlines at the airport. Jet engine aircraft are using kerosene Type Jet A1. The smaller aircraft, which are pushed by piston engines, are using AV-gas, which is similar in it's consistency like super fuel for cars [26].

A certain, safe and reliable storage of fuel is mandatory at the airport. There is normally a specific area needed where the fuel stores can be located. They should neither be too close to the runways nor to close to the terminal areas for safety reasons [27, 28]. Depending on the number of movements and the main destinations, served from this airport, the storage capacity of fuel has to be elaborated. The fuel is stored in big boilers. The boilers have to be refilled either by pipelines, railway tanks, tank trucks, or by a ship supply. Specific pumps will be needed to facilitate the transfer from the oil reservoir to the central tank boilers. Often there are several supply systems installed to have redundant systems operating.
For the service of the aircraft, mainly two possibilities exist:

- Several tank trucks are delivering the fuel to each individual aircraft
- An underground fuel distribution system is installed and at all major gate positions, a specific pump truck is pumping the fuel from the fuel valve connector in the ground to the aircraft wing tank

The monthly or yearly fuel demand at an airport can be roughly calculated from the number of aircraft, departing per day and the standard average trip length from the airport and the aircraft types. The biggest aircraft B747 and A380 may have a maximum fuel capacity of 250 m^3. However, the average fuel demand per aircraft at MUC is in the order of 6–7 m^3 with a daily average of $\sim$2500 m^3/day.

Other sorts of energy are required at the airport for the aircraft service and all the buildings (terminals, park houses, etc.) and runway and taxiway lights, etc. So specific *power stations* are needed at each airport, mainly electricity generators but also pressurized air and hydraulic energy may be required.

Each aircraft being parked at the terminal will need electrical energy, hot or cold air for the cabin and hydraulic power for the systems operation. Each aircraft has an APU system (Auxiliary power Unit), a specific turbine in the aircraft, which can provide different sort of energies on ground. Most of the bigger airports are not allowed to use APU's on the aircraft and are providing ground power, either via a ground power vehicle or via fixed cables close to each aircraft gate!

8.4.8 Business Aspects

Each airport has to cover its expenses by applying service charges/fees to all operating customers like airlines, business jet and private aircraft operators, helicopter operators, etc.

There is no unique fee system for all airports. But as airports are very different in size, attractiveness and national importance, the charges can vary quite largely.

Specific info can be found in the airport economic report [15], issued yearly by ACI.

Charges are fees, paid by airlines for services and facilities provided by airports such as:

- Use of the runway (landing charges)
- Use of the airport infrastructure (parking and boarding bridge charges)
- Use of the terminal building (passenger charges)
- Airport security (security charges)
- Protection of the environment (noise and emission charges)
- Other air navigation services (meteorological and aeronautical information services)

IATA's role is to drive cost reductions and continuous improvements in cost efficiency. Some key facts regarding charges are provided below [29]:

- External campaigns with major airports involving direct consultation and negotiation
- Leading the industry's position on charges issues
- Closer collaboration with local and regional airline associations
- An industry-wide approach with Member airlines
- Incorporating charging principles of non-discrimination, transparency, cost-relationship and consultation with users
- New approaches and strategies for airlines, airports and ANSPs to achieve greater cost efficiency and performance
- Protecting airline interests in cases of commercialization of airports and ANSPs efficiency and performance

Worldwide, the total user charge of infrastructure in 2008 was US$ 64.1 billion representing 11 % of airline revenues. These infrastructure charges form the second largest external cost to airlines after fuel.

Charges for each airport can be found in the airport websides. One example is given in [30] for the airport in Frankfurt Rhein Main. The brochure for fees is about 30 pages, specifying all different charges for each sort of aircraft type, landing fees, passenger fees, parking fees, baggage handling, and other services.

But there are several other opportunities for the airport to generate additional income.

- Most airports are amongst the biggest car park owners and are generating a lot of income via the renting of car space.
- In the terminal area, more and more space is found today to be rented to travel agencies, holiday agencies, magazines, shops, restaurants, car rental services, bank counters and cash machines, etc. …
- All the airlines need space to welcome the passengers, to have the Check-in counters, differentiated by class booking, and also have some space for ticketing and rebooking counters.
- Hotels and conference centres are also a new domain, requested by a lot of companies and business travellers
- The airport is becoming more and more a specific attraction for families, friends to visit the dynamic atmosphere of international travelling flair. The airports are more often therefore to arrange some specific family events and also shopping events

In total, an airport with sufficient passengers per year is one of the partners in the air transport system, who is best placed to generate income and be profitable. Most of the big airports today are generating more income by their additional services compared to the normal passenger and airline fees. This can be seen today by the majority of the airports, who have only little interest to provide and support a quick boarding process. IATA is also seeing this role critically, compared to their own difficult economic situation [29].

The European ACARE initiative was claiming a 15 min period as target to get the passenger from the arrival at the airport on board of the aircraft. However, the business model of most airports is based on a much longer period for the passenger in order to have time and to use all the other offers from the airport shopping malls! A seamless and optimized transfer of the passengers is not only the main interest of an airport!

8.5 Airport Planning—Infrastructure

The airport planning process is very complex and has several constraints to follow, and an ideal planning is rarely happening due to too many restrictions which will have to be considered. On the other hand, all the big cities have already an airport and it is their tendency to enlarge and develop the existing airport then to start with a complete new planning process. Here are only some general remarks to be given. More details can be found in the specialized literature [3, 4, 16, 17] or at the ACI, ICAO and IATA websites.

8.5.1 Airport Planning Process

The first important decision is the selection of the airport location. As the airport should be on one hand close to a major city, the noise and environmental aspects pushes the airport a bit more remote from the city centre and to install a fast link (metro, railway, etc.) between the city centre and the airport. As there may be several possibilities, all relevant regional infrastructure aspects have to be analyzed, meteorological and environmental aspects have to be investigated, legal national/regional planning constraints have to be considered, the air space structure (take-off, landing, holding patterns, approach procedures) have to be defined, socio-economic aspects have to be evaluated and also the potential development has to be considered and integrated. A final decision has to be done, before the detailed master plan can be developed.

The master plan will then have to look at the legal requirements and procedures, to develop the functional concept, integrate all sorts of market studies and to provide some air traffic forecast. Most of the airport master plans (Munich, Kuala Lumpur, Madrid, .) after some years of operation have become obsolete, as the growth potential was under or overestimated. But experience has also shown, so that you cannot start with a big plan based on a 20 year forecast and establish and invest all equipments, you need in the long term. Each airport today is under a permanent development plan and you will hardly see an airport which is not under continuous reconstruction. With the still constant increase of air traffic, the airports have to master their expansion plans and integrate them permanently.
A certain market study is required at the beginning of each airport planning (Fig. 8.7). This market study should include estimations about

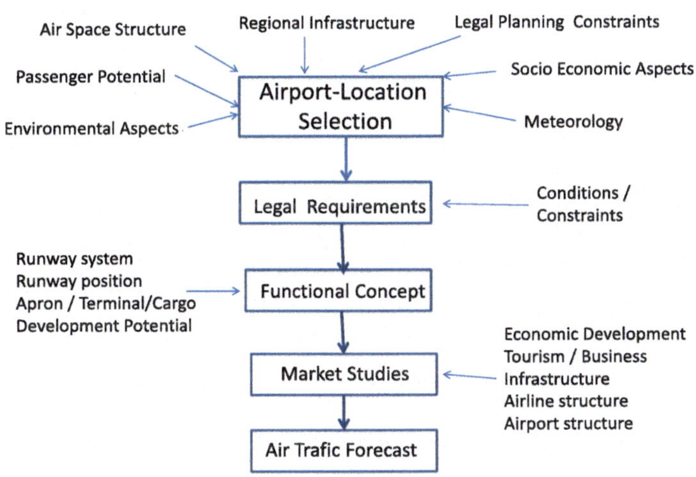

Fig. 8.7 Schematic airport planning process

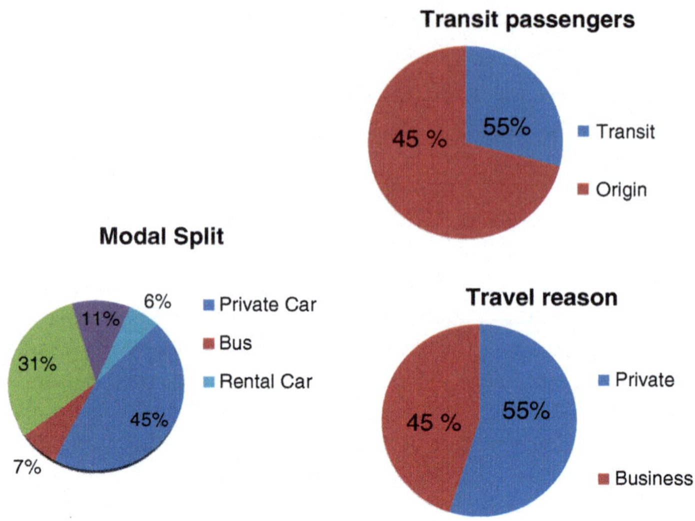

Fig. 8.8 Passenger behavior for airport planning

- The amount of passengers,
- the passenger behavior of how to join the airport,
- the air travel streams
- the expected airlines and route structure.

Figure 8.8 provides some of the influencing factors.

In parallel the basic requirements for an efficient airport design have to be listed. Figure 8.9 shows the main parameters for the initial airport layout. But some assumptions about the future expansion are also required. It is important to secure the additional land around the airport to avoid land speculations and to allow a future expansion in the most important areas like additional runways, another terminal, increase of apron space for aircraft parking and handling.

ICAO and IATA have issued documents with regard to airport planning [1, 2, 11, 31].

ICAO (International Civil Aviation Organization, see Chap. 3) has issued international quality and safety standards. There is a differentiation between "Standards", which are mandatory for all member countries and their airports and "Recommended practices", which are only recommendations.

The ICAO Annex 14 contains several major chapters like

Chapter 2—Aerodrome Reference Code
Chapter 3—Physical Characteristics
Chapter 4—Obstacle Restriction and Removal
Chapter 5—Visual Aids for navigation

Fig. 8.9 Basic airport requirements

Requirements for an Airport:

- Function related and safe operation
- 24-hours operation
- Public Acceptance and Economical operation
- Good accessability by road and rail
- Minimizing of environmental charges (noise, pollution)
- Optimum use and distribution of space/area

Factors for airport design:

- Number and direction of runways
- Number and distribution of taxiways
- Size and form of Apron
- Country geometry of landscape
- Navigation hinderances
- Use of Land within and outside airport
- Meteorology (fog, snow,)
- Size of planned airport system (space for future expansion?)

5.1—Wind direction indicators
5.2—Markings
5.3—Lights
5.4—Signs

Chapter 8—Equipment and Installation
Chapter 9—Emergency and other Services

In the Attachment A is also explained the ACN-PCN method for reporting pavement strength (see Sect. 8.5.3)

Besides ICAO Annex 14 airport planning [19], IATA has also issued an "Airport Development Reference Manual" (ADRM). This ADRM is not compulsory, based on the ICAO document but gives more details about the airline and aircraft specific requirements of the airport. Main points are to deal with the passenger terminals and their Check-In areas, passenger waiting rooms, baggage system, etc.; the apron layout and the link to the public transport system.

8.5.2 Terminal Layout

The basic arrangement of terminals and gate positions can be very different. This depends a bit on the available space, the general weather conditions in the region, some historical development and the vision of the planning team and the owners.

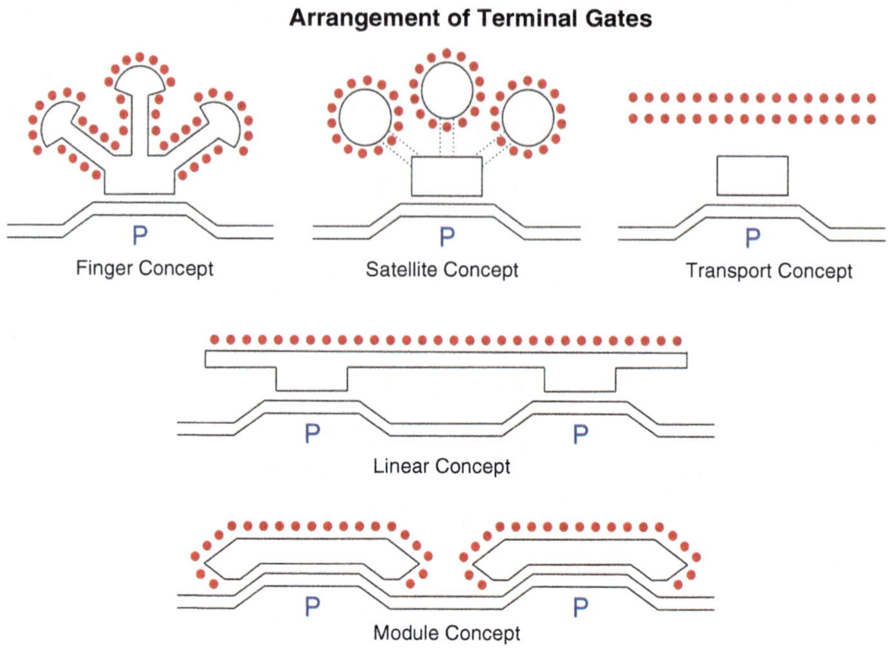

Fig. 8.10 Different concept for the Terminal layout

Figure 8.10 shows several layouts for airports. The Finger concept is realized in Amsterdam, the satellite concept is realized in Paris Charles de Gaulle airport terminal 1, the linear concept is realized in Munich Terminal 2 and Madrid Barajas airport.

The terminal concept is a compromise between aircraft positioning and passenger movements. The finger concept allows the passenger a short way, when changing from one gate to another. For example if a passenger arrives at A and has to move during transit to gate B, his physical way is quite reasonable. Whereas in the "Linear Concept, when he arrives in A and has to transit to B, he has quite a long way to walk. On the other hand, there are some aircraft in the finger concept are a bit constraint, especially when looking at aircraft position B. When this aircraft is ready for departure but another aircraft is just entering to go to gate position C, it has to wait till this aircraft has moved to his position. There is no independent aircraft movement possible. This is the advantage of the linear concept, where aircraft have no limitations (or only very little) for departure and arrival. So a reasonable compromise between passenger comfort (reducing walking times during transit and from security check to remote gates) and aircraft movement flexibility has to be found.

The recent airport terminal designs (Munich Terminal 2, Madrid Terminal 2) seems to favor the linear concept. There is another big argument from the airport side to favor the linear concept. The linear concept is giving ample space for shops, boutiques, restaurants and bars, and people walking to their gate, are offered a variety of opportunities for shopping and relaxing. This side effect is quite interesting for the financial situation of an airport. The long walking distances can be alleviated by the installation of moving belts in the terminal area.

8.5.2.1 Terminal Area, Separating Arrival and Departure Streams

A basic choice for each terminal is needed, how the different streams of passengers for departure, for arrival and for transit have to be directed, guided and separated within the terminal area. Normally the passengers are entering the terminal from the aircraft via a passenger bridges (Sect. 8.5.2.2) and therefore, arriving at the same level like the departing passengers. There are now different possibilities how to separate the streams of arriving and departing passengers within the terminal area. For large airports with a major part of transit passengers, this offers the opportunity to give the arriving passengers the chance to be guided

- either to the baggage area and the exit
- or to the transfer desk or the big boards, where the connecting flight will depart

The arriving passenger stream is then guided to the baggage claim area, customs, passport control and exit. This happens normally on a different level in the terminal area.

Figure 8.11 shows basic concepts, how a separation of arrival and departing passsengers can be achieved by applying different levels in the terminal building for the passenger streams and also the baggage flow (green dashed line). At most medium airports the 2-level concept is applied where departure is normally at the upper level and arrival with all the functions is on the lower level.

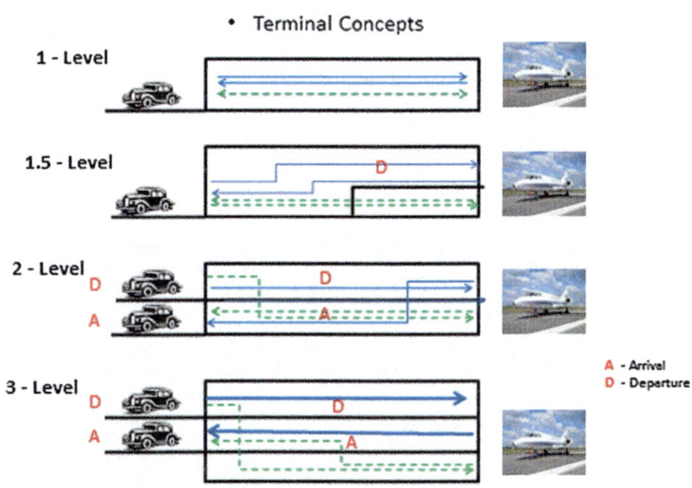

Fig. 8.11 Different terminal concepts for separating arrival and departure

This 2-level concept allows already from the landside (arriving from the city) to direct the cars to the arrival or departure level and reduce the crowd of cars in front of the terminals. More details are given in [4, 17, 31].

8.5.2.2 Passenger Boarding Bridges

A Passenger Boarding Bridge (PBB) is a flexible connection between the terminal building and the aircraft door. There are several names in use for the boarding bridges like jetbridge, loading bridge, airbridge, gatebridge, passenger walkway etc.

The PBB consists of a fixed part at the terminal and the movable part which can be adapted to all heights of different aircraft doors (sill heights, fuel status, loading status of the aircraft, etc.). PBB provide all-weather dry access to the aircraft and enhance the safety and security of terminal operations (Fig. 8.12). They are mostly permanently attached at one end by a pivot to the terminal building and have the ability to swing left or right. The "cab", located at the end of the loading bridge, may be raised or lowered, extended or retracted, and may pivot, in order to be positioned to all different types of aircraft.

PBB provides enhanced access to aircraft for passengers with several types of disabilities. They may board and disembark without climbing stairs or using a specialized wheelchair lift.

Some airports with international gates have two bridges for larger aircraft with multiple entrances. This allows faster boarding and disembarking of larger aircraft. In addition, it is quite common to use one bridge for only passengers in first class and business class, while the other bridge is only for the use of passengers in economy class. With the arrival of the full double-deck airliners such as the Airbus A380 and the new B747-800, most airports have installed new loading bridges in each deck that will have one or more loading bridges to accelerate the turnaround process.

Smaller aircraft are normally parked in the apron area. Passengers will be brought by bus to the aircraft and then have to board the aircraft via a mobile staircase, which is positioned directly at the door and adopted to the sill height of the aircraft. The advantage of this apron position is that two staircases can be placed and passengers can disembark and board the aircraft via front and rear door.

Fig. 8.12 Passenger loading bridges seen here at London Heathrow

8.5.2.3 Terminal Elements

The terminals are an important element for the airport planning: The following functions have to be foreseen:

- Counters for airline ticketing
- Check-In Counters
- Security and passport Control zones
- Waiting rooms and waiting areas in front of each gate
- Counters at the departure gates
- Arrival areas with baggage claim zone and customs
- Baggage handling and distribution system
- Lounges for airlines
- Service areas for children, disabled persons, medical service, religious zones, etc.

Besides the functional needs in the terminal area, the airport is interested to provide a lot of additional space planned for bars, restaurants, shops, snacks, bank counters, health services etc. These areas are becoming an increasing importance, as the airports are realizing that a lot of additional income can be generated by providing sufficient space in the terminal areas, so that passengers can spend money while waiting for their aircraft for departure.

8.5.3 Runways, Taxiways and Aircraft Geometry Codes

8.5.3.1 Runways

In [19] a lot of details about the layout of runways are given. Major basic principles are:

- Parallel runways should be separated by at least 1050 m. This will allow an independent operation on both runways. If the distance is less, the departing and arriving aircraft have to be staggered and this will reduce the capacity of the runway system.
- The runway length depends very much on the different aircraft type and their maximum take-off mass MTOM. If the runway length is 3500–4000 m, nearly all major big long range aircraft can be arriving and departing on this runway.
- The critical design point for the definition of the runway length is normally the aborted take-off case, where the aircraft during take-off with an engine failure close to the point of rotation must still be capable to cancel the take-off and come to a complete stop before reaching the end of the runway. This is a rare case in reality; however the safety considerations are requested in this case as a design and certification point for the aircraft (see Chap. 4).

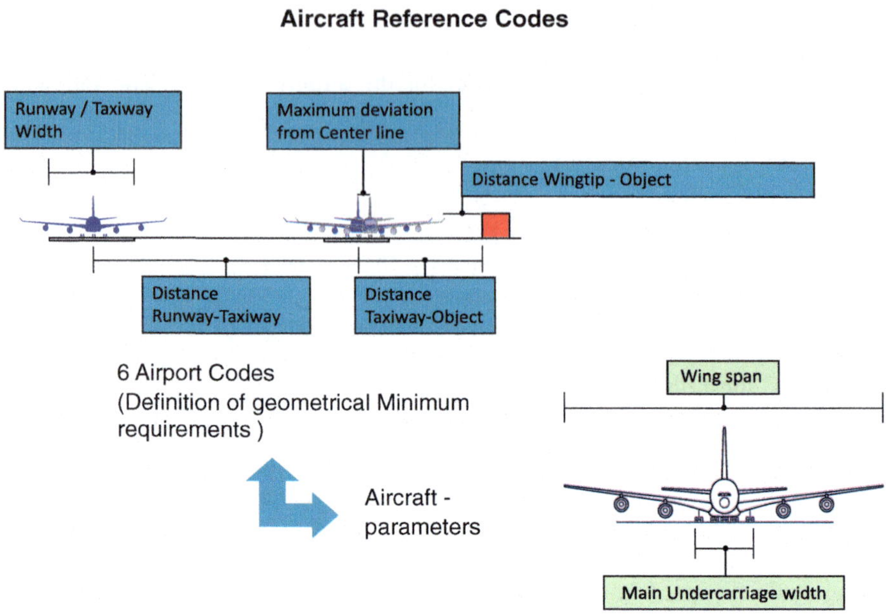

Fig. 8.13 Aircraft characteristics, influencing the airport reference code

- At the end of each runway, there has to be a zone called "stopway and clearway" which is defined in [2, 19] and is needed, if the aircraft has an aborted take-off and has to turn to come back to the terminal and apron area.
- An obstacle free zone at the end of each runway is needed to allow the aircraft, when taking off with an engine failure and a very small climb gradient, not to be obstructed on its climb out phase.
- The construction of a runway has to follow strict rules to obtain a certain PCN value (defined in Sect. 8.5.3) which on the other hand allows all aircraft with a smaller ACN value to use the runway without providing a major impact or damage at this runway.
- The width of a runway should have 60 m to allow big four engine aircraft like B747 and A380 to take-off with full thrust and ensuring that the outer engines are still located during the take-off phase above the paved runway. During take-off with full power, the engines are sucking the maximum amount of air and if the outer engines are not over the paved runway, there is a high risk of sucking some unforeseen elements like stones, small animals like rabbits or others through the engine air intake (Fig. 8.13).

The number of runways, their geometry, the distances between runways is a very important element for each airport and is defining the capacity of an airport. A later addition of another runway is often not possible or it will cost the airport a fortune. Therefore, it is mandatory to foresee already in the basic airport planning an extension of the runway system and protect the necessary space.

8.5.3.2 Runway Loading—Aircraft and Pavement Classification

A very important point in the airport design is the choice of the maximal size of aircraft, the airport is prepared in the long term to accommodate. The dimensioning part is the loading of the pavement of the runway.

ICAO has defined a system, which allows balancing the maximum aircraft weight and the pavement strength of the airport runway. [32, 33]

There are two critical figures defined, which allows to compare aircraft mass and pavement strength:

ACN (Aircraft Classification Number) is calculating the impact of a given aircraft on to the structural pavement of a runway. The necessary aircraft elements, which determine the ACN are:

Aircraft Take-off mass MTOM
Aircraft mass on main undercarriage legs
Wheel geometry
Wheel tyre pressure

PCN (Pavement Code Number) is the figure which describes the quality and loading capability of the runway pavement. The PCN requires the following inputs from the runway pavement construction:

Flexible or rigid Pavement surface quality aspect
Sublayer construction of runway

For a safe operation for big aircraft on an airport runway, the PCN number must be higher than the ACN figure PCN > ACN
The PCN classification number for Munich airport is expressed:
PCN 90/R/A/W/T

- Where the first number is defining the final PCN value
- The second figure is a letter, defining the pavement characteristics itself. The letter is either R or F, R standing for *rigid* (typically concrete upper layer) or *flexible* (typically asphalt).
- The third part is another letter, ranging from A to D, expressing the different sublayers and substructure of a runway, where A is very strong and D is very weak.
- The fourth part is also a letter, typically ranging from W to Z and expressing the maximum tire pressure the pavement can support. W is the highest letter, indicating that the pavement can support all tires of any pressure. The letters X to Z are defining maximum tire pressures between 0.5 and 1.5 MPa.
- The fifth part describes the evaluation methodology. The letter can be T for technical evaluation or U for a physical testing procedure.

So the Munich PCN Number means: load carrying capability of 90 (supports all aircraft with an ACN less than 90!), has a rigid pavement (concrete), has a very strong subgrade structure, has no limit on tire pressure and has been calculated by technical evaluation.

8.5.3.3 Taxiways

The taxiway system is providing the interface from the runway to the apron area. The design should be done in such a way that no conflict or only a minimum of crossings between the departing and the arriving aircraft will take place. This leads to two parallel taxiways and to from the runway and also at the apron area, two parallel guiding lines for the arriving and departing traffic will be important for a simple and efficient airport operation. Figure 8.14 indicates other critical areas at the taxiway system, where specific markings will be necessary to help the pilots of big aircraft to find their way to the terminal area.

The Apron area has to be large enough to accommodate all arriving aircraft. A certain amount of direct passenger bridges (gate bridges) are normally installed directly at the terminal to allow a smooth and easy embarking and disembarking process for the passengers (see Sect. 8.4.2). The standard is to board the passengers

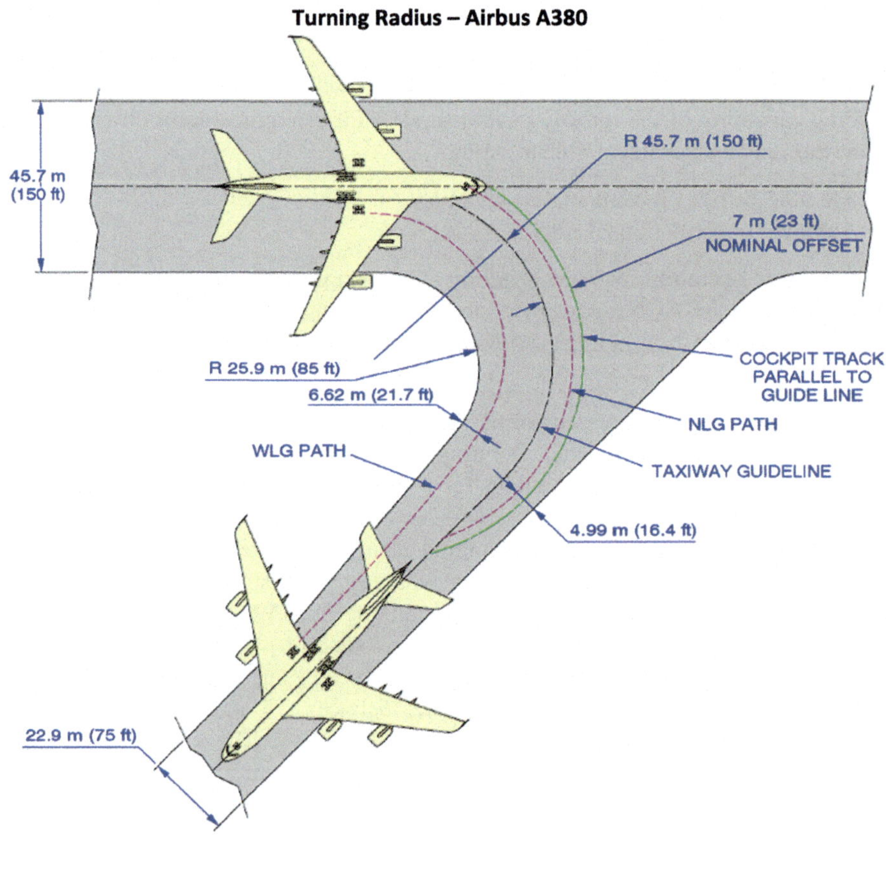

Fig. 8.14 Impact of large aircraft on the taxiways and their layout

from the left side of the aircraft (seen in flight direction!). For bigger aircraft, two or three gate-bridges can be installed and can be used to facilitate and accelerate the boarding and deboarding process. The gate bridges are normally defined in such a way that alternatively two big aircraft or 3–4 smaller aircraft can be handled at the same terminal space.

8.5.3.4 Codes for Aircraft Sizes and Limitations

One important characterization has been issued by ACI, the classification of aircraft sizes and their corresponding aircraft codes. These codes are important, as the gates at the terminal have to be installed in such a way to allow a maximum number of aircraft being placed directly at the terminals. But the aircraft codes are also limited to the maximum dimensions of the aircraft to be accepted at the international airports. Especially, before the A380 entered the market, the airports have established the new category F (Fig. 8.15), which has defined the 80 m by 80 m box as maximum dimension in wing span and fuselage length. Bigger aircraft can not be handled at the airports without major modifications to taxiways, apron areas and terminal access. This limit was a boundary for the development of the A380, where the engineers from Airbus would like to increase the span by some five additional meters.

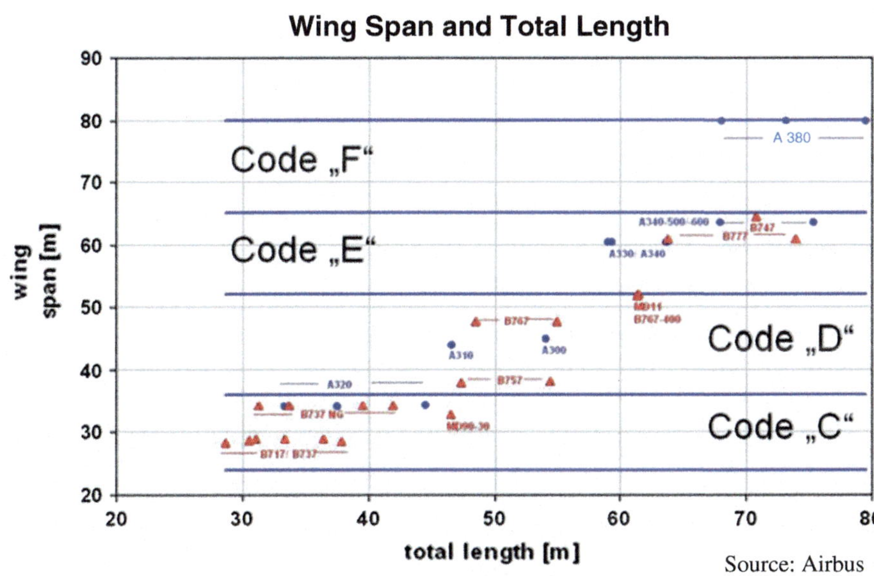

Source: Airbus

Fig. 8.15 Airport codes for the characterization of aircraft sizes

8.5.4 *Planning of Baggage and Cargo Handling*

Passengers baggage and freight/cargo are normally handled at the airport at different places. The baggage of all passengers is normally dropped off at the Check-in counter. It has then to be delivered to the aircraft through an automatic baggage handling system.

Freight, which is often transported also in the cargo hold of passenger aircraft, is handled at the specific freight centre of the airport. The freight centre needs apron space to load specific freighter aircraft. But it needs also a large building, where ULD and pallets can be packed and unpacked and where there is a link to trucks or railway wagons, which will continue to bring the freight to the consignee. This is a mode change from air to road, needed in the system to bring the cargo to the places of destination. In addition, the internet and the globalization is also favoring a worldwide transport system, where most high value goods are transported today by aircraft.

8.5.4.1 Baggage Handling System

There are several requirements for a modern Baggage Handling System (BHS). There are several constraints to be balanced like: time, high reliability, care, registration and storage of baggage. The following tasks have to be fulfilled:

- Departing baggage has to be delivered in time to the aircraft
- Transit baggage has to be transferred to the corresponding flight by respecting the time constraint
- Arriving baggage has to be delivered to the passenger at the baggage claim carroussel as quickly as possible
- All baggage has to be X-ray controlled before departing
- Correct sorting of baggage per flight (Only 1 out of 10.000 bags to be faulty advised)
- Automatic storage and redistribution of all baggage arriving in advance

Figure 8.16 shows a schematic view of a baggage handling system.
Although the primary function of a BHS is the transportation of bags, a typical BHS will have to make sure that a bag gets to the correct location in the airport. The sortation of a bag, i.e. the process of identifying a bag, and the information associated with it, to make a decision on where the bag should be directed within the system, is one of the crucial and critical elements and requires a complex IT system of bag ticketing, bag tracking, and bag control. This IT system will help to support and control the BHS by

- Detection of bag jams
- Volume regulation (to ensure that input points are controlled to avoid overloading)
- Load balancing (to evenly distribute bag volume between sub-systems)

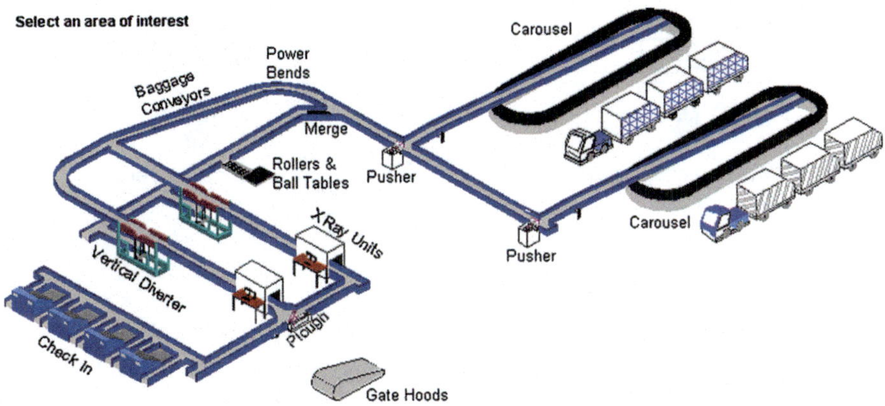

Fig. 8.16 Principle of a baggage handling system BHS

- Bag counting
- Bag tracking
- Redirection of bags wrongly directed

There has also been a breakthrough with "mobile" BHS software where managers of the system can check and correct problems from their mobile phone.

8.5.4.2 Cargo Handling

When defining the planning of a freight handling centre, the following parameters have to be estimated in order to assess the necessary area for the freight centre. These parameters are shown in Fig. 8.17. Some statistical basis values are needed to make the estimation. Typical values for normal freight are 10–15 t/m^2/year. For specific goods (express and big volume) this may be slightly less.

The planning of the freight centre at the airport starts with the basic freight process at the airport. Similar to the passenger departure, shown in Fig. 8.4 the freight process has also a landside and an airside (Fig. 8.17). The freight (good) is arriving at the airport, will be accepted and all the necessary documentation has to be established, including customs declaration etc. The good can either be received as fully consolidated ULD (unit load device, i.e. container or pallet) or as simple parcel. Parcels have then to be consolidated in a ULD within the same destination. When finished, they will be brought to the ULD export side and will be loaded to the aircraft as quickly as possible. Normal time from delivery to the airport and boarding on an aircraft lasts from 2 to 24 h. The incoming freight (Import) will be offloaded from the aircraft and stored in the import storage house. When it is just transit freight, it will go immediately to the export storage place. When an incoming ULD contains several parcels/goods, it will be broken down and the single goods will go to the landside, where some trucks will take them to their final destination.

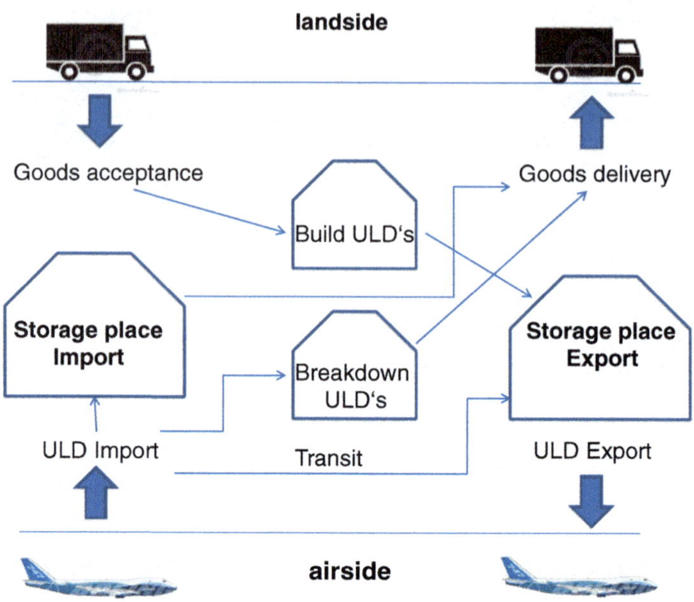

Fig. 8.17 Freight process at airport

In the freight centre, there should also be space to accommodate the administration (customs, security staff, airline staff, freight agents and freight forwarders) (see also [34, 35]).

There are different classes of cargo goods like:

- Standard goods
- Express freight
- Frozen foods
- Perishables
- High Value goods
- Airmail
- Animals
- Hazardous goods

The above mentioned process describes only the process for standard goods. It is clear, that for all specific goods, mentioned above a specific treatment is required. Specific care and specific areas have to be established to handle animals, airmail, frozen food, perishables, high values, express and hazardous goods.

Depending on the overall freight volume, a reasonable area has to be provided on the landside for all trucks arriving on the airport. In a similar manner, the apron at the airside has to be large enough to handle the dedicated freighters and their loading and unloading with all the necessary equipment.

In each large region or continent, there are some specific airports, which have concentrated their business on freight. In the US, Memphis is such an airport, where the Express freight company, FedEX has their home base. A strategic location for an air cargo airport in Europe is Luxembourg. Luxembourg has a very good road and rail connection to all neighboring countries like France, Germany, Belgium and the Netherlands. CargoLux has established as a specialized cargo airline. Luxemburg is now becoming strong competitors from the airports in Cologne and Leipzig, where there is also little passenger traffic, but a strong concentration on freight handling.

8.5.5 Specific Critical Airport Elements

The airport needs a lot of other elements (see Table 8.3) to ensure a proper operation. Some critical and important elements are shortly described below and also the main aspects, which have to be considered in the planning phase.

8.5.5.1 Fuel Storage Centre

Depending on the size of an airport, a fuel infrastructure for a storage, supply, distribution and provision has to be developed. Also a lot of safety aspects have to be considered [27].

Aviation fuel can cause severe environmental damage; all fueling vehicles must carry equipment to control fuel spills. Fire extinguishers must be present at every fueling operation. Airport firefighting forces are specially trained and equipped to handle aviation fuel fires and spills. Aviation fuel must be checked daily and before every flight, for contaminants such as water or dirt.(see also [28]).

In the airport planning process, the fuel storage zone has to be carefully selected. On one side it should be remote to the runways for any aircraft incidents/accidents; on the other hand the fuel area should be connected to the road, rail or a specific pipeline, providing the delivery. Normally, the fuel zone is a protected separate area, where only specific authorized and trained personal has access. The authorities are requesting certain skills from the personal, who are handling the fuel distribution.

The supply of fuel to the airport can be done by

- Pipeline (Best choice!)
- Railway system
- Road with tank lorries
- Ship (if a water system is available in the vicinity)

For big airports several transport means are foreseen to have alternatives, in case of blockages by strikes or other incidents.

For all these systems are some infrastructure needed. Specific pumps to transfer the fuel from the rail tank or tank truck are requested with a certain capacity.

The fuel is normally stored in big tanks. The airport of Munich has five big tanks with a total capacity of 30.000 m,[3] in addition to a direct pipeline to a refinery plant.

The storage has also several specific problems like material of the tank, avoiding water ingestion, filtering, microbiological impacts etc. which can be seen in [26]. The fuel then has to be transported from the fuel storage depot to the aircraft. 2 basic options are available:

- either by a fuel truck (tank lorry)
- Or by an underground pipeline system, where the fuel can be directly pumped at the aircraft position into the aircraft via a pump vehicle.

IATA is providing some recommendation [28] to determine the optimum storage capacity for an airport. their airport is to:

- Understand the airport's current profile and perspective
- Clarify the purposes of storage relevant to that airport
- Quantify the measurable parameters that are applicable to those purposes
- Calculate Optimum Storage
- Review Operational Considerations, Finance & Permissions

8.5.5.2 Rescue and Fire Fighting

The principle objective of an airport rescue and fire-fighting service (RFFS) is "to save lives in the event of an aircraft accident or incident". Rescue and Fire Fighting Services (RFFS) is a special category of fire-fighting that involves the response, hazard mitigation, evacuation and possible rescue of passengers and crew of an aircraft involved in an aerodrome ground emergency (or potentially off aerodrome). The International Civil Aviation Organization (ICAO) defines the requirements for aerodrome Rescue and Fire Fighting Service (RFFS) in Annex 14, Volume 1—Aerodrome Design and Operations [36]. The Civil Aviation Authority of each State inturn publishes the corresponding regulations and guidance for their operators. Modern commercial aircraft can have the capacity to carry several hundred passengers and crew. Therefore, due to the mass casualty potential of an aviation emergency, it is critical that emergency response equipment and personnel arrive at the scene within the minimum possible time. The maximum response time from initial notification until the first vehicle is on scene and spraying fire retardant is defined by State regulation and generally ranges from three to four minutes under conditions of good visibility and uncontaminated surfaces. At large aerodromes, this often means that more than one fire station will be necessary. The timely arrival and the firefighters' initial mission is to protect the aircraft against all hazards—most critically fire—increase the survivability of the passengers and crew on board. Airport firefighters have advanced training in the application of firefighting foams and other agents used to extinguish burning aviation fuel in and around an aircraft. This helps to provide and maintain a path for the evacuating passengers to exit the fire hazard area. Should fire be present within the cabin or encroach upon the cabin from an external fire, the responders must work to control and extinguish those fires as well.

The number and type of firefighting vehicles or appliances based at an airport will be determined by the airport's category. Specialized fire vehicles are required for the RFFS function. The design of these vehicles is predicated on many factors but primarily on speed, water-carrying capacity, off-road performance and agent discharge rates. Since an accident could occur anywhere on airport property, sufficient water and other agents must be carried to contain any fire. This will allow the maximum possibility of a successful evacuation and the best probability of extinguishing or suppressing any post crash fire until additional resources arrive on the scene.

Most airport fire vehicles are equipped with a roof-mounted cannon or nozzle which can shoot fire extinguishing agents at large distances. This allows the fire vehicle to begin extinguishing flames as soon as it closes the scene of the fire.

Munich airport has around 12 fire fighting vehicles with different equipment, 40 other vehicles for rescue, service and coordination and about 30 specifically trained and qualified persons.

ICAO Annex 14 directs that "All rescue and firefighting personnel shall be properly trained to perform their duties in an efficient manner and shall participate in live fire drills commensurate with the types of aircraft and type of firefighting equipment inuse at their aerodrome, including pressure-fed fuel fires". It further states details about the training curriculum and the related topics.

The aircraft manufacturers provide detailed aircraft rescue and firefighting charts for each of their products. Important is here the fact, that for new products and the application of new materials (CFRP on B787 and A350), the aircraft manufacturer are providing evidence and tests that the hazard is similar, bigger or lower, compared to the existing aircraft standard [37].

8.5.5.3 Winter Operation and Aircraft de-Icing

The process of winter operation remains an important part of an airport's operation. Since 2010, when arctic conditions covered the majority of Europe and North America's airports in a blanket of ice and snow, many have come under pressure to clear runways, taxiways and aprons as swiftly as possible to maintain normal flight movements. When an airport has to be closed for one or more than 1 day, these means a huge financial loss for the airport as well as for the airlines. So there is a huge economic interest to keep an aircraft operating even in severe winter conditions.

Planning and preparation are two of the key factors involved with successful winter operations, as well as a strong execution of clearing procedures. Munich, the sixth busiest airport in Europe, relies heavily on keeping planes airborne, so implementing a sound winter operations plan is essential [38, 39].

Due to its operating duties the airport is obliged to remain open during operating hours and he is therefore, responsible for snow removal and e de-icing.

There are two different procedures needed (see [40]):

- The de-icing of aircraft surfaces prior to take-off
- The cleaning(mechanical and partly chemical) of all operational surfaces at the airport

De-icing of Aircraft

In critical weather conditions, (when snow or cold rain is falling) all aircraft have to be de-iced prior to take-off. Specific fluids will be sprayed on the wings and tailplanes of each aircraft to avoid freezing of all liquids(water, snow, rain, ice) on the upper wing surface. If ice would build up on the upper wing and tailplane side, this could change the aerodynamic flow around the lifting surfaces and degrade the performance of the aircraft during the critical take-off- phase. This spraying on the upper lifting surfaces will be done either by a specific crane construction or a spraying vehicle.
Specific fluids are used for De- and Antiicing of aircraft on ground. The AEA has defined two fluids:

- AMS 1424: a Newtonian fluid, SAE Type 1
- AMS 1428: a Non-Newton fluid, SAE Type 2

SAE defines even type 3 and type 4 anti-icing fluids. Details are given in [41–43]

They differ in time to stay on the wing during take-off. A Type 1 fluid is quickly washed away while the aircraft is accelerating for take-off. Type 2 fluids have some thickening agents included, which keep the fluid on the wing up to speeds of 100 knots, before the fluid will be washed away due to the air speed on the upper wing surface. So this type 2 is more often used for large aircraft with high take-off speeds.

Nearly all used de-icing fluids contain some toxic elements (glycol) and are not very good in terms of environmental usage. There is a big interest to recover most of the fluid, which is therefore sprayed at specific stations with some recovery installations.

Cleaning of Operational Surfaces at the Airport

The airport takes care of the mechanical clearing of the snow with ploughs and cutter blowers and also the chemical de-icing of operational surfaces.

There is a specific ice warning system which checks the constant temperature of the ground and air through specific sensors. A fairly precise temperature can be measured in the take-off and landing area of the runway. This information makes it possible to use the ground de-icing chemicals in an extremely environmentally friendly way. Airports that are facing critical winter operation (Northern Europe,

Northern America, Russia and airports at higher altitudes or in the mountains, see [38]) are operating variety of vehicles for winter operation like snow ploughs, snow blowing machines, snow cutters and runway liquid de-icers, snow milling machines, runway friction testing machines and other specialized equipment. Airport Munich is operating more than 50 specialized vehicles for winter operation.

In [39] it is reported that Frankfurt Airport has expanded its winter services capability by adding more manpower and equipment to its snow-clearing and de-icing fleet.

A snow clearing convoy is led by a command vehicle followed by as many as 14 runway sweepers, 2 snow blowers, upto three spreading vehicles and a second command vehicle at the end.

The additional equipment means that Frankfurt Airport now operates a combined fleet of six snow-removal convoys: two large convoys of vehicles and four smaller convoys—comprising a total of 243 vehicles and pieces of equipment for winter operations.

Frankfurt's snow team has also been significantly expanded—growing from 180 staff to a total of 450 personnel.

This move will allow three shifts to work in round-the-clock snow removal operations at one of the world's ten busiest airports.

Beyond the runways and apron areas, Fraport's winter services team is also responsible for about 105 km of roads, 450,000 m^2 of car parking areas, and about 120 km of sidewalks around the airport city.

This report highlights the economic importance of a huge airport to keep also during critical winter conditions, the operation at a very high performance level.

8.5.5.4 Other Airport Services

The airport is normally an independant small city in it's region. A lot of additional service functionalities have to be provided like a *medical service station* with permanent staff.

It could happen that a passenger in an aircraft may have a heart attack and the pilot decides to land at the next the next possible airport then a "First medical aid" service is required.

Most airports provide rooms for religious ceremonies and contemplation.

Relaxation areas are also often provided.

Section 8.3.8 is providing other elements for the economic development of an airport.

More details about these subjects can be found in the specific literature [3, 4, 16, 17, 19, 44].

References

1. IATA Airline policies: http://www.iata.org/whatwedo/workgroups/Pages/governments-policy-makers.aspx. Accessed 29 Nov2014
2. ACI airport policies: http://www.aci.aero/Airport-Service-Quality/About-ASQ. Accessed 29 Nov 2014
3. Ashford, N. et al.: Airport Engineering, Toronto (1992)
4. Ashford, N. et al.: Airport Operations, 3rd edn. Toronto (2013)
5. Airport codes defined in http://www.world-airport-codes.com/. Accessed 29 Nov 2014
6. Munich Airport figures: http://www.munich-airport.de/en/company/facts/index.jsp. Accessed 29 Nov 2014
7. Auckland Airport economic study: http://www.aucklandairport.co.nz/~/media/Files/Corporate/AIAL%20EIA%20Report%202021%20and%202031%20final%20291010.pdf. Accessed 20 Nov 2014
8. Airport Economic Sustainability for North Central Texas, 2011 in: http://www.nctcog.org/trans/aviation/plan/EconomicSustainabilityReport.pdf. Accessed 26 June 2012
9. Cherry, J.: Aeroports de Montreal, The Economic importance of airports; under ICAO. http://de.scribd.com/doc/56621753/ICAO. Accessed 28 Nov 2014
10. Europ. Commission; ADES report: http://www.espon.eu/export/sites/default/Documents/Projects/TargetedAnalyses/ADES/DFR/01_ADES_DFR.pdf. Accessed 28 Nov 2014
11. ACI Policies and Recommended practices: http://www.aci.aero/Media/aci/file/Publications/2009/ACI_Policies_and_Recommended_Practices_seventh_edition_FINAL_v2.pdf. Accessed 29 Nov 2014
12. Airport safety aspects in http://www.aci.aero/Services/APEX-in-Safety. Accessed 30 Nov 2014
13. Riga Airport; security charges. http://www.riga-airport.com/en/main/passengers/useful-information/aviation-security-requirements/security-charge. Accessed 28 Nov 2014
14. Airport as shopping event: http://www.changiairport.com/at-changi/events-and-promotions. Accessed 28 Nov 2014
15. Airports economics reports in http://www.aci.aero/Data-Centre. Accessed 29 Nov 2014
16. ACI, Airport Economics in http://www.aci.aero/News/Releases/Most-Recent/2013/04/25/ACI-Releases-its-2012-Airport-Economics-Report, Accessed 28 Nov 2014
17. Mensen, H.: Planung, Anlage und Betrieb von Flugplätzen. Springer, Berlin. ISBN 978-3-540-68106-9 (German edition)
18. de Neufville, R., Odoni, A., Belobaba, P.: Airport systems (e-Book) McGraw-Hill. ISBN 9780071770590
19. Richter, T.: Simulationsmethodik zur Effizienz-und Komfortbewertung von Menschenflussprozessen in Verkehrsflugzeugen, PhD thesis at LLT; TU Munich (2007) (in German)
20. Gaffal, R.: Modell zur nachhaltigen Schadstoffreduktion an Flughäfen; PhD thesis at LLT; TU Munich (2010) (in German)
21. Baggage handling system provider: http://www.daifukuwebb.com/Products. Accessed 29 Nov 2014
22. Baggage handling system provider: http://www.beumergroup.com/en/products/airport-baggage-handling-systems/. Accessed 28 Nov 2014
23. Airport baggage handling systems: http://www.herbertsystems.co.uk/products. Accessed 28 Nov 2014
24. IATA C2K process in http://www.iata.org/whatwedo/cargo/cargo2000/Documents/c2k-presentation-part-1-c2k-purpose-and-benefits-v16-2012-05-02.pdf. Accessed 28 Nov 2014
25. Unit Load Devices, in IATA, ULD Technical Manual, 2010, Montreal Canada, IATA
26. Exxon aviation fuels: http://www.exxonmobil.com/aviation/productsandservices_aviation-fuels_jeta-a1.aspx. Accessed 30 Nov 2014

27. FAA Bulletin on Aircraft fuel storage, handling, training, and dispensing on airport: http://www.faa.gov/documentLibrary/media/Advisory_Circular/150_5230_4b.pdf. Accessed 29 Nov 2014
28. IATA recommendation on fuel storage on airports: http://www.iata.org/pressroom/facts_figures/fact_sheets/Documents/fuel-fact-sheet.pdf. Accessed 30 Nov 2014
29. IATA, Airport Competition, IATA Economics Briefing No 11, Nov 2013, http://www.iata.org/whatwedo/Documents/economics/airport-competition.pdf. Accessed 29 Nov 2014
30. Airport Frankfurt/Main, Business partners, in http://www.fraport.com/content/fraport/en/misc/binaer/our-expertise/aviation-services/airport-charges-2014/jcr:content.file/140801_entgelte-charges_08-2014.pdf. Accessed 29 Nov 2014
31. Wells, A., Young, S.: Airport Planning and Management. McGraw-Hill, New York (2004)
32. FAA: Airport Pavement Design and Evaluation Advisory Circular; AC 150/5320-6D
33. FAA: The ACN—PCN System. http://www.faa.gov/documentLibrary/media/Advisory_Circular/150_5320_6e.pdf. Accessed 29 Nov 2014
34. Munich airport Cargo Center. http://www.munich-airport.de/en/micro/cargo/index.jsp. Accessed 30 Nov 2014
35. Air freight handling process: http://air-cargo-how-it-works.blogspot.fr/2011/01/air-transport.htm. Accessed 29 Nov 2014
36. Fire fighting services. §9.2 in ICAO Annex 14, vol. I. Aerodrome Design and Operations. http://www.skybrary.aero/index.php/Category:Fire_Smoke_and_Fumes. Accessed 29 Nov 2014
37. B787 aircraft rescue and fire fighting: http://www.boeing.com/assets/pdf/commercial/airports/faqs/787_composite_arff_data.pdf. Accessed 29 Nov 2014
38. Winter operation at airport Zurich: http://www.zurich-airport.com/~/media/FlughafenZH/Dokumente/Business_und_Partner/Flugbetrieb/Revision_13_2014_2015_snow_comm_011114_v1_0.pdf. Accessed 29 Nov 2014
39. Munich winter service: http://www.munich-airport.de/en/general/presse/pm/2013/q4/pm61/index.jsp. Accessed 29 Nov 2014
40. AEA, Recommendations for De-icing/Anti-Icing of Aircraft on the Ground in http://www.aea.be/component/attachments/attachments.html?id=97&task=download. Accessed 30 Nov 2014
41. Aircraft deicing fluids: http://en.wikipedia.org/wiki/Deicing_fluid. Accessed 29 Nov 2014
42. Tanner, C.: The effect of wing leading edge contamination on the stall characteristics of aircraft. SAE Technical Paper 2007-01-3286 (2007)
43. SAE Standards: Fluid, Aircraft De-icing/Anti-icing, SAE Types II,II,IV, SAE AMS1428. http://papers.sae.org/
44. Ashford, N., Mumayiz, S., Wright, P.: Airport Engineering. Wiley, New York. ISBN 9780470398555
45. Thales Group: Airport infrastructure Security—Towards Global Security. https://www.thalesgroup.com/en/worldwide/security/what-we-do/critical-infrastructure/airports-ports. Accessed 28 Nov 2014
46. ICAO: Aerodrome Design Manual, Montreal (2002)

Chapter 9
Air Navigation Services

Abstract In this chapter, the principle elements of ANS in terms of infrastructures, organizations, and processes as well as the main technical principles of the systems used are introduced. These characteristics are required to complete the view on the ATS and to show how ANSP is capable to fulfil its key responsibilities. First, the main organizational structures are introduced and then the set up of air space structures is discussed. In the third section, the most relevant navigation systems are presented with their general characteristics and features. The control and separation strategies are discussed in the fourth section, finally followed by a short introduction to navigation fees.

The Air Navigation Services (ANS) are an essential part of the entire Air Transport System (ATS) where airport, airline and air navigation service provider (ANSP) are working close together as major stakeholders with different objectives and interests. Following the system engineering systematic and hierarchy as described in Chap. 1, the ANS is one substructure of the entire ATS. Because ATS has been used also as an abbreviation for Air Traffic Services (ATS) it has to be noted, that this abbreviation is used for Air Transport System in this book.

The key responsibility of ANSP is to ensure seamless, safe and cost-efficient air transport flow of all aircraft on the airport and in the airspace.

Communication, Navigation and Surveillance (CNS) are key technologies, which enable an efficient Air Traffic Management (ATM) among the stakeholders. However, only if the information about aircraft actual and predicted position is rationally processed and quickly distributed and shared among ANSP, airport and airline, better efficiency in terms of higher punctuality and infrastructural utilization as well as environmental indicators in terms of less emissions and lower noise impact can be achieved [1].

D. Schmitt and V. Gollnick, *Air Transport System*,
DOI 10.1007/978-3-7091-1880-1_9

9.1 Principles of Operation—The Role of the Air Navigation Services

The ANSP is the main stakeholder for ATM. It has to organize, monitor and control the air traffic in order to ensure safe and efficient flying.

As public services they are governmental institutions, which are sometimes managed according to private law (e.g. German DFS, Swiss Sky Guide). Their responsibility covers the provision and operation of the ANS as described by Fig. 9.1, [2].

Looking at Fig. 9.1 acquisition and information handling in form of management can be identified as key driving technologies, which influence the future progress in ATM.

Three technology areas are of crucial relevance:

- Management and organization of *information* flow
- Networked *communication* systems which allow worldwide exchange of information between aircraft and ground stations as well as among aircraft only
- Systems to determine the *position* of aircraft either by onboard measurement or surveillance.

For this purpose the ANSP as the air traffic control entity (ATC) has to control the air traffic flow and to manage the use of the airspace as shown in Fig. 9.2.

Summarizing all these activities ATM is defined as the dynamic, integrated management of air traffic and air space, including air traffic services, air space management and air traffic flow management. ATM is intended to provide all necessary to inform (ATS), organize and coordinate (ASM) and control (ATFM) the aircraft before and during their flight through the airspace.

Also the provision of flight plan information and approval before and during flight is one of the main tasks of the ANSP. Proposed flight plans are sent to the Central Flow Management Unit (CFMU) of Eurocontrol, which approves these proposals or gives advice for adaptions based on their global knowledge about the actual aircraft flow conditions in Europe. Before flight, for example weather

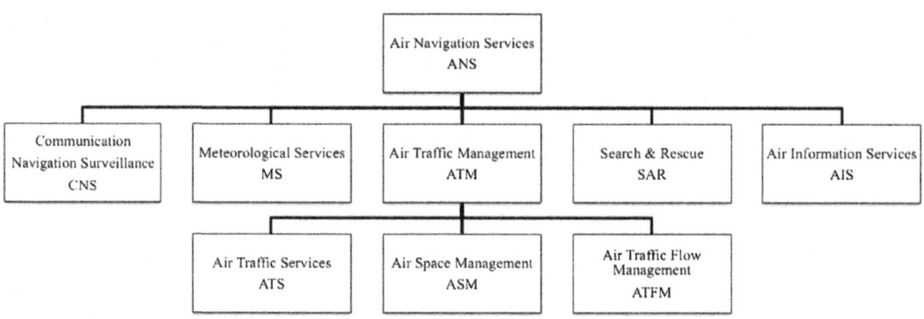

Fig. 9.1 Air navigation services

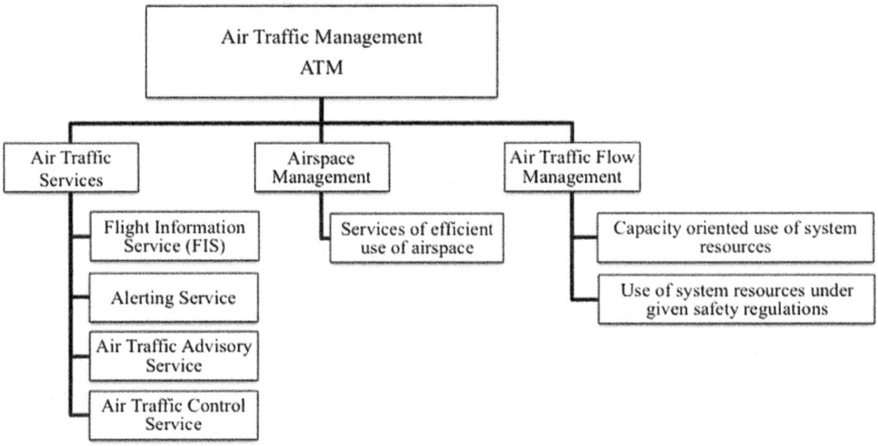

Fig. 9.2 Definition and structure of air traffic management, [2]

information as well as the availability of air routes, sectors and airspaces is provided by ANSP as part of Air traffic services and Flight Information Services (FIS).

Further, during flight the air traffic controller organize and guide the aircraft when they are passing various air space sectors. Here also the capacity-driven allocation of routes and aircraft through sectors is done.

In order to ensure flight safety the ANSP defines and controls horizontal and vertical as well as timely minimum separation between controlled aircraft. The absolute distances between the aircraft are depending on the principle flight rules to be applied, which are distinguished between Visual Flight Rules (VFR) and Instrument Flight Rules (IFR). For visual flight the principle "see and avoid" applies. While for VFR minimum horizontal and vertical line of sight are required as well as minimum lowest cloud levels, IFR is performed assuming no visual orientation is given, but only cockpit navigation aids are given. Flying according to IFR therefore requires special additional equipment like the artificial horizon and radio communication aids, which in addition need to provide a high reliability.

At controlled airports ATC is responsible for giving take-off and landing allowances as well as it controls the airfield movements of the aircraft.

To fulfil all these tasks various regulations are set up by ICAO, which require to be transferred to the national level by the national air transport authorities, see Chap. 4 and [3–5]. Internationally agreed flight procedures and CNS systems are required on ground and in the air to ensure worldwide safe and efficient operations.

The ICAO document 4444 "Rules of the air and air traffic services" provides the set of regulations about structures, procedures and required systems to establish the ATM worldwide in a harmonized and very similar way, [6]. It is important to note at this point, that ATM is mainly composed of processes and activities, which are defined by commonly agreed formal rules. This is different from other processes, which are based on best practices or company internal rules. Therefore, ATM rules as described in the ICAO document have a semi regulatory character.

Table 9.1 Procedures for air navigation services, [6]

• Part 2: General provisions
• Part 3: Area control services
• Part 4: Approach control service
• Part 5: Procedures for aerodrome control services
• Part 6: ATS surveillance services
• Part 7: Flight information and alerting services
• Part 8: Coordination
• Part 9: Air traffic services messages
• Part 10: Phraseologies
• Part 11: Controller—pilot data link communications (CPDLC)

Table 9.1 gives an overview about the most important chapters, which address the ATS:

In order to provide an understanding of the way of working of the ATS within the context of the ATS it should be noted here, that in Part 2 of the ICAO Doc. 4444 responsibilities for the provision of air traffic control services and information are defined to provide clear roles. Further operating processes like the setup of flight plans as well as the way of changing between VFR and IFR flight rules or the control procedures for air traffic flow are defined.

Part 3 gives advice for the separation of air traffic and especially how air traffic shall be separated vertically and horizontally. At last due to the further improvement of navigation accuracy Part 3 provides guidelines for the reduction of separation minima in order to safely extend airspace efficiency.

The way, how aircraft are guided and managed during approach to and departing from airports is in the scope of Part 4.

Part 5 deals with the critical operation on airports including take-off and landing.

Air traffic surveillance especially based on primary and secondary radar systems is described in Part 6. Here, the technical specification of the required radar systems including identification is given. Further the procedures how the radar systems shall be used, e.g. during approach and landing are described.

Part 7 gives advice how flight information and alert information shall be given and in which way it has to be transmitted.

The way, flight information transmission is performed in a coordinated manner, is provided by Part 8. These procedures are crucial because in ATM various service providers are involved like en route radar control and terminal area control.

Parts 9 and 10 give definitions about the contents and the phrases to be used in communication.

Part 11 addresses the special case of communication between air traffic controller and cockpit crew via data link systems. In this particular case, the data structures and wordings are different to verbal communication and need clear definitions. Also for radio communication defined terminology and phrases are specified to ensure clarity and identity in the communication channel. Those clear definitions are mandatory to ensure safety since especially verbal communication can be disturbed by signal noise but also lingual differences.

The ICAO Doc.4444 provides comprehensive information and rules about guidance and communication procedures to ensure safe aircraft operation in normal and adverse condition, these rules need to be transferred into national orders.

9.2 Airspace Structures

The airspace around the world is structured in a very similar way and normally the horizontal extension is oriented along the geographical country borders. Vertically there is no upper limitation. However for air navigation service purposes the airspace is vertically organized in an upper and lower airspace. The limits are defined on national level, e.g. in Germany the upper airspace begins at 24.500ft, also called flight level FL 245, while the lower airspace is below 24.500ft down to the ground (GND). In both airspace sections Flight Information Regions (FIR) are defined, which are characterized by special rules. For the lower airspace the regions are called FIR, while Upper Information Regions (UIR) are the corresponding definitions for the upper airspace.

As shown in Fig. 9.3 the upper limits of various airspaces are varying and decreasing the closer the airspace is to the Terminal Control Area (TMA).

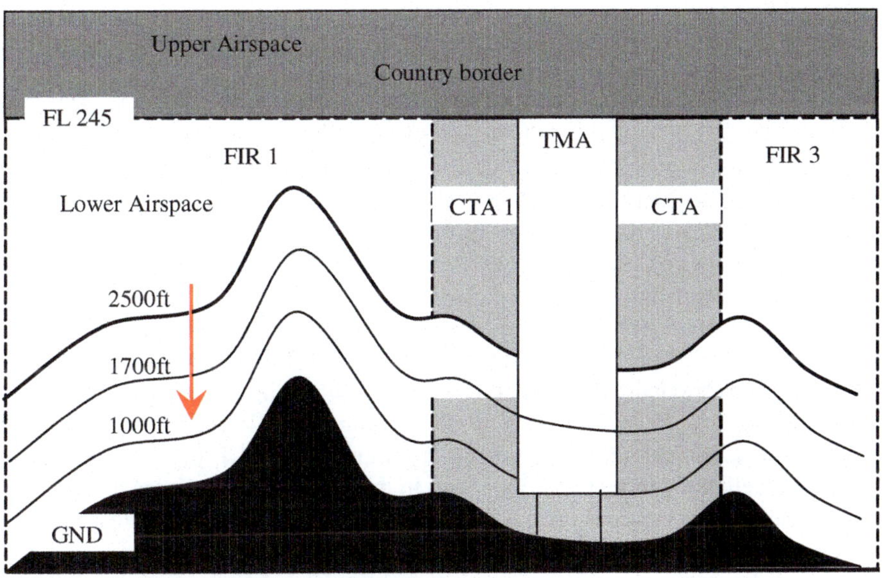

Fig. 9.3 Principle vertical airspace structure setup

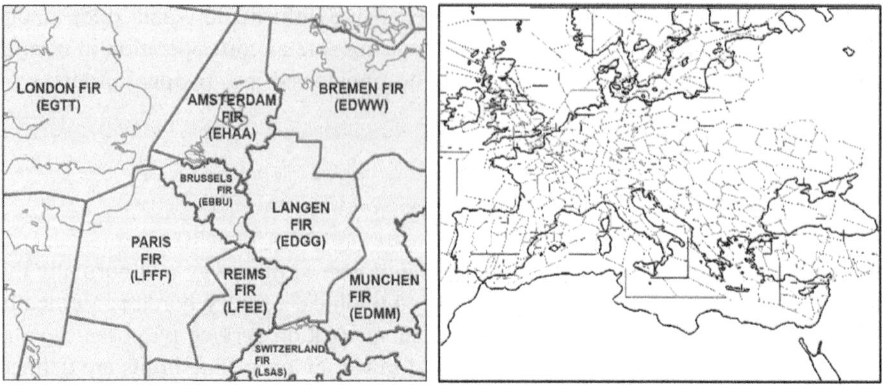

Fig. 9.4 Flight information regions in Germany and European sectors

The principle set up of airspace structures in a horizontal representation is shown in the Fig. 9.4, where the German flight information regions (FIR, left) and European sectors (right) are presented.

While the lower airspace has clear lower and upper boundaries, the upper airspace has only a limit for the controlled regions, which are established in Germany up to flight level FL 660. Above this flight level no controlled airspace and associated rules are established. However, this upper level may be different for various countries. The Control Area (CTA), which is a vertical and horizontal definition of a supervised area, and is segmented into radar sector responsibilities, which are supervised by an air traffic controller team of the ANSP.

A special area within the FIR around controlled airports is called Control Zone (CTR), which is oriented along the runway directions, and TMAs.

In order to ensure safe operation in such high traffic density areas the maximum speed is limited to 250 kts, in airspace category C for VFR flights below 10000ft and in category D also for IFR flights further the minimum visual sight is to be 5 km for Visual Meteorological Conditions (VMC). To enter such a controlled area, an explicit allowance from ATC is requested.

Terminal Control Areas are especially supervised regions for approaching and departing aircraft. Last according to ICAO-SARP (ICAO Standards And Recommended Practices) airspace classes A-G (A = maximum supervision, G = no supervision) are defined according to the level of ANSP supervision. A TMA typically has a conical vertical geometry from ground to FL100.

As a guideline it can be mentioned, that in the vertical and horizontal vicinity of airports the level of supervision and therefore the level of classified airspace sectors is higher, than in the airport far field regions.

Further also dangerous areas and temporarily restricted areas are defined. Those areas are typically military areas above fire and training ranges. But also within the airspace restricted areas are defined e.g. for military air-to-air fights.

Flight Level or Altitude	France	Germany	Italy	Spain	UK
Upper Limit CAS	FL660	FL660	FL460	FL460	FL660
FL245 - FL460					B
FL205 - FL245	A		A	C	
FL195 - FL205		C			
FL150 - FL195	D				
FL130* - FL150					G
FL95* - FL130*		C | E	G	G	
3000ft* - FL95*	G	E			
SFC - 3000ft*		G			

Fig. 9.5 Comparison of different airspace type flight level settings [7]

Although the general classification of airspaces and sectors is worldwide the same, each country defines its own flight levels, where the airspace types are placed.

As shown in Fig. 9.5, e.g. in Germany no airspace types A and B are defined. Type G reaches much higher altitudes in France, Italy, Spain and especially Great Britain. Sector G requires permanent line of sight to the ground, which is hard to achieve at higher altitudes under inclement weather conditions. Further no permanent radio communication listening mode is requested and also flying in clouds is not allowed, since VFR rules apply.

On the other hand airspace type C requires permanent radio communication listening. A horizontal line of sight of 8 km is requested at flight level above FL100 according to VFR flight rules. Below FL100 at least 5 km are requested. 1000ft vertical distance from clouds and 1500 m horizontal distances are required.

Regarding ATS air traffic control and traffic information is provided. Air traffic operating under VFR and IFR is completely separated, because different procedures underlie. Also different equipment is required.

The operational structures and procedures around airports require the definition of Standard Instrument Departures (SID) and Standard Terminal Arrival Routes (STAR) at this point.

Both procedures are defined to organize a staggered departure and approach to the airport. Also balance between air traffic load on inhabitants and noise impact shall be achieved. Therefore, the definition of the SID and STAR are to be developed by the airport and ANSP and these are approved by the ANSP. In the following figure an example of a SID for Frankfurt airport is presented.

In Fig. 9.6 aircraft leave Frankfurt on two different ways both ending up at Dinkelsbuehl (DKB) NDB (bottom right, see Sect. 9.5.1.1).

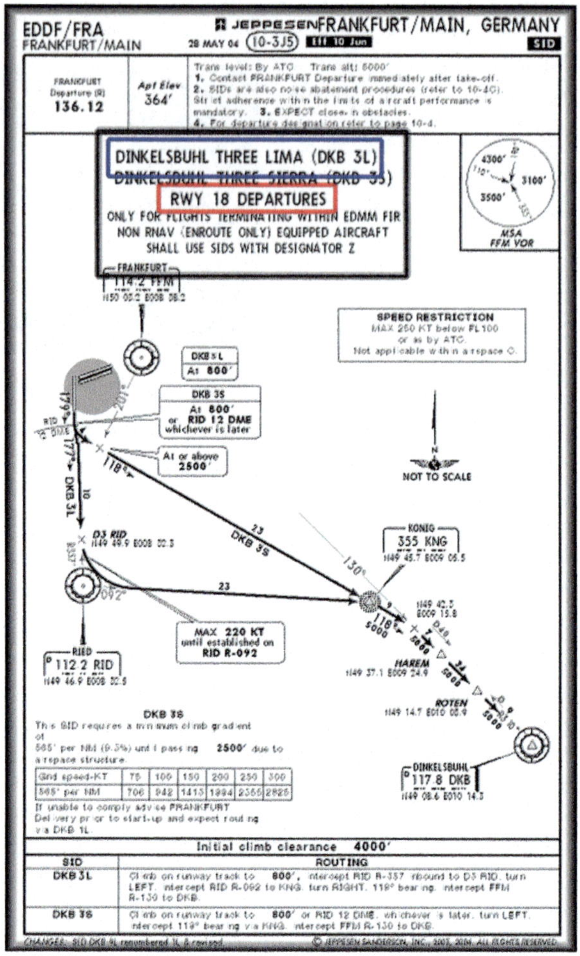

Fig. 9.6 Frankfurt standard instrument departure (SID)

Similar rules become visible if STAR are considered. Taking Hamburg Arrivals as an example as shown in Fig. 9.7, one can identify four circular areas (IAF, NOLGO, RARUP, BOGMU), where aircraft have to approach to before they go to the airport.

The STAR chart, Fig. 9.7 shows two NDB for instrumental orientation for the aircraft with the related frequencies. Aircraft pilots are obliged to follow these routes under IFR to approach to Hamburg.

From ATS point of view it is important to understand, that the airspace is structured according to clear routes. Further for each controlled airport specific operational procedures exist to organize operations in a safe way. These various operational procedures and some others introduced later require special equipment onboard the aircraft and at the ANSP.

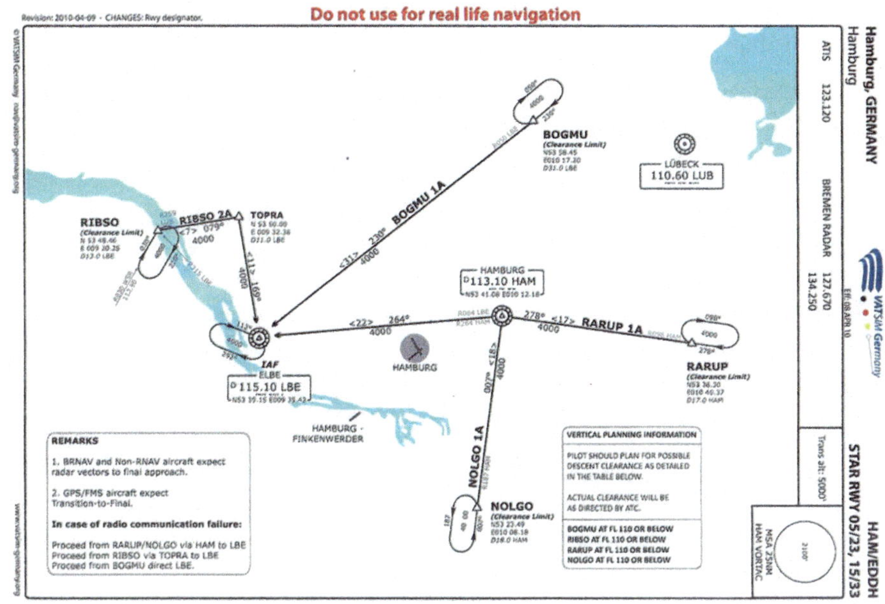

Fig. 9.7 Hamburg Standard Terminal Arrival Routes (STAR) [8]

9.3 Airspace and Airport Capacity

Capacity in air transport is the capability of a subsystem, e.g. an airport runway system, to handle a certain amount of aircraft in a given time window [9].

It is an issue of the transport flow, that the frequency of vehicle movements and the capacity of the rail, road, air networks as well as air spaces, airports and railway station become the essential design parameter.

Taking this principle into account, establishing a requested amount of people′s mobility in terms of passenger-kilometre (Pkm) will lead to transport systems of either high frequency low payload vehicle operations or low frequency high payload vehicles.

While the first provides more individuality and flexibility the later can offer more efficiency regarding energy effort and environmental compatibility, because for the same energy effort more people can be moved. For the development of future transport systems a trade off is to be made between this two fundamental approaches, which never exist solely but emphasis has to be given to more global objectives people want.

Runway capacity is therefore depending on various influences like

- Number of runways
- Runway dependency
- Amount and position of taxi ways

- Weather conditions
- Aircraft mix

Airport capacity in general depends on other parameters, as shown in Chap. 8.

Based on these characteristics the maximum capacity of an airport or airspace is variable. Capacity is related to the demand of aircraft movements. This is different from the passenger demand for transport capacity on certain lags or flows. This passenger demand can be fulfiled by a certain amount of aircraft with a particular seat capacity. But also the frequency of flights is a measure to increase the transport capacity.

In ATM the demand describes the request of airlines to operate a certain amount of aircraft on given routes at selected airports at certain time windows. For example in the early morning phase between 6:30 and 9:00 h there is typically a high demand for flights to bring people to business locations. A similar situation is given in the afternoon between 16:00 and 19:00 h when many people travel back. In between there might be a much lower request for air travel opportunities at a given airport. Therefore spare capacity is available. Capacity of an airport is mainly related to the throughput of airport runways. It is defined as the maximum amount of aircraft, which can take-off and land within an hour.

When the demand for flights is higher than the defined service capacity, unacceptable delay rates arise.

In order to make air transport as attractive as possible it is required to reduce the delay to a minimum but not necessarily to zero. As long as a delay can be forecasted it can be managed in either way. Therefore in the last years the term "planned delay" derived from railway transport has been proposed also for aviation, [10]. Since CNS systems like precise navigation or satellite-based communication allow more and more for real time information transfer it will become feasible to predict delays in the actual tactical ATC planning and to react as early as possible to that.

Eurocontrol has developed Key Performance Areas (KPA), where capacity has been identified as one factor of ATM success, [11]. As a KPA, capacity has been defined as the ability of the system to cope with air traffic demand.

Airspace capacity covers any individual or aggregated volume of airspace. It relates to the throughput of that volume per unit of time for a given safety level. Network capacity is concerned with the overall network throughput taking into account the network effect of the airspace and airport capacity as a function of traffic demand patterns.

Further an increase of network capacity by 15 % at full scope implementation across the network has been defined as related Key Performance Indicator (KPI). This is associated with an upper airspace sector capacity increase by up to 15 % and lower airspace sector capacity growth by up to 10 %.

To make air transport as attractive as possible capacity is not needed at its maximum, but should enable punctuality, predictability, reliability but also safety.

The required average total capacity C_{tot} of a subsystem in the air transport infrastructures should fulfil the demand and provide some spare capacity to compensate a limited amount of unexpected events within the considered time window:

$$C_{\text{tot}} = C_{\text{Demand}} + C_{\text{Spare}} \tag{9.1}$$

It is very difficult to provide fixed values for the required spare capacity, because it is a tradeoff between the economical effort and the achievement of fluency of air transportation. As a rule of thumb C_{Spare} should be calculated as 5–10 % of C_{Demand}.

9.4 Aircraft Separation

In order to ensure safe flight operation, there are binding rules for separating aircraft from each other, see [6] part III "Area Control Services". Additionally also horizontal evasion rules are defined, which are very similar to those known from maritime sailing and car driving.

Vertical separation also called altitude separation is based on flight levels, which are defined by barometric pressure height according to the ICAO standard atmosphere. Because the barometric pressure varies depending on local weather conditions two different procedures apply. For en route flight above a determined transition height of 5000ft every barometric altimeter of an aircraft is set to the 1013.25 hPa reference pressure.

Although the use of relative pressure altitudes referencing to this more theoretical value implies significant absolute failures compared to the actual local barometric pressure, the procedure is quite robust and safe, because all aircraft in a local region have the same failure in its altitude measurement, but the relative failure and uncertainty among the aircraft altitude measurement is very small.

If the real absolute pressure altitude would be constantly used, aircraft might reduce their altitude relatively to the ground, which may cause flight into the ground or collision with aircraft at different altitudes above ground, Fig. 9.8.

In general a vertical separation of 1000ft shall be kept. Exceptions are allowed to apply smaller separations of 500ft with the introduction of Reduced Vertical Separation Minima (RVSM). This is possible, if highly accurate position navigation

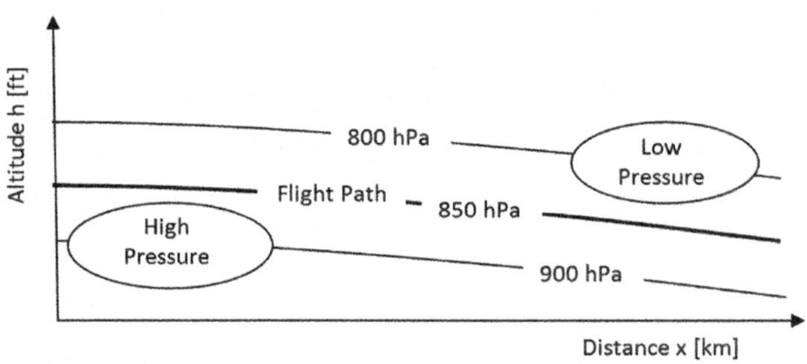

Fig. 9.8 Absolute altitude variation depending on continuous pressure level flight

and surveillance aids are used, like integrated navigation onboard systems, see Sect. 9.5.2.

For the landing phase at a transition level of typically 5000ft ± 500ft between last flight level and the transition altitude the reference pressure is set to local airfield pressure to ensure, that the aircraft will not fly at a pressure -driven sink rate as shown in Fig. 9.8. Flying at constant pressure altitude would drive the aircraft to the ground, which causes severe safety risks, especially if a flight is going from high to low pressure areas and airport field elevation is above SL. Due to the introduction of Required Vertical Separation Minima (RVSM) in any case the vertical separation of 1000ft has still to be kept at altitudes above 5000ft. If no more precise Required Navigation Performance (RNP) of the navigation systems is given or RVSM operations are applied, 2000ft vertical separation has to be kept in the upper air space above 29000ft.

In the case of horizontal separation two different situations must be considered. First, generally aircraft shall follow each other in a time or range depending distance of at least 5 min or 20 nm if no radar surveillance is available.

Second, for wake vortex situations especially at ground proximity during approach and departure operations the aircraft mass is becoming the driving factor. Due to the pressure distribution on the wing at the aircraft wing tips wake vortices are induced containing much energy and cause heavy turbulences for subsequent aircraft. Figure 9.9 shows the principle effect of wake vortex situations. The aircraft approaching nearly perpendicular from the left will experience a significant pitch down at the first wake and counter rotating pitch up at the right wake. A lighter aircraft flying directly behind a bigger one, will be pressed down or rotated by the departing wakes.

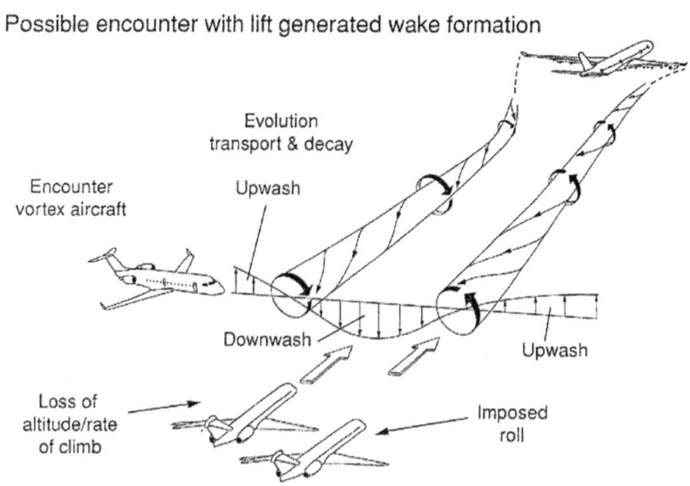

Fig. 9.9 Possible encounter with lift generated wake formation [7]

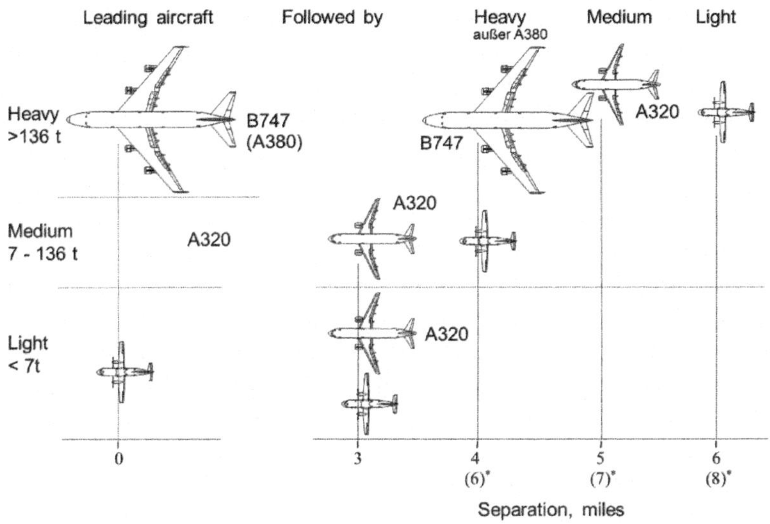

Fig. 9.10 Wake vortex separation minima between fixed wing aircraft [12]

For safety reasons horizontal separation is prescribed to protect aircraft flying on the same flight level too close to each other. In the following Fig. 9.10 required horizontal separation minima for fixed wing aircraft on approach to land, or on take-off into initial climb are presented.

The required distance depends on the maximum take-off weight of the leading aircraft, because it determines the magnitude of the wake vortex circulation intensity. In this context the introduction of very heavy aircraft of the A380 class, it is actually in discussion to introduce a new separation category. On the other hand there are observations that the wake vortex intensity of the A380 seems to be in the same order as the B747. Further data gathering is needed to justify new rules. In principle the required separation distance will increase the greater the weight difference between the leading aircraft and the following one is, because the energy content of wake vortexes increases directly with the weight of an aircraft.

9.5 Flight Guidance Systems

Communication Navigation and Surveillance (CNS) systems are required to handle traffic flows safe and efficient. Relevant KPIs are defined in ACARE, e.g. increased punctuality, capacity and security but also reduced emissions, [13]. Only regarding CO_2 emissions ATM is expected to provide the potential of 8–12 % CO_2 emissions reduction. Here navigation systems can provide some improvements like

- Shorter tracks due to higher accurate navigation performance
- Increased onboard en route and approach accuracy independent from ground systems

- Stable landing performance with more robust and reliable ground systems regarding maintainability/calibration and electronic perturbations.
- Assured integrity and availability
- High resistance to interferences and perturbations.

These features shall be accomplished either with increased performance and safety at the actual cost level or with reduced acquisition and operating cost at the actual performance level.

Space-Based Augmentation Systems (SBAS) and Micro Electro Mechanical Systems (MEMS) are technology trends in communication and navigation, which provide significant potentials to achieve the objectives mentioned before. Especially, these technologies are also appropriate to use available and finally limited capacities in air space and at airports at its best.

9.5.1 Navigation Systems

According to DIN 13312 navigation is every measure (observation, measurement and analysis) which determines a geographic location and/or the movement of an object or vehicle, [14].

There are various principles, which enable navigation. In this section, the most relevant navigation systems are introduced with their general functions and main characteristics.

Aside from fundamental visual navigation, in aviation navigation systems can be distinguished between radio navigation, inertial navigation, satellite navigation and so-called integrated navigation systems. The following table gives an overview about the related systems from functional point of view (Table 9.2):

Table 9.2 Various navigation system types [2, 15]

Navigation principle	Navigation aid	Abbreviation
Radio navigation	Non-directional beacon, automated direction indicator	NDB/ADF
	VHF omnidirectional (radio) range	VOR
	Distance measurement equipment	DME
	Instrument landing system	ILS
Inertial navigation	Inertial navigation system (stabilized gyroscopic inertial platform)	INS
	Inertial reference system (captivated "Strapdown" platforms	IRS
Satellite navigation	Global navigation satellite system (GPS, GALILEO, GLONASS etc.)	GNSS
Integrated navigation	Flight management system	FMS
	Area navigation (integrated radio-and satellite-based navigation)	RNAV
	Precision navigation	PRNav

Table 9.3 Required navigation performance for area navigation [16]

Flight phase	Dimension	Accuracy	Type	Integrity
Oceanic/remote	2D	10 nm	RNAV10	10^{-5} [h]
		4 nm	RNP4	
En route/continental/terminal	2D	5 nm	RNAV5	
		2 nm	RNAV2	
		1 nm	RNP1	
Non-precision Approach	2D	0.3 nm	RNP APCH	10^{-5} [h]
	3D	0.1–0.3 m	RNP AR	10^{-7} [h]
		16.0 m horiz. 4.0 m vert.	LPV	10^{-7} [h]
Landing	3D	3.6 m horiz. 1.0 m vert.	GLS CAT I & II	
			GLS CAT III	10^{-9} [h]
Ground surveillance Guidance	2D	6 m		
		0.5 m		10^{-5} [h]

The required performances of the navigation systems depend on the operational phase. They are summarized as requirements in the RTCA DO-236B standard for area navigation, published by the Radio Technical Commission for Aeronautics, [16] (Table 9.3):

In addition to these functional and technical requirements like range, resolution or accuracy also equipment manufacturing cost and operating cost in terms of equipment mass and power consumption are further constraints. The latter are especially interesting for the design of future aircraft, while the first class of requirements will affect the efficiency of ATM and the aircraft too.

From ATS point of view it is essential to know about the principle functions and characteristics of the various systems to develop future concepts and to assess the potentials. For more technical information regarding detailed design and optimization of those systems the referenced literature is recommended, while in the following the descriptions are limited to the physical principles and limitations or disadvantages necessary to understand the application in the ATS [3].

9.5.1.1 Radio Navigation Systems

Radio navigation systems are subsystems in the ANS substructure of the ATS. The most relevant systems are introduced with its main features in the following sections.

Distance Measurement Equipment

The Distance Measurement Equipment (DME) is used since the early 1950s of the twentieth century, standardized by ICAO. It provides the information about the transverse distance between the aircraft and the radio ground station.

Its physical principle is based on runtime measurement, where the relative slope distance between the DME ground station and the aircraft is calculated,

$$R = c \cdot \frac{t}{2} \tag{9.2}$$

where c = 299.792.458 m/s is the speed of light and t is the runtime between sending out the signal from the aircraft and receiving the response from the DME ground station. In practice there is a system specific time shift of 50 μs introduced to the runtime signal.

The transmitter unit on board of the aircraft, called interrogator, sends out the measurement signal, which is received and processed by the ground station (transponder). At the end the slope distance is measured, which implies some error concerning the intended horizontal distance measurement.

$$d = \sqrt{R^2 - h^2} \tag{9.3}$$

The more the aircraft approaches to the DME ground station, the more the measurement error increases, i.e. when the aircraft is directly above the station, the measurement provides the height above the station, Fig. 9.11.

Consequently, the error in distance measurement increases from 0.047 % at 360 km to nearly 20 % at 20 km.

The accuracy of the DME is therefore limited by the combination of flight level and relative distance. Further the accuracy of the radio measurement is in the range of ±450 m related to target ranges of 200 nm. Due to the interrogation and respond

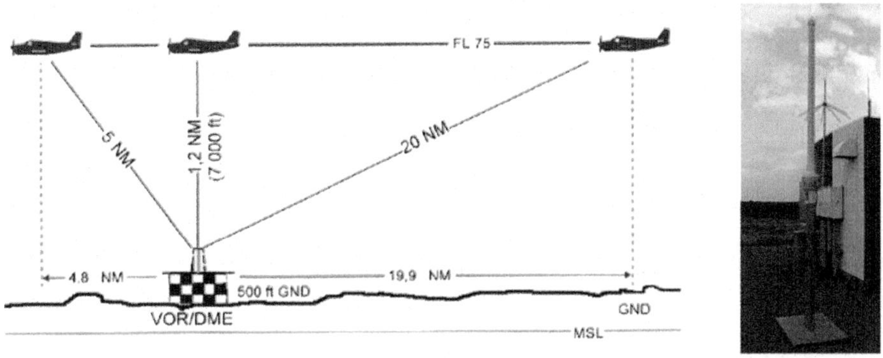

Fig. 9.11 DME distance and height error relation [2]

principle of the system the maximum capacity of a single DME station is limited to 360 aircraft theoretically. In practice a lower amount of about 250 aircraft can use the station.

DME stations are very often combined with VOR stations, which provide the corresponding relative bearing or directional information. Providing both type of information the precise aircraft position can be determined.

Very High Frequency Omnidirectional Radio Range

The Very High Frequency Omnidirectional Radio Range (VOR) provides a relative bearing information to the aircraft. Typically VOR stations are positioned along major traffic routs, and offer a measurement range of about 130 nm. Like DME stations also VOR stations are identified by an individual code, which is associated to the radio frequency of the station. The pilot has to set this code or frequency in the cockpit, when he wants to use the VOR station.

VOR stations are also placed within the vicinity or directly on the site of airports, where their range is limited to approximately 25 nm. The accuracy of the system is about ±5.2° with conventional technologies. Using a Doppler-VOR the accuracy can be improved to 1°.

Setting a fixed phase difference corresponding to the requested track allows course tracking to a VOR station, Fig. 9.12.

In the cockpit specific displayed information either in an integrated display or in a separate device Course Deviation Indicator (CDI) like below is provided, Fig. 9.13.

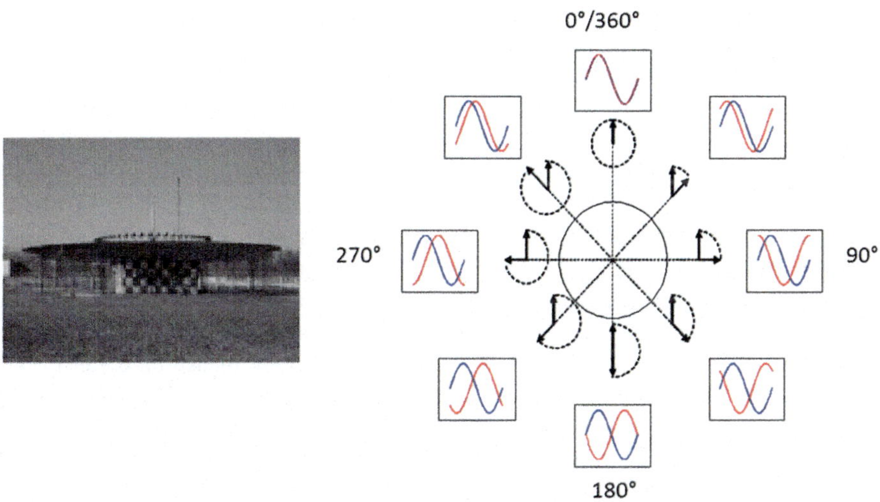

Fig. 9.12 VOR phase difference measurement for relative bearing and VOR antenna [2]

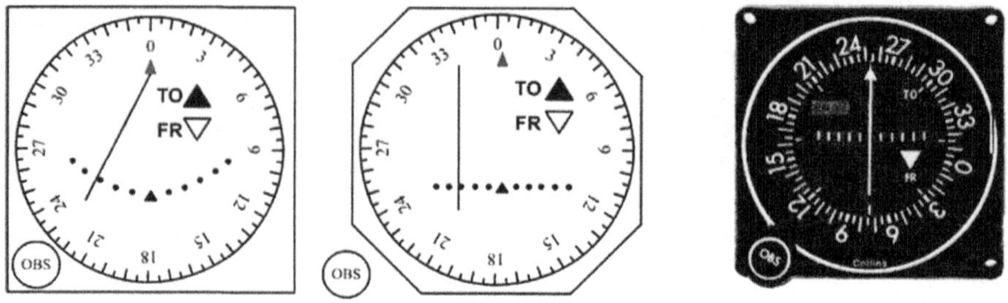

Fig. 9.13 VOR cockpit indication and course deviation indication (CDI) [2]

The CDI also indicates, whether the aircraft flies To (TO) or From (FR) the VOR station. Very close to the VOR station also this radio navigation device has an error cone, which is called "cone of silence". Right above the station the onboard receiver does not get any signal information.

Non-Directional Beacon (NDB)

A third radio navigation aid to be introduced is the Non-Directional Beacon (NDB), which is widely used since the 1930s of the last century. In combination with the onboard device called "Automatic Direction Finder" (ADF) this system also provides directional information. The operational range of the ground station is between 25–150 nm. NDBs are often used in the vicinity of airports, where they are widely used by airliners for cross-check or as "locators" for pre-visual clearance to an airport. Since NDB provides only horizontal directional information it is used as a Non-Precision-Approach device. Because it is not providing the direction, which the aircraft is approaching to and it is sensitive to interferences, the NDB cannot be used as a primary navigation device. Only in conjunction with the ADF installed on aircraft accurate directional information can be determined.

Radio Navigation Errors and Deficiencies

From ATS point of view not only the physical principles of the different radio navigation systems but also the deficiencies and errors need to be considered for their right operation and assessment. Here it should be mentioned, that the previously addressed "cone of silence" close to the navigation ground station as well as electromagnetic interferences and losses cause and increase measurement error and reduce navigation accuracy. Also reflections of the signals by mountains, hills, buildings and water called multipathing reduce the reliability of the received position information. Multipathing in this context describes the effect that radio waves colliding with flat planes maybe multiplied and develop additional travel ways and signals.

9.5.1.2 Satellite-Based Navigation

In the 1970s, the American military services developed a satellite-based navigation system called Global Navigation Satellite System (GNSS) of Global Positioning System (GPS). This system became fully operational with 24 satellites in the 1990s. Similar systems have been established in Russia, called GLONASS and China, where it is named COMPASS. In Europe the GALILEO system is under development, which is intended to enter into service around 2015. All systems follow the same setup consisting of

- a *space segment*, representing the satellites
- a *ground segment*, used for controlling and supervising the satellites as well as for data transmission
- a *user segment*, which is represented by different kind of civil and military users.

Four satellites each are operating on at least six nearly circular trajectories at about 20200 km altitude. The resulting 24 satellites are required to ensure a worldwide coverage over 24 h, Fig. 9.14.

Also four different satellites are required to provide the necessary runtime information to calculate the position parameters latitude, longitude, altitude and time. The fundamental satellite navigation formula is used to calculate the *Pseudo Range*. The term *Pseudo Range* describes the fact that the measured distances between the aircraft and the satellites differ from the true distances by a constant factor. This constant deviation is to be determined and corrected to calculate the true distance.

Fig. 9.14 General GPS satellite arrangement and pseudo range measurement

The American GNSS provides a highly accurate military signal and a less precise open civil signal. The European GALILEO system will offer a free of charge open service and chargeable commercial services, which shall be used for safety relevant flight navigation and other commercial data transmissions. Therefore, this system offers a higher bandwidth of 500 bit/s and higher data security.

If a high precision ground reference position is available, for example through a ground station close to an airport runway the accuracy can be improved significantly. Because of the accurately known position of the station, this information can be used to reduce the satellite runtime signal error.

This method, also known as Ground-Based Augmentation System (GBAS) or Differential Global Positioning System (DGPS) has been used several times to improve the position calculation, when the less accurate civil GPS signal is used. GBAS is today the most promising way to achieve a level of accuracy and integrity to replace Instrumental Landing Systems (ILS), see Sect. 9.5.1.3.

In a similar way, so-called Airborne-Based Augmentation System (ABAS) have been developed using Receiver Autonomous Integrity Monitoring (RAIM) or Aircraft Integrity Monitoring (AIM). RAIM uses internal monitoring algorithms to supervise the GPS receiver onboard of an aircraft. Those algorithms check regularly the correct functioning of the receiver components. The AIM uses onboard sensors like inertial sensors or radio navigation systems, to compare the GPS signals and to check the correctness and accuracy as major characteristics for the integrity of GPS.

Space-Based Augmentation Systems (SBAS) represent a third type of supplementary systems, which have been installed by the United States (Wide Area Augmentation System, WAAS), Japan (Multifunctional Satellite Augmentation System, MSAS), India (GPS Aided Geo Augmented Navigation, GAGAN) and Europe (European Geostationary Navigation Overlay System, EGNOS). Those systems are actually used to support the existing ILS CAT I approaches with at least 550 m horizontal vision. Actual research is also looking for improvements to reach CAT II and III capabilities.

When SBAS or GBAS are used for aircraft landing they provide the advantage, that approaches are no longer limited to straight flight paths of 3-3.2° descent angle, which are the limitations of current ILS. In addition also curved and optimized approaches may become possible, which are no longer limited by the navigation systems, but the structural layout of the aircraft must be adapted concerning additional loads. In conjunction with high precision short-term navigation devices like Inertial Navigation Systems (INS) more dynamic and flexible approaches and departures will become possible.

9.5.1.3 Instrumental Landing Systems and Landing Minima

Since the fifties of the last century the ILS has been established as the standard landing system to enable aircraft landings also in degraded visual environments.

According to ICAO three categories have been defined to classify the visual conditions for landings:

- **Category I (CAT I)**: A decision height not lower than 200 feet (61 m) above touchdown zone elevation and with either a visibility not less than 800 m or 2400 ft or a runway visual range not less than 550 m (1,800 ft) on a runway with touchdown zone and runway centerline lighting are required for a precision instrument landing .
- **Category II (CAT II)**: Here a decision height lower than 200 feet (61 m) above touchdown zone elevation but not lower than 100 feet (30 m), and a runway visual range not less than 350 m (1,150 ft) (ICAO and FAA) or 300 m (980 ft) (JAA) are at least necessary for a precision instrument approach and landing.
- **Category III (CAT III)** is subdivided into three sections:
  - Category III A—A decision height lower than 100 feet (30 m) above touchdown zone elevation, or no decision height (alert height); and a runway visual range not less than 200 m (660 ft).
  - Category III B—A decision height lower than 50 feet (15 m) above touchdown zone elevation, or no decision height (alert height); a runway visual range less than 200 m (660 ft) but not less than 50 m (160 ft) (ICAO and FAA) or 75 m (246 ft) (JAA).
  - Category III C—Zero visual range is given requiring guidance to taxi in zero visibility as well.

All visual landing minima requested for CAT III landings imply no sufficient outside view to the pilots. Therefore, a highly accurate automatic landing system like an ILS is required. The compliance with the CAT I-III requirements is given by the calibration of an ILS.

The instrument landing system is the most distributed landing system on airports around the world. It is composed of a localizer and a glideslope guidance system, Fig. 9.15.

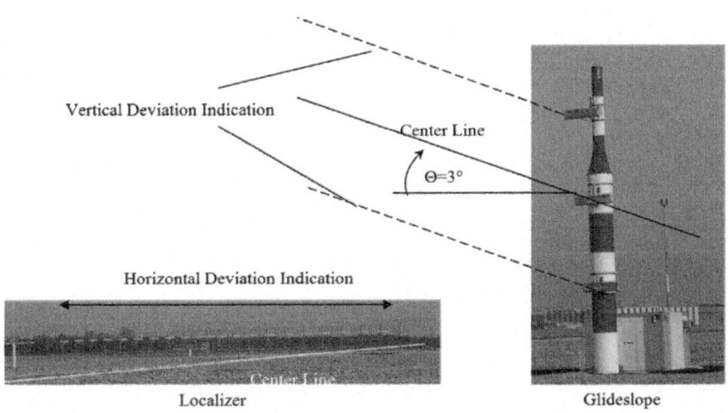

Fig. 9.15 Instrument landing system localizer and Glideslope

Providing a centerline 3° fixed ideal glideslope the ILS system also provides information to the cockpit, if the aircraft is above or below the glideslope. Further also lateral deviations are indicated. Because only a fixed final straight approach is possible the alignment of the landing aircraft will appear around 20 nm away from the runway threshold. From overall ATS perspective this limitation is of disadvantage because the different approach trajectories cover a wider area around the airport, which may cause noise complaints of the surrounding population. If more flexible Standard Arrival Routes (STAR) could be realized, the affected urban area can be reduced, but the noise intensity may be concentrated to smaller areas. Looking at this example tradeoffs are to be made to find the most appropriate landing system.

9.5.1.4 Inertial Navigation

Onboard navigation subsystems are to provide two main functions. First, they have to deliver the critical control parameter like air data, attitudes, angular rates and acceleration. Second, the aircraft positioning information like position, time reference and speed is needed to allow more accurate and safer aircraft guidance.

Gyroscopes and accelerometers are known as inertial sensors because they are representing the property to resist a change in momentum, [17]. This principle is used to sense angular and linear motion. Due to this principle gyroscopes and accelerometers are essential as well for automatic flight control systems (FCS) as for spatial reference in navigation.

Today different technical solutions for either gyroscopes or accelerometers are available. Gyroscopes and accelerometers are integrated on INS or Attitude and Heading Reference Systems (AHRS).

Different requirements for state and position measurements exist from system point of view.

The selection of adequate gyroscopes and accelerometers but also INS and AHRS is mainly based on the type of application and the related requirements. Following *Collinson* requirements according to Table 9.4 can be formulated, [18].

Especially, the scale factor is different for applications in FCS and INS, because the precision required for navigation measurements as performed with INS is much higher to achieve an overall position accuracy of 1 nm/h. Therefore the required drift is 500 times lower, than for the FCS. The FCS needs high dynamic short-term responses and only short-term accuracy to measure the aircraft state.

Table 9.4 Accuracy requirements for inertial sensors depending on application

Sensor	Accuracy requirement	FCS	Strapdown INS
Gyroscope	Scale factor	0.5 %	0.001 % (10 ppm)
	Zero offset/rate uncertainty	1°/min	0.01°/h
Accelerometer	Scale factor	0.5 %	0.01 % (100 ppm)
	Zero offset/rate uncertainty	5×10^{-3}g	5×10^{-3} g (50 μg)

Gyroscopes

While formerly gyroscopes were mainly built as angular momentum gyros to date most of the gyroscopes are ring laser gyros (RLG). But also the Fibre Optical Gyro (FOG) has reached a high level of accuracy, so that this principle is often used for AHRS due to cost reasons and less mechanical complexity [17]. Both concepts use the time measurement principle to sense angular rates. Splitting a light beam the light running clockwise and counter clockwise through a path needs different time if the optical wire of the gyro is rotating due to external excitation.

This principle is called "Sagnac Effect". Because this effect is very low for low rates of rotation some kind of amplification is needed, to sense and measure also small rotations accurately. Compared to the FOG a RLG has the advantage of a very high reliability of 60.000 h MTBF (Mean Time Between Failures) and a very low drift. The weight of such an optical gyro is in the order of 450 g requiring 7.5 W.

Accelerometers

The translational movement of a vehicle is measured using the Newton's principle by accelerometers. Over the years different types beginning with typical mass-spring devices and ending up with solid state accelerometers nowadays have been developed. Like for the gyro systems the trade off between required accuracy on the one hand and low manufacturing cost has to be made. Here simple spring-mass devices provide low cost but also low accuracy [17].

9.5.2 Future Trends in Navigation

Based on the principle navigation systems described previously today integrated navigation systems are used to provide high performance navigation in terms of accuracy, reliability, weight and energy consumption. Integrated navigation merges the positive characteristics of single navigation devices and tries to compensate their disadvantages [3].

Because radio navigation systems as described before are ground-based systems they are typically limited in their range and show increasing inaccuracies in position determination with increasing distance.

While the radio navigation systems mentioned before have been used for navigation over land some radio navigation systems like LORAN or OMEGA were used in the past for area navigation over sea. Those systems have been superseded by satellite-based navigation system as introduced in Sect. 9.5.1.2.

In Table 9.5 the major characteristics of the different principle navigations systems are summarized.

Although satellite and inertial navigation provide accurate 3D position both, inertial systems offer the big advantage of short-term high dynamic accuracy, while

Table 9.5 Comparison of various navigation aids [18]

Navigation aid	NDB/ADF	VOR	DME	GNSS	INS	Doppler
Information	Direction to ground station	Direction to ground station	Slope distance to ground station	3D-position	3D-position	2D-position
				3D speed	3D speed	3D speed
					A/C state	
Range [NM/km]	200/370.4	200/370.4	200/370.4	10.799/ 20.000	n.a.	n.a
Max. user	Unlimited	Unlimited	200	Unlimited	Unlimited	Unlimited
Precision	1…5°	2°	0.1NM	100-300 m	1.5–2NM	0.5–1 % of flightpath
Reliability	Good	Very good	Very good	Very good	Fair to good	Fair
Basis	On ground	On ground	On ground	On ground	Aircraft	aircraft

satellite navigation shows high precision long-term accuracy. Consequently today integrated navigations are developed, which combine the advantages of those single sensors in a complementary way to improve to overall navigation performance.

Despite the required accuracy and area coverage for future navigation systems it is mandatory to provide guaranteed integrity and resistance to interferences and corruption. In this context today it is required, that air- and space born area navigation systems (B-RNAV) provide not more than ±5 NM position deviation in 95 % of the flight time, [16]. For precision area navigation (P-RNAV) not more than ±1 NM position deviation is allowed in 95 % of the flight time. Potential causes for those deviations are given by sensor drift, abnormal behavior, atmospheric disturbances and reliability characteristics of the individual sensor systems.

For landing such integrated navigation systems must provide accuracy in horizontal position of 3.6 and 1.0 m vertically. These accuracies need to be fulfiled at an integrity level of 10^{-7} (CAT I-II) to 10^{-9} (CAT III) to ensure sufficient reliability and safety.

In conjunction with satellite-based area navigation systems, which provide a highly accurate position, also improvements in inertial navigation systems are addressed. Reductions in weight and energy consumption of about 40 % are envisaged if MEMS technologies (MEMS) are used. These improvements are associated with increases in accuracy and reliability, which might be doubled compared to 2010, [19].

9.5.3 Air Transport Surveillance

In order to ensure a maximum level of air safety, which is one of the major ANSP tasks, Radar (Radio Detecting And Ranging) systems are used to control the airspace. For this reason, worldwide radar stations are distributed to observe the airspace and to control the air traffic flow. Nevertheless radar surveillance is limited to over ground areas as well as coast areas, due to the limited radar system tracking range. As an example, the radar stations in Germany are presented in Fig. 9.16.

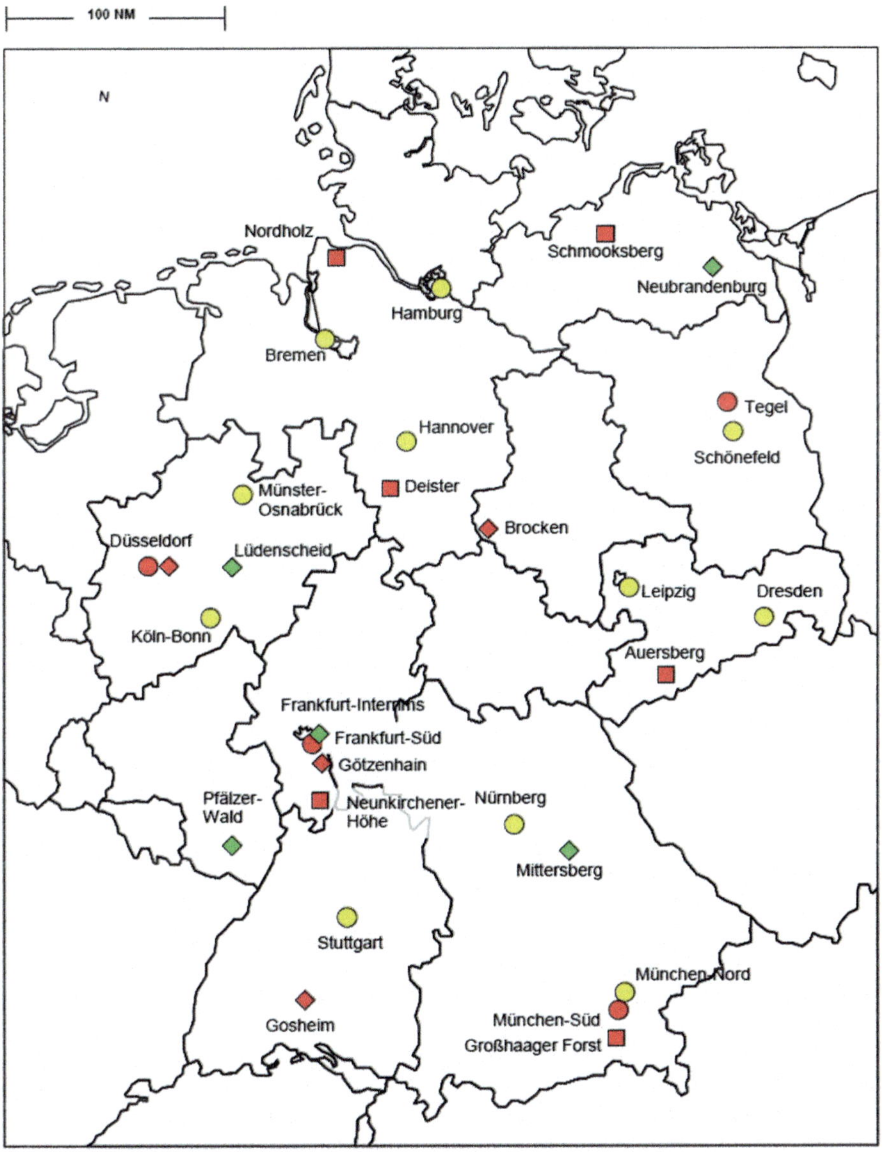

Fig. 9.16 Overview about the radar stations in Germany, [DFS]

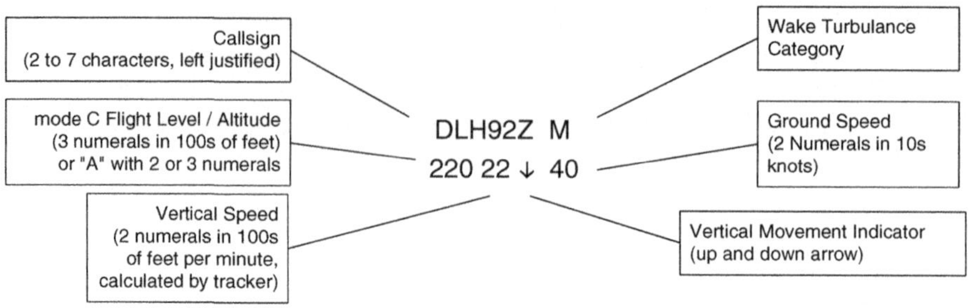

Fig. 9.17 Typical aircraft label on a radar display [20]

As one can see the various stations are distributed along to the most relevant coordinated airports in Germany. In the middle there is only a reduced coverage by stations at the Brocken mountain and Erfurt to ensure en route tracking.

There are two different principles of RADAR systems used, the autonomous or non-cooperative Primary Surveillance Radar (PSR) and the cooperative Secondary Surveillance Radar system (SSR), which is based on a bidirectional communication between a ground station and the air vehicle.

Both systems are used in parallel and the provided information is merged and displayed at the air traffic controller station. Here the aircraft position and additional objects like navigation stations, runways of sector boundaries are presented to provide best situation awareness to the controller.

In the background of such an ATC controller display modern Radar Data Processing Systems (RDPS) are operated to process all incoming information. They calculate the forecasted aircraft position and place the symbol at the most probably expected position. In this context, the RDPS calculates on the one hand the measured aircraft trajectory, which is called "plot" and adds to this past time information the expected next position, which is called "track". Tacking the SSR-code data for the calculation of the aircraft track this information is correlated with the Flight Data Processing System (FDPS), which contains the flight plans of aircraft. The entire information is covered by the so-called aircraft label, as shown in the next figure.

In Fig. 9.17 the different information is presented, which is associated to he relevant aircraft.

9.5.3.1 Physical Characteristics of Radar Systems

Radar operates in a similar manner as the DME measuring the runtime of an emitted signal. While the DME only analyses the runtime itself, radar systems also use the reflected energy content of the signal. The reflection characteristics of objects can be used not only for detection but also for classification and identification of

Table 9.6 Radar reflection characteristics of different aircraft [2]

Object	Max. cross section σ [m²]	Min. cross section σ [m²]
General aviation aircraft	10	5
Business jet	20	10
Transport aircraft	100	20
Fighter aircraft	5	0.3

σ = radar cross section, A = Area, λ = wave length, a = diameter

objects. This capability makes radar systems attractive for surveillance and identification applications on ground as well as onboard of aircraft. The following Table 9.6 provides typical reflection area characteristics of selected objects.

By its nature the radar cross section σ describes the relation between the electromagnetic power reflected by the target to the power received by the target. For flat metal plates the radar cross section is not only depending on the area A but also on the wave length λ. The active reflective radar cross section is depending on the geometric size and form of the respective object and its material and surface characteristics.

Due to the physical refraction of the radio waves from thinner upper atmospheric layers to the thicker ones, radio waves follow roughly the curvature of the earth. As a consequence the maximum detection range of radar systems is much longer than the detection range of optical systems, which allows the so-called "over the horizon targeting", [21]. Additionally also the ionosphere at 80–400 km has good reflection characteristics, which allows to extend the detection and transmission range of radar systems significantly, [2, 21, 22].

Further one can show, that the energy content of electromagnetic waves decreases by the quadratic power of the distance in vacuum. In real atmosphere, friction and scatter at atmospheric particles lead to further losses of wave energy. These effects a partially compensated by collimating antennas which focus and concentrate sent and received signals as shown in Fig. 9.18.

Further collimation provides a better azimuthal resolution, which is introduced in the next section.

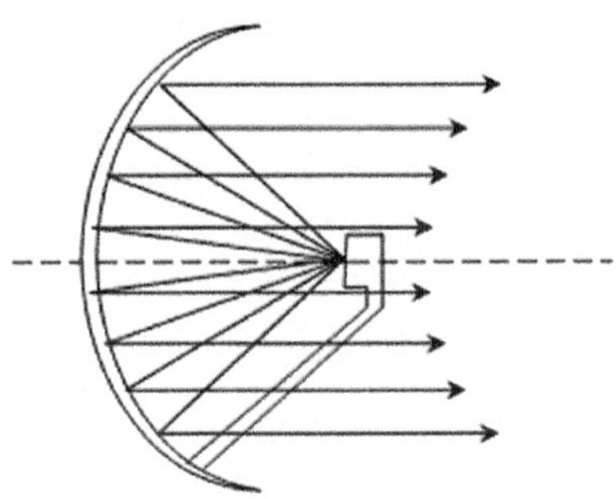

Fig. 9.18 Collimation principle of radar beams

9.5.3.2 Primary Surveillance Radar

Primary radar systems (PSR) are active unidirectional surveillance systems, which send out electromagnetic signals, which are passively reflected by moving and fixed objects, Fig. 9.19.

Due to the passive reaction of the objects only their position is identified by signal runtime measurements.

$$R = c \cdot \frac{\Delta t}{2} \tag{9.4}$$

Considering the speed of light of $c = 2.99792458 \times 10^8$ [m/s] for signal transmission, the angular distance R between the radar ground station and the object can be easily calculated by measuring the runtime of signal towards and backwards to the object:

Primary radar systems, working in a pulse mode instead of a continuous wave mode, are designed and used for different applications in air traffic surveillance. Depending on the required range and associated transmission power the following PSR systems are applied:

Airport Surface Movement Detection Equipment (ASDE) used for airfield and runway surveillance at the airport with a range of about *2 nautical miles* and *50 kilowatt* pulse power typically at wave length 0.9–2 cm (K-Band).

Route Surveillance Radar (RSR) is usually used for observation of air traffic control areas, which are at altitudes of about 50000–70000ft. Due to a typical range of 120–150 NM, its transmission power is much higher at 1–5 MW using wave length of 23 cm (L-Band).

Precision Approach Radar (PAR) running in the X-Band is used for supervising the precision approaches and landings on the glide path. Sometimes it is also used for so-called "Ground Controlled Approaches, GCA", where the controller gives verbal commands to the pilot to stay on the right glide path.

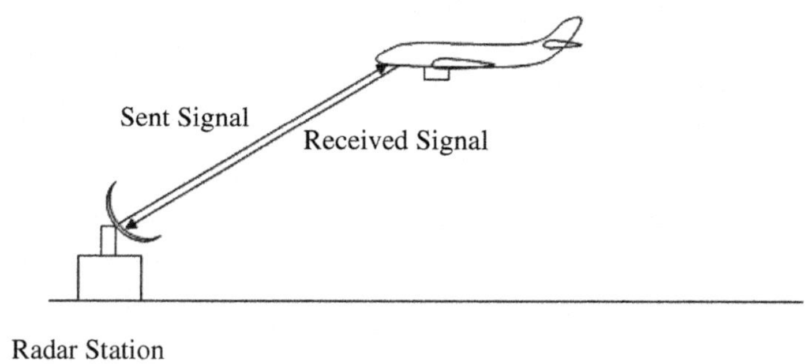

Fig. 9.19 Runtime measurement principle of primary radar systems

In order to achieve a requested resolution and precision of the target detection different wave length are used. The closer the air vehicle approaches to the airport the more precisely the measurement must be and the shorter the wave length is. At the airport various systems are used for different surveillance activities to prevent from interferences, [21, 22].

Since the main purpose of a radar system is the detection and tracking of objects, the radial and azimuthal resolution are of crucial relevance.

The radial resolution defines the minimum distance between two objects, where these objects can be distinguished and detected.

Beside these performance characteristics the radar equation gives a lot of inside about the performance of a radar system and its major design parameters:

$$P_R = P_T \cdot \frac{G^2 \cdot \sigma \cdot \lambda}{(4 \cdot \pi)^3} \cdot \frac{1}{R^4} \cdot \frac{1}{L} \tag{9.5}$$

Looking at the radar equation it indicates, that the received power P_R is directly affected by the radar cross section of the target object.

The size of the antenna, which is covered by the antenna gain G and the selected wave length are the relevant radar system design parameters, which define the system performance. It is important to note, that for physical reasons the transmitted power P_T is reduced by the fourth power of the range R, which is a strong loss. At last further losses L caused by atmospheric damping have to be mentioned. Such losses heavily depend on the weather situation like rain, snow, light or heavy rain. The damping characteristics of these occurrences differ with frequency in general. Therefore the operational function of the radar system defines its layout, e.g. a weather radar system needs to detect different reflection characteristics of various weather conditions, while this information is not useful for aerial surveillance, where air vehicle are to be tracked.

Another example showing the relevant knowledge about radar principles is the design of high resolution long range radar systems. Both requirements are strong contradictory, because increasing frequency will increase the resolution but the range will decrease if the transmission power is kept constant. This conflict might be considered if for future ATS concepts more automation is considered and automatic sensing systems will support pilots and controller in supervising the flight track. It is also an issue if radar systems will be used for bird crowd detection to prevent aircraft from potential bird strikes especially in the vicinity of airports. Finally radar systems are to be designed to suppress fixed objects and clutter from ground to ensure accurate detection of moving objects.

The way, how these detailed physical issues are realized are subject of electronic radar system design and not subject of this book. Therefore specific literature like *Barton* 2005 is recommended, [21].

9.5.3.3 Secondary Surveillance Radar

Secondary radar systems (SRS) which are also called Air Traffic Control Radar Beacon Systems (ATCRBS) are cooperative surveillance systems, where the aircraft transponder actively responds to interrogator signals of the ground station. Consequently this system works only if the aircraft is equipped with such a transponder. While the primary radar system provides the track information (direction and relative distance to the PSR ground station) of the air vehicle the SRS delivers encrypted the identity of the air vehicle, which is called Mode A and its barometric flight level information (Mode C) in 100 feet resolution. These are important information for the air traffic controller to guide the aircraft en route as well especially during approach and landing.
Compared to the PSR system the SRS shows some differences:

- The SRS can only be used with onboard transponders. If an air vehicle is not equipped, e.g. many general aviation aircraft or sailing planes, no information is available.
- Due to the active reply of the onboard transponder no clutter, i.e. mal information is possible
- Additionally the active reply reduces the signal energy loss to $1/R^2$ compared to the 2 ways $1/R^4$ energy decrease of the PSR.

Some kind of "garbeling" (i.e. overlap reply signals of two aircraft being close together within the resolution of the radar beam) makes a clear identification and signal processing impossible.

The amount of Ident-Codes is limited to 4096, which have to be allocated manually and dynamically to the aircraft by the air traffic controller.

In order to overcome some of the deficiencies of the ATCRBS the Mode-S secondary radar system has been developed by the MIT Lincoln Laboratory, which plays today a paramount role in the ATS in different ways [23]. Two interrogator modes are installed named "All Call" and "Roll Call". Typically the ground systems requests alternatively in both modes. With the "All Call" mode all aircraft equipped either with classical or Mode-S transponder within range are addressed. The classical transponder reply with the Mode A and Mode C information while the Mode-S transponder additionally provides their individual, fixed ICAO-address. In this case, the transponder is directly addressed in the so-called "Roll Call" in the next interrogation cycle.

With the progress of digital data protocols, more and dedicated information can be exchanged. Those details can be read in, e.g. [22–24]. For the scope of the ATS the type of information is of importance, which is transmitted by a Mode-S transponder (Table 9.7):

The implementation of further aircraft state information into the protocol like heading, indicated airspeed, climb/sink rate, etc. is actually under discussion. As

Table 9.7 Comparison of ATCRBS and mode-S information content

Information	Mode-S transponder	Classical ATCRBS transponder
24 bit aircraft address	X	–
Aircraft identification	X	–
SSR mode 3/A	X	X
SSR mode C	X	X
Aircraft status (on ground/in flight)	X	–
Avoidance devices	X	–

shown in Table 9.7 the Mode-S-SRS data protocol provides not only a unique automatic identification of the aircraft but also additional ATM relevant information like avoidance advices within the Terrain Alerting and Warning System (TAWS) see also Sect. 9.7.3. Because this information is of crucial relevance for safety Mode-S systems are installed with two antennas and receivers to ensure spatial diversification and to avoid shading effects.

9.6 Communication Systems

Oral radio communication has played a paramount role in aviation over decades. Also to date it is an elementary communication media, which is now supported by data communication or so-called data link systems. This section gives a brief overview about the relevant systems and their major characteristics.

9.6.1 Voice Radio Communication

Pilots and air traffic controller traditionally use radio communication systems in High Frequency (HF) and Very High Frequency (VHF). Due to an increasing lag of frequency resources the bandwidth of usable frequencies is more and more limited. Additionally the wave length of the VHF is limited in its range to about 150–200 nm, because a line of sight is required between transponder and receiver. For a rough estimation there is a relation between the range of the radio R in [nm] and the related flight level of the aircraft h [ft]:

$$R = 1.225 \cdot \sqrt{h} \qquad [\mathrm{m}] \tag{9.6}$$

For a 10000 ft flight level, the achievable communication range is about 122.5 nm as an example. Another advantage of VHF radios is their ability to be used for direction finding, which is an additional safety feature. The controller can use the radio communication to identify a certain aircraft by voice and direction. Due to the range limitation of VHF radio systems also HF radio systems are used especially for oversea cruise. Like the radar systems HF radios allow the so-called "over the horizon" ranging, because they take advantage of the ionospheric reflection. This physical advantage is useful for communication over oceans or sparsely populated regions.

9.6.2 Data Link Communication

There are various data link systems, which have been developed up to date. They use the VHF and HF radio transmission for operation.

High Frequency Data Link (HFDL) uses HF radio communication for data transmission. It is appropriate for aircraft as a long range data link, which are not equipped with SatCom. Additionally HFDL is valuable redundancy for SatCom around polar regions, where SatCom has reduced coverage and performance.

VHF data link (VDL) has been defined in four modes using VHF radio communication frequencies for data transmission. The modes define whether data and voice are transmitted point to point or broadcast.

SatCom is a further data link subnet for ground—onboard communication. The key benefit is the worldwide coverage and accessibility. Based on different satellite systems it will become the backbone for the future integrated ATM System.

Mode-S previously introduced as secondary radar and identification system is also used as a data link. Using four different modes up- and downlink communication and also point to point communication between aircraft is supported.

Data links also provide service functions like transmission of maintenance information to speed up turn around or ground time or transmission of passenger check-in or rebooking data. They will become a valuable functional feature of future aircraft.

9.7 Integrated Air Traffic Management and Control Systems

In order to extend air transport capacities and to improve efficiency in terms of punctuality and energy effort the ANSP and related industries develop integrated CNS-systems, which combine the performances of the individual systems. The Aeronautical Telecommunication Network (ATN) specified by ICAO integrates several data link technologies into a comprehensive network to combine the air-to-ground (Downlink), ground-to-air (Uplink) and air-to air communications, [25]. For this purpose the data links mentioned before are integrated to ATM/C systems.

Since the 1990s of the last century mainly driven by military applications digital data links using the features of VHF and HF transmission have been developed to improve communication between aircraft and ATC.

In civil aviation the Aircraft Communication Addressing and Reporting System (ACARS) has been developed. Due to the line of sight restriction a ground-based receiver/transmitter network is used to exchange all data. Thus ACARS can be used only over land mainly. The services offered are aircraft system health data transmission or passenger booking details as examples.

The development of satellite-based navigation and communication systems has brought out the Automatic Dependent Surveillance-Broadcast (ADS-B) as a non-commercial successor of ACARS. It is a general purpose data-link-system using satellites as relay stations, which provide a worldwide coverage. Also ADS-B is now on the way to be used for additional services like those mentioned for ACARS. This information will be used for future collision avoidance concepts superseding the actual TCAS, when the position information will be broadcasted to other participants of the air traffic. Additionally, the so-called Position-Velocity-Time-Vector (PVT-vector) is transmitted by ADS-B, which contains the actual geodetic height, the speed components air data, ground speed and some more. An architectural overview about the ADS-B system setup is given in Fig. 9.20.

If this information is broadcasted between all air traffic participants collision avoidance can be significantly improved, which contributes to enhanced safety in air traffic and also provides potential for increased capacities in the airspace. Also areas without radar surveillance can be controlled by air traffic control. All participants can also monitor the track of each aircraft in its region and provision can be made. Last ADS-B information is useful for airfield surface movements control, which is in the focus of the Advanced Surface Movements Guidance and Control System (A-SMGCS), which is currently short implementation on airports.

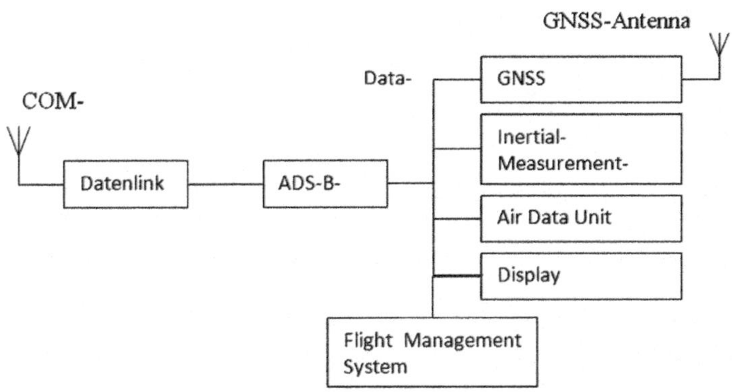

Fig. 9.20 System architecture of automatic dependent surveillance—broadcast, ADS-B

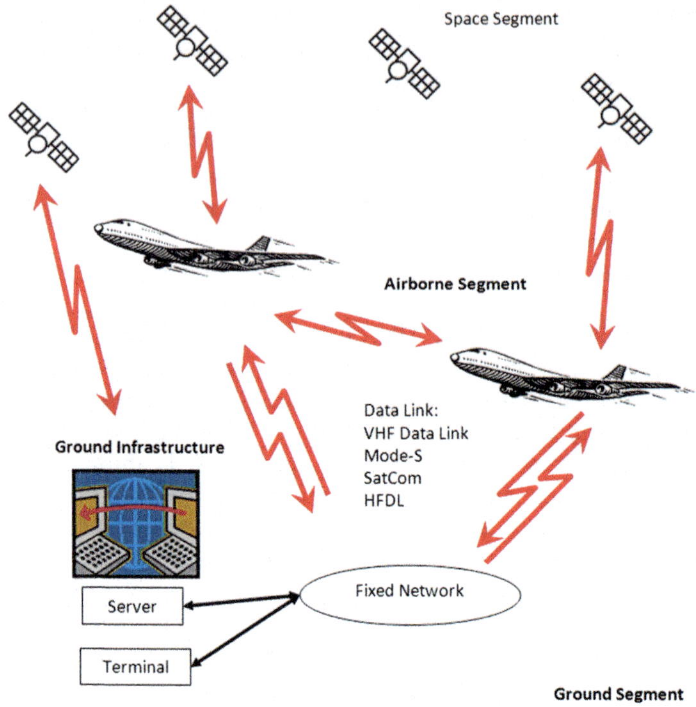

Fig. 9.21 Principle concept for Aeronautical Telecommunication Network (ATN)

Radar, navigation and data link systems are merged and integrated to ATM/C systems with the major objective to make data and information available to any air transport participant and any time. This is the main progress in ATM, guidance, navigation and control systems. For this purpose the concept of ATN has been developed, Fig. 9.21.

ATN will be installed to provide various services:

- Next generation collision avoidance system based on ADS-B (Automatic Dependent Surveillance—Broadcast)
- Controller—Pilot data link communications (CPDLC) replacing standard information communication by automatic data transfer to relieve VHF oral communication
- Traffic Information Service—Broadcast (TIS-B) providing uplink air traffic situation information Flight Information Services—Broadcast (FIS-B) providing weather, departure and arrival information
- Ground-Based Augmentation System to improve satellite navigation-based position identification by providing ground correction data.

9.7.1 Multilateration (MLAT)

Multilateration is a well-known method for position measurement, which has been used for long time with long range navigation systems like LORAN or OMEGA. Today the principle of multilateration, which is based on the so-called "time difference of arrival" (TDOA) is using data link signals like those of VDL, HFDL, Mode-S, or ADS-B. At least three ground-based reference stations are required, which calculate the time differences of the arriving data link signals. As a result 2 or 3D positioning is achieved, using existing onboard equipment in a different way.

This principle is applied for A-SMGCS for airport movement surveillance and control as well as it can be used also for Wide Area Multilateration (WAM) surveillance at airport terminal areas where no radar surveillance is provided.

At last WAM can be used for en route airspace surveillance especially over sea and difficult topography where no ground-based radar can be operated.

9.7.2 Airborne Collision Avoidance Systems

In order to install provisions to avoid "Mid Air Collisions" the ICAO has established regulations for Airborne Collision Avoidance Systems (ACAS), [25]. Technical solutions for those ACAS are realized as "Traffic Alert and Collision Avoidance Systems, TCAS". Two systems have been introduced and all aircraft with more than 5.7 t to takeoff mass or 19 passengers have to provide a TCAS II system, which not only gives warnings about aircraft in the vicinity ("Traffic Advisory") but it also provides recommendations for vertical evasion maneuvers ("Resolution Advisory").

Technically those systems are based on the Mode-S transponder functions, where the course and altitude information of the responding aircraft are used by the TCAS system to track aircraft in the vicinity [23]. Separate displays provide this information and evasion recommendations to the cockpit crew.

9.7.3 Terrain Awareness and Warning System

Like ACAS mainly enforced by FAA a Terrain Awareness and Warning System (TAWS) has been defined to prevent aircraft from Controlled Flight Into Terrain (CFIT), which is one of the major causes for aircraft accidents. Those systems use onboard information like radar and pressure altitude, vertical and horizontal speed, as well as glide path deviation of an ILS and landing gear and flapping settings. Up to seven different modes are available to create warning, e.g. about exceeded sink rates, glide path deviation and also shear winds. In an extended version today also synthetic topographic data bases are used in conjunction with GNSS to provide better situation awareness to the cockpit crew. TAWS also called Ground Proximity Warning System (GPWS) act in a similar way as ACAS. The main task of the cockpit and cockpit devices is to provide interfaces to select the different functions and to display the information to the cockpit crew.

9.7.4 *Interfaces Between ATM and Aircraft*

In the previous sections the most relevant up to date systems for guidance navigation and control of air traffic have been introduced as well as the organizational setup of ATM. In order to demonstrate the interfaces between all this elements Fig. 9.22 provides an overview about the principle architecture of these elements.

Ground and air side of ATM cooperate through the provision, receipt and exchange of information provided by the various ground-based communication and navigation systems. Air Navigation Services provide and operate the different navigation systems like ILS, VOR and DME. They also provide radio communication systems like VHF and HF. The Air Traffic Controller and the cockpit crew use this information including weather forecast to coordinate and update the flight plan including the arrival and departure procedures.

Maintenance Repair Overhaul (MRO) services based on data links are used by airlines and MRO companies.

Navigation information is mainly used by the flight management system (FMS) to calculate the flight path and the related aircraft performance [26]. This information is further used to feed the FCS, which automatically controls the rudder and flaps of the aircraft. For this purpose also the onboard inertial and air data sensors are used. The traffic alert and warning system (TAWS) uses this information as well and includes also the information of the cooperative secondary radar to calculate potential mid air collision situations.

The integration and interfacing of the ground and air side of the ATM infrastructures and processes is based on the cooperative principle, Fig. 9.22. This principle worked well during the last decades for civil aviation. During the early years of the twenty-first century unmanned air vehicles or systems called UAV or UAS became more and more relevant for military missions but also for aviation in general. The integration of those systems regarding coordinated navigation, communication and surveillance is a major challenge of research and development for the next decades. UAV/S are typically used for reconnaissance and surveillance missions. Especially military and industrial reconnaissance missions are intended, not to be detected and therefore are non-cooperative. In the future procedures and requirements need to be developed to deal also with those non-cooperative systems in the airspace.

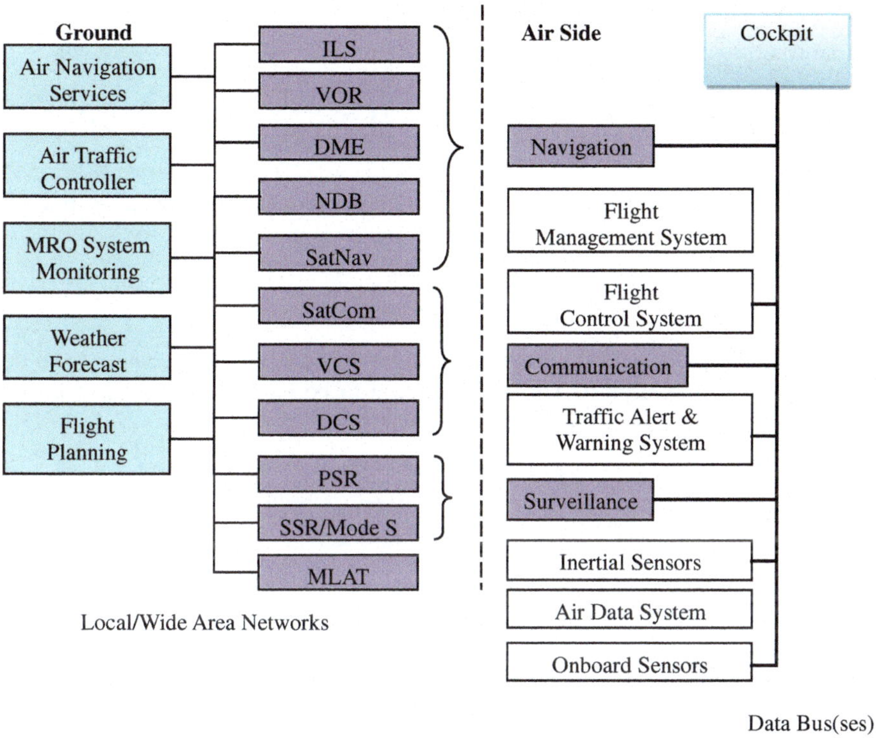

Fig. 9.22 Integration and interfacing air traffic management ground and air side

9.8 Navigation Fees

Each airline has to pay navigation fees for the ANS provided en route as well as during approach and departure on an airport for each individual flight. To introduce the principles the procedures of Eurocontrol and the German Air Navigation Service (DFS) are used as an example.

Terminal charges are levied by the ANS for providing services and facilities for aircraft during take-off and landing at German airports. Route charges are collected for en route ANS and facilities, which are used by aircraft in the airspace of the FIR of the Federal Republic of Germany.

9.8.1 Take-off and Landing Charges

Approach and departure, as well as repeated touch and goes, count as one flight [27, 28]. The counting unit is the departure.

Table 9.8 Examples of terminal navigation charges in Germany [28]

Aircraft type	Cessna	LR35	B737	A320
m_{TOM} [t]	0.7	45.51	58.0	75.5
Terminal navigation charge [€]	8.13	8.3	180.42	212.93

For aircraft with a maximum take-off weight beyond 2 tons the following applies:

$$r = t \cdot p \tag{9.7}$$

The TakeOff and Landing charges r are computed by the service value t and the mass factor p of the aircraft. The actual charge to be paid is calculated using the weight factor in the case of terminal services.

$$p = \sqrt{\frac{m_{TOM}}{50}} \qquad \text{with } m_{TOM} \text{ in [tons]} \tag{9.8}$$

Using the maximum take-off mass (m_{TOM}) as a reference parameter might be questionable but there are some aspects, which let it assume to be representative. First, the m_{TOM} is a direct indicator for the maximum passenger and cargo capacity of an aircraft. The resulting airline revenue can be used as a reference for the landing fees. Further the abrasion and load of the airfield and runway is directly affected by m_{TOM}. Also the runway capacity is indirectly affected by the aircraft mass, which drives the intensity of wake vortices and the resulting separation minima. The resulting maintenance cost to keep the runway operational can be directly linked to the aircraft m_{TOM}. At last the specific fuel consumption (sfc) and the resulting CO_2 emissions are also proportional. Therefore, the m_{TOM} is also an indicator for the environmental impact of the individual aircraft.

As an example based on the navigation fees of the German Air Navigation Service (DFS) for 2010 of 162.54 € per unit rate the navigation charges typical aircraft types are calculated as follows (Table 9.8):

In Germany charges are levied for the provision of ANS for arriving and departing aircraft at the airports of Berlin (Tegel, Schönefeld), Bremen, Dresden, Düsseldorf, Erfurt, Frankfurt/Main, Hamburg, Hannover, Köln/Bonn, Leipzig/Halle, München, Münster/Osnabrück, Nürnberg, Saarbrücken and Stuttgart.

9.8.2 *En Route Charges*

For each flight in controlled airspace through the airspace of a state navigation fees are calculated on various input parameters [27]:

$$r = u_i \cdot N \tag{9.9}$$

Where u_i defines the unit rate of the individual service, N is representing the amount of the services.

$$N = d \cdot p \tag{9.10}$$

It is remarkable to highlight, that the fees a directly depending on the route length d of the relevant section, which is 1/100 of the orthodromous distance (great circle) between the departure airport or the entry point of the relevant airspace and the first arrival airport or the exit point of the sector. With p as the representation of the aircraft mass:

$$p = \sqrt{\frac{m_{\mathrm{TOM}}}{50}} \qquad \text{with } m_{\mathrm{TOM}} \text{ in [tons]} \tag{9.11}$$

The entry and exit points of the sectors are given in the aviation manuals. Following these calculations, the overall fees are composed of

$$r_{\mathrm{tot}} = \sum\nolimits_{i=1}^{n} r_i \tag{9.12}$$

where n gives the amount of sectors, which are crossed in the respective country.

Taking MTOM as a baseline for navigation charges could be of relevant disadvantage for short range aircraft, which have a nominal higher m_{TOM} like it is in real operations on very short distances. About 80 % of all short range flights cover only 1/3 of the design range and therefore related m_{TOM}.

References

1. Fricke, M., Hüttig, G.: Flight Guidance. Lecture Notes, Institute for Aeronautics and Astronautics, Technical University Berlin (2005)
2. Gollnick, V.: Introduction to Air Traffic Management and Flight Guidance Systems. Lecture at the Technical University of Hamburg-Harburg, 4th edn. (2012)
3. Gollnick, V.: The Air Transportation System. Lecture at the Technical University of Hamburg-Harburg, 6th edn. (2013)
4. ICAO: Rules of the air. Annex 2, Montreal (2010)
5. ICAO: International Civil Aviation Organization. www.icao.int (2011). Accessed 18 Oct 2011
6. ICAO: Procedures for Air Navigation Services—Rules of the Air Traffic Services. ICAO Doc.4444-RAC50 1, 17th edn. www.icao.int (2007). Accessed 21 Nov 2013
7. Gollnick, V., Stumpf, E., Szodruch, J.: ATS beyond 2020. Forum The Green Air Transportation System, Bonn, Germany, 31 Oct 2007
8. Vatsim: Standard terminal arrival route Hamburgs. http://edww.vatsim-germany.org/charts/eddh/eddh_star_2010_04_09.pdf (2013). Accessed 10 Feb 2013
9. Ashford, N., Stanton, H.P.M., Moore, C.A.: Airport Operations, 2nd edn. Wiley, New York (1997)

10. Kösters, D.: Regular delays in aviation—definition and application of a quality measure for airport coordination. Ph.D. thesis, Institute for Transportation, RWTH Aachen, German (2010)
11. Eurocontrol: ATM-airport-performance-report. https://www.eurocontrol.int/sites/default/files/attachments/200912-atm-airport-performance_report.pdf (2013). Accessed 31 Oct 2013
12. Skybrary: An electronic repository of safety knowledge related to ATM and aviation safety in general. http://www.skybrary.aero/index.php/Main_Page (2013). Accessed 17 Feb 2013
13. ACARE: European aeronautics: vision for 2020. www.acare4europe.org/docs/Vision%202020.pdf (2001). Accessed 27 Feb 2011
14. DIN: Navigation- Concepts, Abbreviations, Letter Symbols, Graphical Symbols. DIN13312, Beuth publishing, German (2005)
15. Moir, I., Seabridge, A.: Civil Avionics Systems. AIAA Education Series, 1st edn. AIAA, Reston Virginia (2002)
16. RTCA: Minimum Aviation System Performance Standard—Required Navigation Performance For Area Navigation, RTCA DO-236B
17. Lawrence, A.: Modern Inertial Navigation Technology 1st edn. Springer, New York (1998)
18. Collinson, R.P.G.: Introduction to Avionics, 1st edn. Chapman & Hall, England (1996)
19. Mary, M., Clemencaeau, P.J., Bouniol, P.: Technological trends for future navigation systems. In: 27th International Congress of the Aeronautical Sciences, Nice (2010)
20. Hassa, O.: Personal notes. German Air Navigation Services (DFS), Oct 2012
21. Barton, D.: Radar Systems Analysis and Modeling. Artech House, Boston (2005)
22. ICAO: Surveillance Radar and Collision Avoidance Systems. Aeronautical Telecommunications, Annex 10, International Civil Aviation Organization, Montreal (2007)
23. Eurocontrol: Principles of Mode S Operation and Interrogator Codes. European Organization for the Safety of Air Navigation, Bruxelles (2003)
24. Flühr, H.: Avionics and Air Traffic Management 1st edn. Springer, German (2009)
25. ICAO: Aeronautical Telecommunications. Annex 10 to the Convention on International Civil Aviation, Volume IV: Surveillance Radar and Collision Avoidance Systems, International Civil Aviation Organization, Montreal (2012)
26. Moir, I., Seabridge, A.: Aircraft Systems: Mechanical, Electrical And Avionics Subsystems Integration. AIAA Education Series, 1st edn. AIAA, Reston Virginia (2012)
27. Eurocontrol: BasicUnitRates. http://www.eurocontrol.int/crco/public/standard_page/basic_unit_rates.html (2010). Accessed 13 Oct 2010
28. DFS: air navigation services terminal charges. German Air Navigation Service Provider, Issue January 10th 2010. http://www.dfs.de/dfs/internet_2008/module/unternehmen_dfs/englisch/about_dfs/business/charges/index.html (2010). Accessed 13 Oct 2010
29. ICAO: Aeronautical Telecommunications. Annex 10 to the Convention on International Civil Aviation, Volume III: Communication Systems, International Civil Aviation Organization, Montreal (2007)
30. Mensen, H.: Aviation Manual, 1st edn. Springer, German (2003)

Chapter 10
Environmental Aspects of Air Transport

Abstract This chapter introduces environmental impacts of aviation. Starting with principle considerations, major emphasis is given to explain the physical effects of CO_2, NO_x, and contrails. Further, measurement methods to classify the environmental compatibility of engines are introduced including measures to improve the environmental compatibility. Aircraft noise as another major aspect is considered starting with a physical and mathematical explanation. Also, various sound metrices are introduced to provide an understanding of the sound impact. Regulatory requirements as well as measurement methods and criteria are discussed. Lastly, measures to reduce noise sources are discussed.

Aviation and the environment are cross-interacting. On the one hand, aviation produces emissions and noise affecting people and climate. On the other hand, weather conditions like rain, snow, ice, storm, thunder storm, wind and turbulence but also volcanic contamination of the atmosphere are influencing the flight of an aircraft and could lead to operational restrictions. It is not only the flight dynamics in terms of safety and comfort of an aircraft, which is affected by wind, gust and turbulences, icing or volcanic contamination reduce the aerodynamic performance of the wings. Further, also the engines and sensors can be influenced by those atmospheric conditions. For example, the sensor detection range of laser sensors is significantly reduced, if atmospheric humidity is increasing in terms of fog, rain, or snow [1].

For the development of future air transport systems, it is of paramount importance to understand the principle characteristics how emissions affect the climate. Therefore, the chemical and physical as well as the time-dependent behaviour of emissions on the short- and long-term perspective are introduced briefly in Sect. 10.2. Additionally, principle technical and operational solutions to reduce the impact of air transport on its environment are discussed.

Further, it is necessary to know in which way aviation noise is created and how it influences human life. These aspects are described in Sect. 10.3 including some measures, which reduce the generation of noise at its source as well as procedures decreasing the noise impact on people.

D. Schmitt and V. Gollnick, *Air Transport System*,
DOI 10.1007/978-3-7091-1880-1_10

10.1 Introduction

The aircraft itself as a technical system incorporates all technical features and performances, which may have an effect on noise impact and emissions affecting the atmosphere. But climb and landing performance of an aircraft also has an impact on the required runway length and therefore on land use and also local air quality and noise impact.

It is the responsibility and interest of the aircraft manufacturer to develop aircraft, which are attractive and accepted by the airline and people on the one hand and fulfil ecological regulatory conditions and are also economically competitive on the other hand [7].

The airport's contribution to sustainable air transport covers a much wider range of aspects, starting with the energy effort for its buildings regarding heating, air conditioning and lighting. But also ground handling services, airfield lighting are issues to be addressed for sustainable air transport contribution. Also land use, herbicides and pesticides, which are used on airports to clear the airfield are elements of the environmental impact. But also the use of low pollutant deicing liquids as well as low emissions and noise during taxiing is important from airport's and airline's point of view.

From airline perspective, it is further essential to operate the aircraft fleet economically competitive while fulfilling the legal conditions and being attractive and accepted by the people. Also in this case ecological compatibility needs to be realized during operation on ground, especially regarding handling of waste and wastewater. In flight fuel consumption and navigation fees as well as crew cost are the main cost driving factors from airline's perspective, which need to be minimized. Especially, fuel burn and navigation fees have a direct correlation to environmental compatibility in terms of emissions and also noise.

Guiding the aircraft safely through the airspace the air navigation service provider can support sustainable air transport especially with respect to noise abating approaches and departures but also by realizing shortest flight regimes. The ANSP are therefore interested in navigation, communication and guidance systems, which allow for 4D precision flight guidance with minimum detours and minimized delays.

This brief overview shows various opportunities and interests of the main stakeholders of air transport to contribute to a sustainable air transport system.

Noise of the aircraft and its operation is seriously affecting the people as an additional environmental impact. Therefore, as described in Chap. 4 fulfillment of minimized-noise requirements and also proof of minimum engines emissions are prerequisites for aircraft and engine certification.

According to Fig. 10.1 the, environmental impact of aviation is not only limited to those two aspects. During operational life maintenance and service activities like deicing or engine washing create pollutants too.

At the end of the life cycle of an aircraft or any other technical system in air transport recycling becomes more and more a relevant issue. When technical

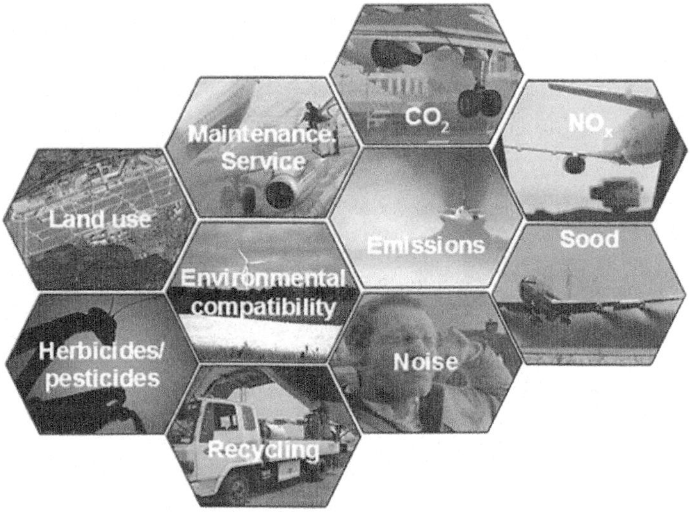

Fig. 10.1 Environmental aspects of aviation

systems reach their end of life, dismounting, separation and sustainable reuse of materials have to be considered.

10.2 Air Transport Emissions Impact on the Climate

The impact of anthropogenic emissions is a sensitive matter with many uncertainties and also a lot of political interests. There is no doubt that man-made emissions contribute to climate change.

Since the 1990s of the last century, the Intergovernmental Panel on Climate Change (IPCC), which is the United Nations climate panel, reports continuously about the scientific knowledge on man-made climate impact [3, 8]. It has been observed, that the global average temperature raised about 0.8 °C during the last 150 years and due to the significant amount of especially CO_2 emissions the IPCC prognoses further 2 °C increase during the next 50 years. Further climate sciences have recently discovered and documented in the latest IPCC report that aviation contrails may have a more relevant impact than CO_2 [9].

There are also other investigations which relate the man-made emissions during the last 150 years since industrialization to the very long lasting climate behaviour [10]. As a conclusion those analyses conclude that man-made emission contribute less, than stated by the IPCC and the direct and indirect effects of the sun, take a larger share of the actual temperature increase.

Due to the very complex mechanisms of the climate system, it is very difficult to come to a final assessment. It is therefore strongly recommended to look carefully at different perspectives and analysis.

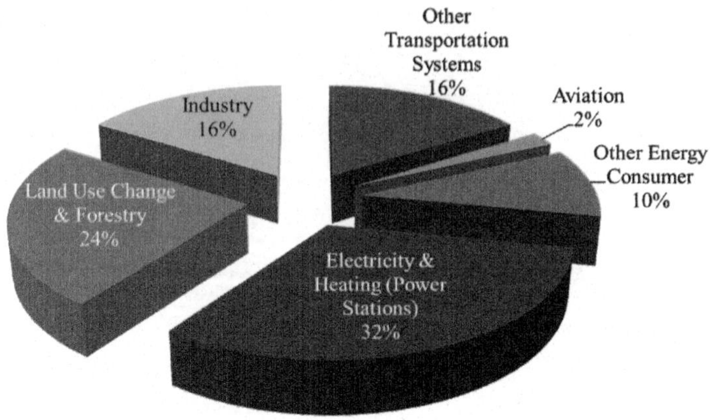

Fig. 10.2 Share of man-made CO_2 emissions [11]

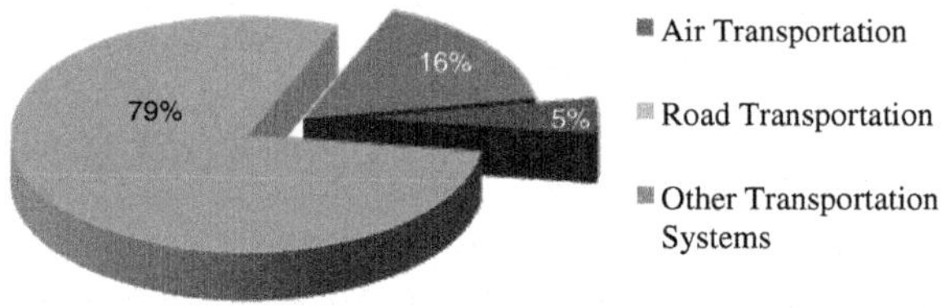

Fig. 10.3 Share of air transport NO_x emissions

The major part of man-made CO_2 is caused by electricity generation at power stations and heating, shown in Fig. 10.2. Further, agricultural and forestall land use contributes about 24 % of the man-made CO_2. Transport and aviation together provide 18 % CO_2, where aviation shares about 2 %.

Therefore, transport or mobility as a major driver for prosperity and welfare is also a significant contributor to CO_2 emissions.

Although the portion of aviation seems to be fairly small, it needs to be considered carefully, because these emissions are occurring in unique form at high altitudes during cruise conditions at 10,000–12,000 m approximately.

A similar situation is given for the contribution of air transport to the overall NO_x emissions of transport, Fig. 10.3. Also for NO_x emissions, a share of about 5 % for air transport seems to be fairly small.

But also in case of NO_x emissions, the occurrence at high altitudes makes the effects on the atmosphere unique as shown in Sect. 10.2.3.

For the understanding of the atmospheric impact of emissions, their dynamic geographic and time-depending behaviour is of paramount importance.

CO_2 emissions cause a slow but long-term increase of temperature. The maximum temperature raise caused by CO_2 is achieved later than 35 years after the pulse

Table 10.1 Life time of emissions in the atmosphere [12]

Emission	Life time
Carbon dioxide (CO_2)	50–200 years
Methane (CH_4)	8–10 years
Ozone (O_3)	Some months
Water steam (H_2O)	Some weeks
Nitrogen oxide (NO_x)	Some weeks
Cirrus, contrails	Up to some weeks

emission, while all other increases are reduced by more than 50 % at this point. This is what makes the impact of CO_2 unique.

While ozone, contrails and water lead to a short-term increase of temperature, ozone in the primary mode and hydrocarbons cause a mid-term decrease of the temperature, which turns to zero change after about 100 years approximately.

As an order of magnitude, Table 10.1 provides an overview about the life time of different emissions in the atmosphere:

Referring to the aforementioned unclear climate impact assessment, especially for CO_2, the long-term dynamics of the climate and especially the influence of the sun must be taken into account to judge the CO_2 [10]. Although the long-term impact of CO_2 is out of question it seems, that it is much lower than actually levelled by IPCC. Further natural cooling phases originated by the sun might overcompensate the CO_2 warming impact.

Nevertheless, Table 10.1 highlights that also unsuspicious emissions like water steam and contrails have a relevant life time, where they can impact the climate. Further, Methane set free, e.g. through defreeze of Siberian ground or in deep sea, must be heavily considered for climate impact. In the previous analysis, it has been shown that the share of aviation of CO_2 and NO_x emission is fairly low on the one hand, but especially the long-term effect of CO_2 on global warming on the other is relevant. In Fig. 10.4, these effects are still visible, but road transport clearly dominates CO_2 and Ozone creation, while aviation is the unique contributor to the creation of contrails and cirrus clouds.

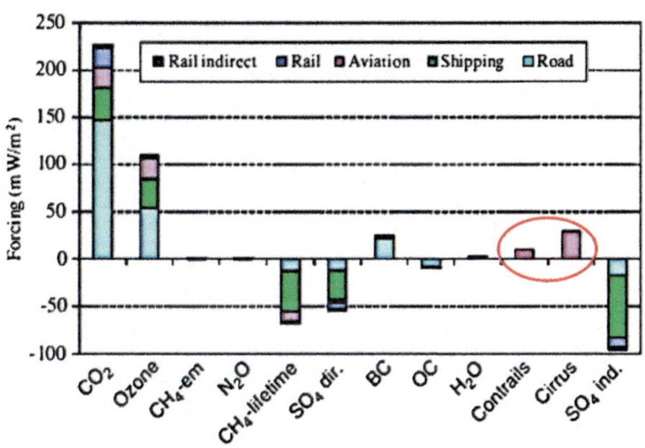

Fig. 10.4 Stakeholder relevance on the environmental impact of air transport [13]

It will be shown later on in this section that contrails and cirrus clouds are of major interest for the global warming of the atmosphere (Fig. 10.4).

The previous general descriptions have shown that it is mandatory to consider the environmental impact of future air transport developments on the atmospheric consistency and the global climate warming, especially due to the high level of uncertainty and controversial discussions.

10.2.1 Aircraft Emissions

Aircraft emissions are generated by engines. Nevertheless, the required thrust to be delivered by the engines is directly depending on the aircraft weight and the aerodynamic drag. Therefore, weight and drag, as shown in Sect. 10.2.4 indirectly affect aircraft emissions. Today, turbine engines dominate the world aircraft fleet as turbo-propeller or turbo-fan engines. Piston engines are used in general aviation aircraft and play only a minor role. For aircraft cruise conditions, Fig. 10.5 presents the principle combustion process and products of a turbine engine.

The chemical combustion process of a turbine engine in cruise condition based on 1000 g of kerosene and 3400 g of oxygen results in 1240 g water steam and 3150 g carbon dioxide mainly. Both components have a major impact on the radiative forcing (RF) (Fig. 10.4). Although constituting minor share nitrogen oxides, carbon monoxide, unburned HC, sulphides and sood are further combustion products, which need to be considered.

Therefore, in a second step some characteristics of the engine thermodynamics are to be investigated.

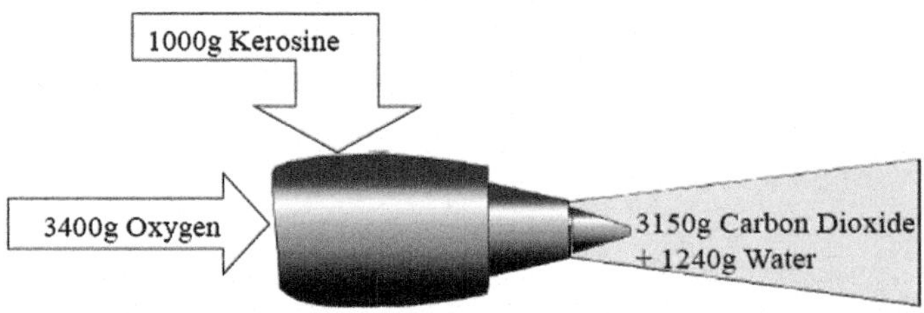

Further combustion products:

6-20 g	Nitrogen oxides (NO_x)	0,1g	Lubrication oil
1,5-10g	Carbon monoxide (CO)	0,03g	OH
0,2-3g	Unburned (UHC/HC)	0,01-0,03g	Sood (C)
0,8—1,0g	Sulfid (SO_2)	0,01g	Micronutrients
0,15g	Nitrous Oxide (N_2O)	0,0001g	Methane (CH_4)

Fig. 10.5 Principle chemical engine process and its products in aircraft cruise condition

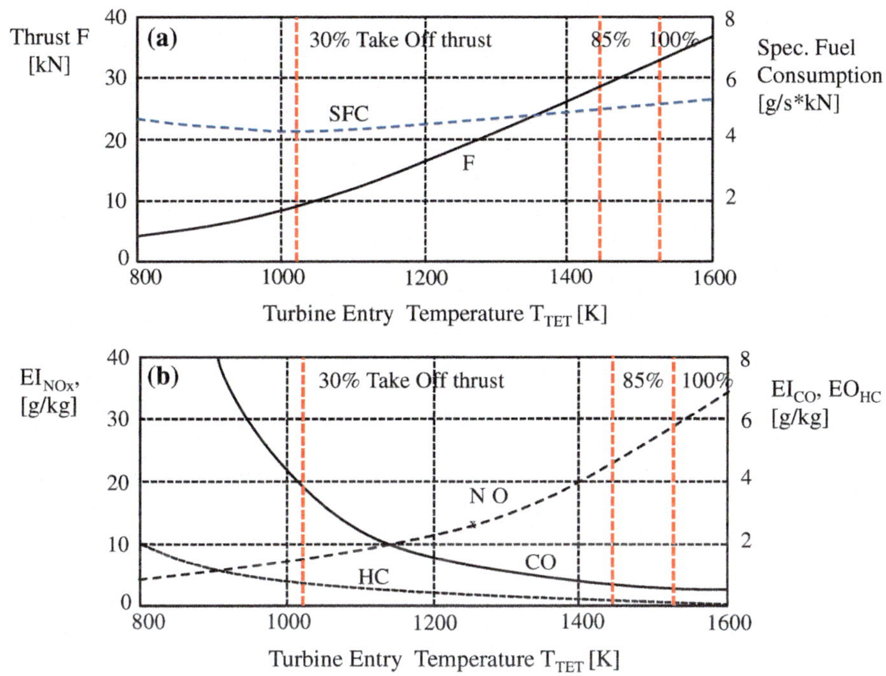

Fig. 10.6 Turbine entry turbine impact on engine thrust (**a**) and emissions (**b**)

Looking at Fig. 10.6, the engine thrust and the resulting emissions depending on the turbine entry temperature are presented. Considering the Turbine Entry Temperature (T_{ET}) as an indicator for the engine power setting, one can see in the upper figure that the resulting thrust is following quadratic or nearly proportional behaviour in a first approximation. Further the specific fuel consumption shows a minimum over a wider range of T_{ET} while it is increasing at lower and higher T_{ET}.

Looking at the lower Fig. 10.6b, the more thrust is required, e.g. during takeoff, the higher the turbine entry temperature needs to be, the higher the nitrogen oxide generation will be in the combustion chamber, while HC and carbon monoxide decrease. In this context, emphasis should be put to the fact, that the absolute values for the emission indices (EI) for NO_x are five times the value for CO and HC.

In parallel, also the specific fuel burn will slightly increase, which results in proportional increase of carbon dioxide creation. Consequently, during takeoff and climb those emissions are of major importance, which will take place in the vicinity of the airport and at lower altitudes.

In cruise condition, the typical thrust level is depending on the required thrust to achieve an appropriate lift to weight ratio of 1. Typically, thrust is set in the range of 60–80 % of the takeoff thrust. Also at 11,000 m or 33,000 ft flight level altitude significant carbon dioxide and nitrogen oxides are therefore emitted.

For the assessment of the environmental impact of aircraft, the effect of these emissions at those altitudes has to be discussed separately.

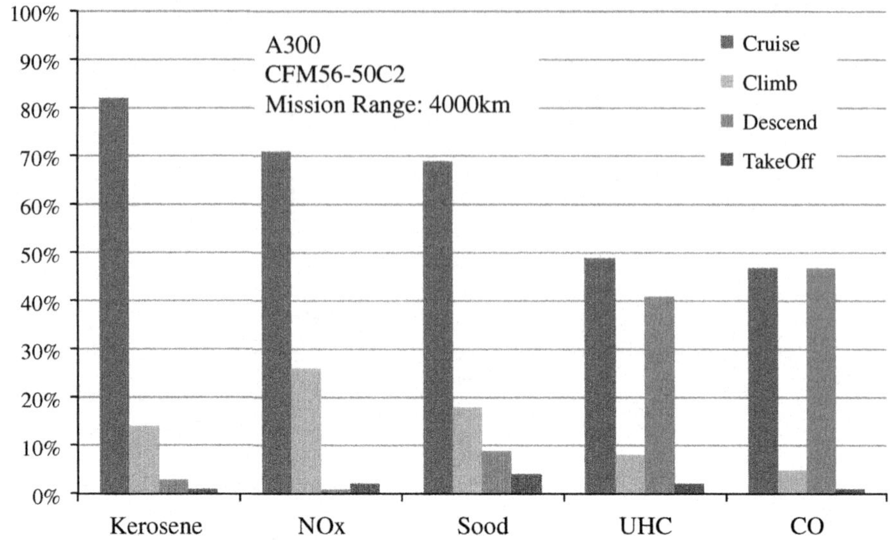

Fig. 10.7 Flight phase depending emissions [14]

Figure 10.7 provides an overview about the portion of different emissions in major flight phases.

Cruise condition is dominating across all types of emissions. NO_x and sood are also relevant during climb, while UHC and CO are significant during descent, where the engine is operated in part-load conditions. Consequently, the impact of those emissions at cruise altitudes of about 10,000–12,000 m has to be considered more in detail.

10.2.2 Physical Principles of the Atmosphere

While the atmosphere covers the entire surrounding of the earth at different levels from troposphere to exosphere, as shown in Chap. 5, climate describes regional atmospheric conditions within the troposphere and stratosphere mainly, which are in general constant but vary within certain ranges of temperature, pressure and humidity over the year.

Weather itself is the description of local conditions of temperature, pressure, density and humidity, which cause certain conditions like dry sunny situations, rain, wind, thunder storms, snow, fog, etc.

All processes and dynamics in the atmosphere are initiated by solar radiation, which varies with the eccentric and asymmetric rotation of the earth. These effects also cause cyclic long-term changes, which result in variations of the earth climate [10].

Based on solar radiation, absorption and reflection of the earth surface and various parts of the atmosphere, the thermodynamic balance is the key indicator of

Table 10.2 Consistence of the atmosphere (main components only) [11]

Element	Portion (volume %)
Permanent elements	
Nitrogen	78 %
Oxygen	21 %
Inert gases	<1 %
Temporary elements	
Aerosoles	Diverse
Argon	
Water vapour	Diverse
Carbon dioxide	>355 ppm
Ozone	Diverse
Nitrous oxide	0.35 ppm
Hydrogen	Diverse

the climate impact of any man-made system. These are affected by trace gases, like ozone, carbon dioxide or nitrogen oxide.

Concerning the climate impact of aviation consistency of the atmosphere is to be addressed briefly to understand the "opportunities" of chemical reactions. In Table 10.2, the main permanent and temporary elements composing the gaseous atmosphere are listed.

While more than 99 % of the atmospheric volume components are permanent, especially the temporary existing elements are of paramount relevance for the radiation and energy balance of the atmosphere.

Since the concentration of natural carbon dioxide is more or less homogenous around the world, the distribution and concentration of water vapour and ozone is temporarily and locally very inhomogeneous [11].

The physical effects of these trace gases with respect to the radiation and energetic balance will be discussed in the following sections because they are main components of the aircraft combustion exhaust.

The so-called "greenhouse effect" describes the mechanism of the atmosphere to absorb infrared radiation and therefore it acts as a thermal radiator, which reflects a significant part of the received energy back to the earth. Water vapour, CO_2, ozone and clouds mainly absorb this infrared radiation. This isolating attribute of the atmosphere has a natural and an anthropogenic/man-made part. The greenhouse effect is a natural and mandatory process, which is vital part of the atmosphere.

10.2.2.1 Carbon Dioxide

Carbon dioxide creation during fuel burn as mentioned in Sect. 10.2.1 is three times proportional to the fuel burn. Additionally, CO_2 has one of the highest and increasing portions in the atmosphere (Table 10.2).

The more air traffic is growing the more CO_2 will be created. Figure 10.8 provides an example of the CO_2 production per flight and seat indicating the relevance of CO_2.

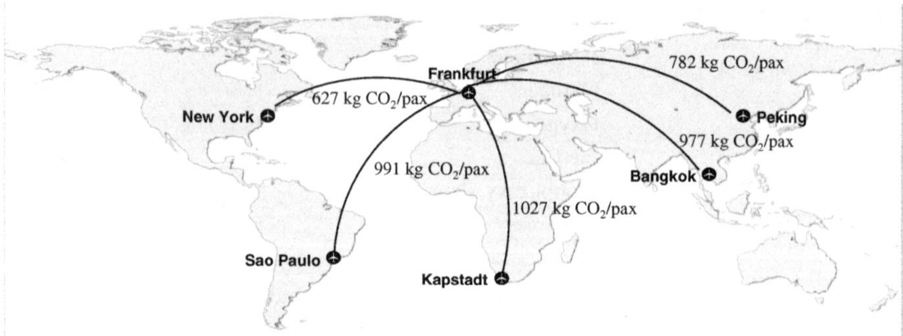

Fig. 10.8 Example of average carbon dioxide creation per flight and seat on selected routes

During a flight between Frankfurt and Cape Town for example, an A330-200 or B777-200 type aircraft with 76 % load factor and 350 seats maximum capacity, will produce about 272,118 kg CO_2 per flight. This results in individual CO_2 emissions of 1027 kg CO_2/pax.

Taking the origin-destination (OD) pair Hamburg—Munich as an extreme short range contrast, 7720 kg CO_2 are emitted considering a typical load factor of 76 % and a A320 type aircraft of 150 passenger as maximum capacity. In this case 67 kg CO_2/pax are produced.

Although in both cases only cruise flight length and conditions are taken into account, both figures provide a rough estimate about the CO_2 emission characteristics of aviation.

To get the global picture one has to consider the worldwide flight tracks in a year. For example, in 2013 about 450 billion passenger kilometres have been produced. Taking the average between long and short range CO_2 emissions as rough orientation air transport produces around 650 billion kg of CO_2 per year. This corresponds to the share of about 1.6–2.4 % of global aviation CO_2 emissions, as mentioned in Sect. 10.1.

While the natural carbon dioxide distribution is nearly homogenous, the CO_2 distribution generated by the air traffic is mainly located in the northern hemisphere. As a result, the warming effect of the tropospheric CO_2 emissions is also in this area, leading more to a global distortion of the overall energy balance of the atmosphere, which might cause more windy conditions due to the temperature and therefore atmospheric energy difference too. Plants and oceans acting as CO_2 sinks can compensate this effect only partly, especially because most of the wooden areas in the world are not below the major flight tracks of the northern hemisphere.

As CO_2 is fairly inert, the atmosphere needs long time to remove emitted CO_2 through natural washout processes, resulting in an average lifetime of 50–200 years. Due to this long lifetime and its resulting equal dispersion in the atmosphere, the locus of CO_2 emission is irrelevant for the global warming effect.

Consequently, it is absolutely mandatory to reduce the tropospheric CO_2 emissions of air transport.

10.2.2.2 Nitrogen Oxide

Nitrogen Oxide (NO_x) emissions from air traffic have an impact on ozone (increase) and methane (decrease), both being important greenhouse gases. O_3 and CH_4 affect the earth climate by the same principle as CO_2 through absorption and remission of outgoing infrared radiation, leading to a temperature rise in the troposphere.

The lifetime of CH_4 is about 10 years, which makes it well mixed over the atmosphere, whereas O_3 is a chemically reactive gas, leading to comparably short lifetimes between days and weeks in the troposphere. The impact of NO_x emissions and its impact on O_3 is thus sensitive to the emission region and altitude, with larger impacts on ozone at lower latitudes but also higher altitudes [15]. As aviation reduces the methane concentration in the atmosphere, the aviation radiative forcing is negative and therefore beneficial, counteracting partially the positive radiative forcing from Ozone production.

Depending on the pressure and temperature conditions in the atmospheric levels, nitrogen oxide either increases or decreases the ozone density. In the stratosphere, the ozone concentration is up to six times higher (i.e. 18×10^4 mbar) than in lower regions of the troposphere (Fig. 10.9). At this level of the atmosphere, ozone is acting as a UV filter, preventing us from too much sun energy. In this region, i.e. above 11,000 m when reaching the tropopause, the existence of nitrogen oxide reduces the ozone density.

At lower levels in the troposphere, the natural density of ozone is much lower. Here, due to the higher temperatures and air density the occurrence of nitrogen oxide leads to additional ozone production.

Higher ozone concentrations at the earth surface can lead to health problems like reduced human performance and breathe symptoms due to the toxic impact (Fig. 9.9).

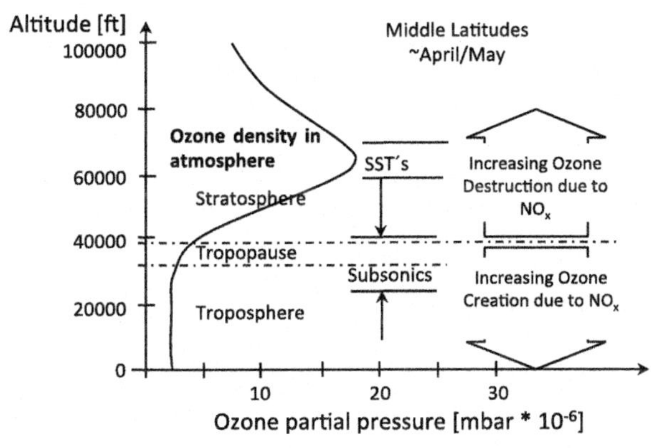

Fig. 10.9 Principle increase and decrease of ozone caused by nitrogen oxide

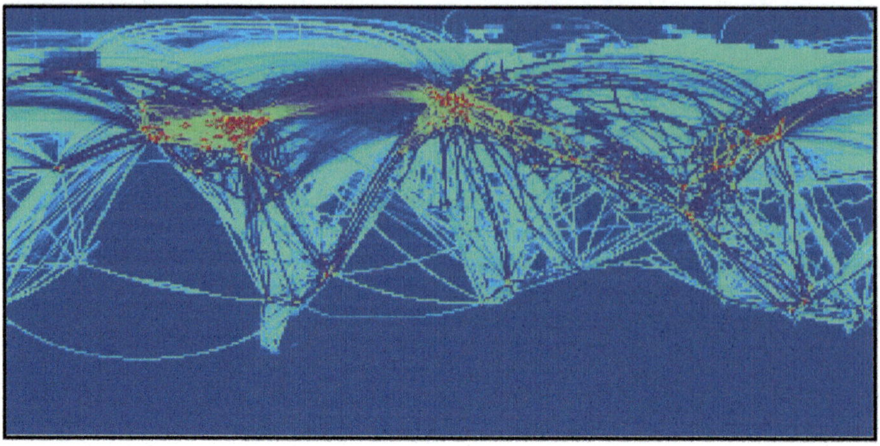

Fig. 10.10 Worldwide distribution of nitrogen oxide concentrations [16]

The global relevance of nitrogen oxide emissions becomes visible, if the worldwide distributions are considered (Fig. 10.10).

More than 90 % of the nitrogen oxides are measured on the northern hemisphere. Obviously, there seems to be no exchange between the northern and southern hemisphere too.

Keeping in mind, that also nearly 90 % of the worldwide air traffic happens on the northern hemisphere a significant reduction of the nitrogen oxide is vital in order to prevent the northern hemisphere from negative ozone effects, as shown in Fig. 10.10.

Nitrogen oxide is generated during high temperature combustion of the aircraft engine as described in Sect. 10.2.1. This engine and flight state occurs typically during the takeoff phase but also at cruise condition, when the engine is operated at higher power settings and temperatures, which is due to the thrust lost caused by the reduced air density at higher altitudes.

Relating these physical effects in the atmosphere to the thermodynamic engine combustion processes, it becomes obvious that the effect of nitrogen production is adverse in any flight phase. Therefore, the ACARE Vision 2020 requirement of 80 % nitrogen oxide reduction becomes logical and mandatory [2].

10.2.2.3 Contrails

Since couple of years scientists more and more discover that contrails may also have a remarkable impact on the climate [9, 17]. Contrails are visible line clouds, consisting of ice particles that form in the exhaust plume of aircraft if the ambient air is cold enough.

Engine emissions, like water steam, aerosols/sood, unburnt and burnt hydrocarbons create under certain temperature, humidity and density conditions contrails, which will develop to cirrus clouds over the time.

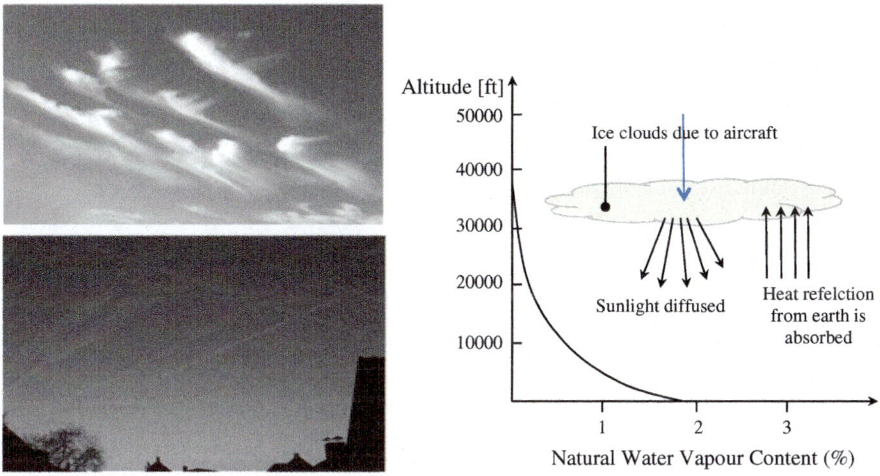

Fig. 10.11 Contrail formation depending on water vapour content

Those additional clouds, which contain also a lot of ice particles, absorb heat reflections from earth and diffuse the sunlight from space to earth, Fig. 10.11.

These effects occur typically at altitudes of around 10,000 m (30,000 ft), which are typical aircraft cruise conditions, especially for long-range flights.

Referring to *Roedel* cirrus and cirrostratus clouds, which exist at altitudes between 8500 and 12,000 m typically, absorb approximately between 16 and 32 % of the relative heat [11]. This significant absorption ratio leads to an increase of the atmospheric temperature.

The emission of aerosols has a direct climate impact through the reflection of incoming radiation by sulphates, leading to a negative radiative forcing, but also through absorption of incoming radiation by soot particles and other particle matter, which results in a positive RF. It is further assumed that emitted aerosols have also an indirect climate impact by serving as condensation nuclei and enhancing the formation of condensation trails (contrails).

Contrail will form if the mixture of exhaust gas and ambient air transiently reaches saturation with respect to liquid water during the plume expansion. Whether this condition will be reached or not is described by the *Schmidt-Appleman* criterion [11].

According to this criterion as shown in Fig. 10.12, contrails form and persist when the isobaric mixing line ends in an ice supersaturated state, otherwise they dissolve quickly, Fig. 10.12. The two solid curves represent the saturation with respect to liquid water (upper curve) and ice (lower curve), respectively. The phase trajectory of the mixture between exhaust gases and ambient air is displayed as dashed curve. The tangent (dotted line) to the water saturation curve marks the warmest temperatures for which contrail formation is possible.

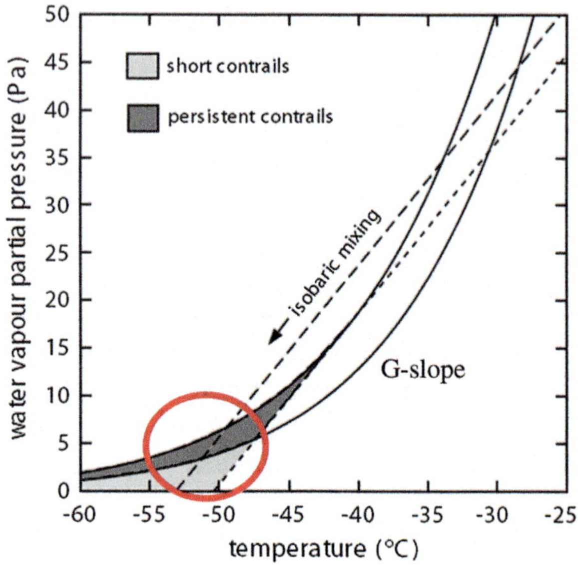

Fig. 10.12 Contrail formation according to Schmidt-Appleman criterion

Persisting contrails can grow by the uptake of ambient water vapour until the ice crystals fall due to their increasing weight into lower and warmer altitudes where they evaporate as soon as they enter an unsaturated state. Contrails can further transform into contrail-cirrus clouds, which look natural but would not exist without prior formation of contrails. The climate impact of contrails and induced contrail-cirrus cloudiness will be substantial if the current assumptions about their RF are found to be adequate, Fig. 10.14.

Summarizing the effects of CO_2, NO_x, sood and water creating contrails Fig. 10.13, elaborated by Lee et al., shows how these emissions also created by aviation end up in contributions to climate change [18].

It is to be recognized, that especially CO_2 and NO_x emissions contribute directly to RF. As a result of recent research in that field of physics of the atmosphere it came out, that the indirect effect of emissions like water, sood, sulphides and hydrocarbons enforce the creation of cloudy conditions, which seem to have a more significant impact on global warming, than CO_2 and NO_x alone [13, 19].

Figure 10.14 displays the resulting RF components, which are caused by the different perturbations of the atmosphere through air transport. The bars indicate the actual best estimates and associated levels of uncertainty for the various contributions of emissions and their secondary effects like contrails to RF. It is remarkable, that the level of confidence associated to the impact of contrails, aerosols and sood is relatively low compared to the impact of NO_x. Nevertheless, the average contribution of contrails of 0.033 W/m^2 is assumed to be higher than the one of CO_2 0.028 W/m^2.

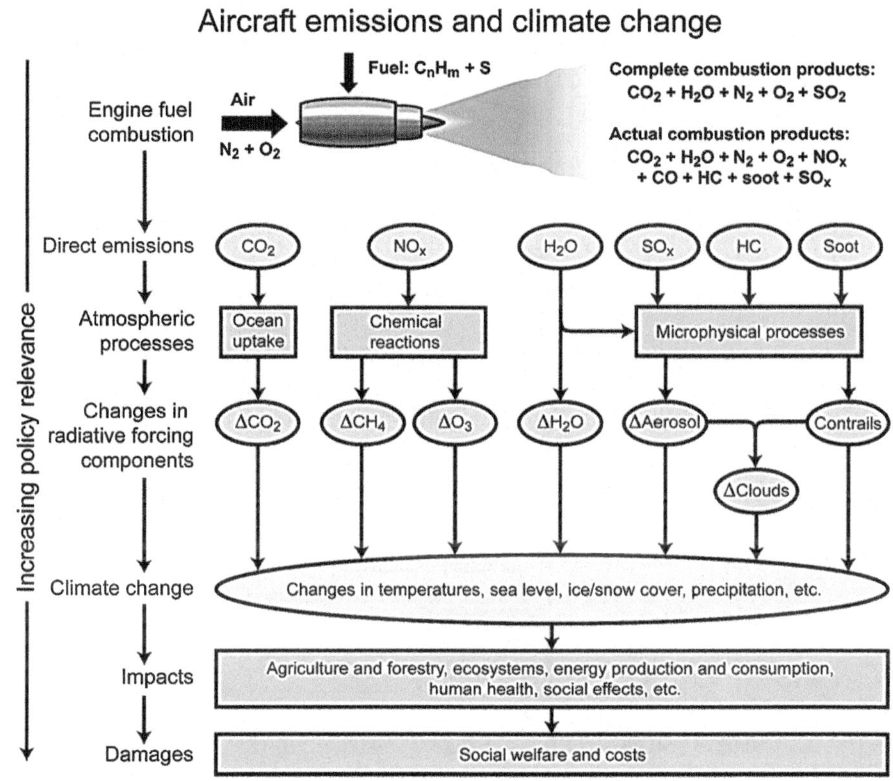

Fig. 10.13 Aircraft emissions, their interaction with the atmosphere and resulting climate impact [18]

It is also visible that RF induced by air transport is caused to a large extent by non-CO_2 effects. Thus it is insufficient to limit the analysis of climate impact of air transport to CO_2 emissions only.

Any technology assessment with respect to climate reduction potential, which focusses on CO_2 only, is likely to point into the wrong direction. Moreover, the authors of the IPCC advisory report (AR4) believe that any assessment should be performed in a most comprehensive way, including RF of all relevant emitted compounds and expressing the climate impact in a reasonable climate metric [21].

In addition, it must be noted again, that it is still not clarified by research how far natural mid to long-term dynamics originated by the sun are superposed by these anthropogenic effects. Much more research is required since the current scientific understanding is not detailed enough.

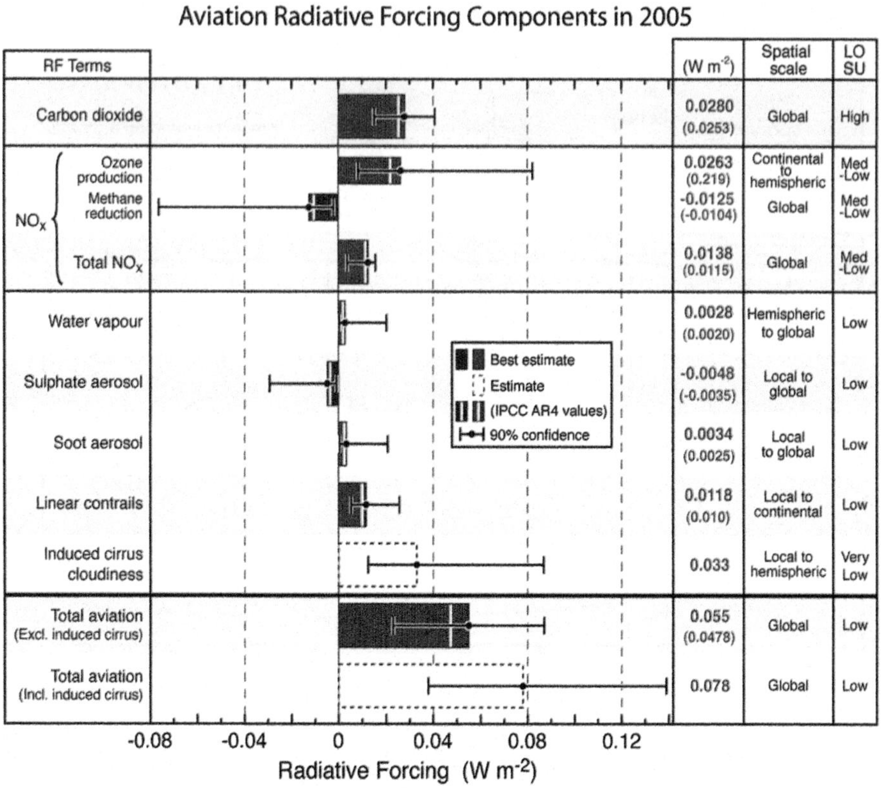

*LoSu=Level of Scientific Understanding

Fig. 10.14 Radiative forcing components from global aviation as evaluated from preindustrial times until 2005 [20]

10.2.3 Emission Impact Assessment in Air Transport

Based on the brief introduction on the emission behaviour of aero engines and the effects of certain emissions on the atmosphere, in this section measures are described to quantify aircraft emissions and global environmental impact.

10.2.3.1 Regulatory Measures to acquire Aircraft Emissions

In order to cope with the public need to acquire the aircraft emission performance, the International Civil Aviation Organization (ICAO) in 1982 defined the first time a method to measure the engine emissions [22].

ICAO Landing-TakeOff-Cycle		
Operating Conditions	Thrust level	Time [min.]
Taxi Out	7% max	19
Take-off	100% max	0,7
Climb	85% max	2,2
Approach	30% max	4,0
Taxi In	7% max	7,0

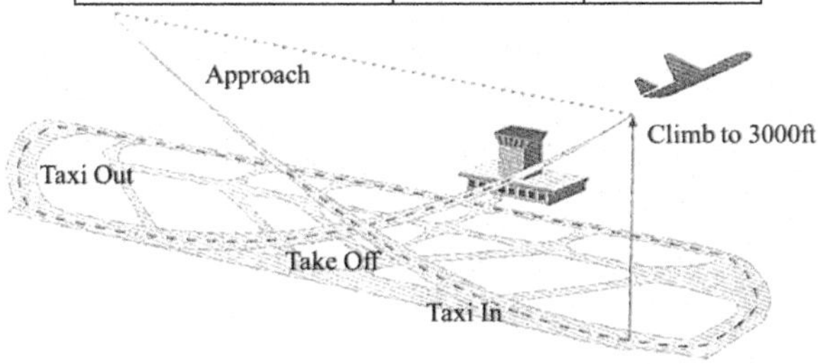

Fig. 10.15 ICAO Annex16 landing-takeoff-cycle for engine emissions determination

Although the emissions are measured at sea level on an engine test stand a comparative assessment is possible. As a reference, a landing and takeoff cycle (LTO) is defined as described in Fig. 10.15.

This method is representative to determine engine emissions in the vicinity of an airport. The emissions generated at flight level at higher altitudes are not covered, because a flight phase is not part of the measurement cycle. Covering also at cruise conditions is rather complicated because they vary much in altitude, length and step climbs.

The standard applies to engines with more than 26.7 kN installed thrust which are manufactured on or after 1 January 1983. For hydro carbons and carbon monoxide, only engines manufactured after 1 January 1986 are considered.

In order to provide an insight into the calculation, some examples for the determination of the emission metrices, which are called EI are given. Practically on the engine test rig, the engine is operated according to the cycle mentioned in Fig. 10.15 and the emissions are collected from the exhaust. Further, referring to the ICAO Annex 16 Vol. II standard, e.g. the allowed smoke number is determined as follows:

$$\mathrm{SN} = D_p/F_\infty \leq 83.6 \cdot (F_\infty)^{-0.274} \text{ or } 50, \text{ whichever is lower} \tag{10.1}$$

Here D_p/F_∞ describes the mass, in grams (D_p), of any pollutant emitted during the reference landing and takeoff (LTO) cycle, divided by the rated output (F_∞) of the engine.

For hydro carbons (HC) and Carbon Oxide (CO) the relevant number shall fulfil:

$$\begin{aligned} \mathrm{HC} &= D_p/F_\infty \leq 19.6 \\ \mathrm{CO} &= D_p/F_\infty \leq 118 \end{aligned} \tag{10.2}$$

For nitrogen oxides, the determination of the allowed emissions is slightly different, distinguishing between first individual production model on or before 31st December 1995 and manufacture date before December 31st 1999. Also the overall pressure ratio is used to define the regulatory limits. For example, for engines with a maximum rated thrust of more than 89.0 kN, like the A380, Rolls Royce Trent 970 with 314–340 kN:

$$NO_x = D_p/F_\infty \leq 19 + 1.6 \cdot \pi_\infty \tag{10.3}$$

Here the pressure ratio (π_∞) expresses the ratio of the mean total pressure at the last compressor discharge plane of the compressor to the mean total pressure at the compressor entry plane when the engine is developing takeoff thrust rating in ISA sea-level static conditions.

In Fig. 10.16 exemplary engine data are shown indicating the relative distance to the ICAO limits.

As shown before, the allowed emission limits are depending on the engine thrust, which has to be taken into account.

While HC and CO emissions are far below the ICAO limits for all different engine sizes, NO_x emissions are at 80–90 % of the limit. Further, the examples also show the younger an engine type is the lower the NO_x emissions are.

This approach is used for certification purposes and determination of airport fees. For design and modelling tasks, the Boeing Fuel Flow Method (BFFM) first published in 1996 by *Baughcum* is used. Generally, this method determines the engine emissions from the calculated fuel flow using various experienced corrections [14].

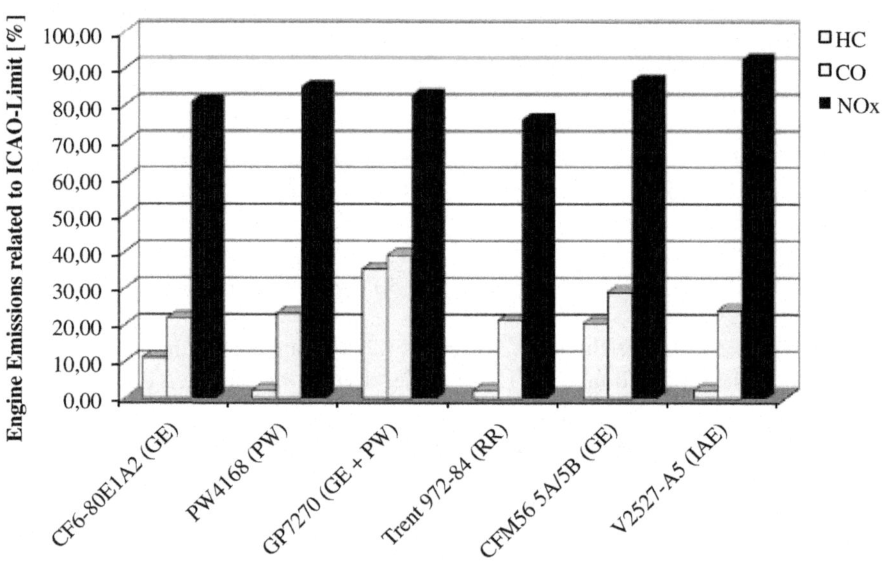

Fig. 10.16 Typical engine emission performances fulfilling ICAO Annex 16 emission limits

10.2.3.2 Climate Impact Metrices

Despite the quantification of engine emissions, it is necessary to develop metrics, which allow quantifying the impact of those emissions on the development of the climate.

Global Warming Potential (GWP) and Radiative Forcing Index (RFI) are well-known metrices in this context.

Radiative Forcing defined in watts per square metre has been developed as a criterion to quantify absolute or relative to CO_2 the net change of radiation compared to the preindustrial period before 1860. It is therefore based on the changes in emission concentrations compared to the past. While RF is well suited for a look back from today, it is not able to forecast the climate impact of emissions for the future. Further emissions have different lifetimes in the atmosphere, which has a significant impact but cannot be reflected by the RF criterion.

In order to cope with this aspect the *Global Warming Potential* (GWP) has been developed. GWP describes the warming potential of an emission mainly related to the power of CO_2 over a certain time period, typically 20, 50 or 100 years. Unfortunately, GWP does not provide any information about the temperature changes of the climate over a certain period.

Studies have shown that metrices like GWPs or RFI are misleading and not appropriate for purpose [23]. Here *Sausen* and *Schumann* developed a linear response model and *Grewe* and *Stenke* presented a way, how to estimate the climate impact in terms of changes in the near-surface air temperature [23, 27]. In those recent publications also the change of global *Average Temperature Response* (ATR) has been elaborated as an appropriate measure to especially address the cumulated effects of aviation emissions [24, 25].

$$\mathrm{ATR}_H = \frac{1}{H}\int_t^{t+H} \Delta T\ (t)\mathrm{d}t \tag{10.4}$$

The presented metric integrates the surface temperature change $\Delta T(t)$ (expressed in Kelvins) over a chosen time period H (for this study 100 years) considering thus impact of short-lived (e.g. contrails) and long-lived (e.g. CO_2) forcing agents in appropriate way.

Trying to merge these climate impact metrics with environmental goals of aviation as they have been described in Chaps. 1 and 11 lead to some problems. Goals formulated as percentage reductions of CO_2 or NO_x as done by ACARE do not cover the real climate impact. Therefore, *Schumann* has postulated, to revise these goals in order to cover the cumulated impact of all emissions as well as showing the real climate reaction [26].

As an example, how such an approach might work, *Koch* has developed an approach to assess in which way different flight profiles of a fleet in terms of flight altitude but also different aircraft designs can contribute to an improved climate compatibility [20].

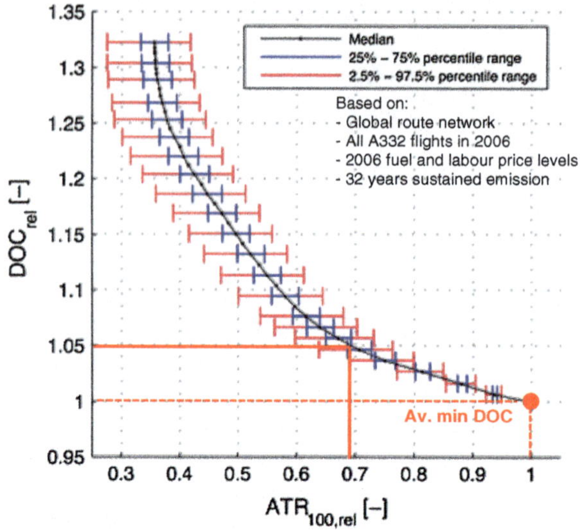

Fig. 10.17 ATR reduction potential and DOC impact of A330 world fleet driven by lower flight altitudes and reduced speed

Figure 10.17 demonstrates the potential of about 30 % damping of temperature raise (ATR) compared to the actual (2013) situation if aircraft are operated at initial cruise altitudes of about 8000 m at Mach numbers of about Ma = 0.72 [20]. This climate impact improvement is achieved at a cost increase of 5 % in DOC driven by increased fuel consumptions caused by drag increase at lower altitudes. On the other hand, at those altitudes the emissions cause less atmospheric and climate impact. All this investigations have been done using existing unmodified aircraft. If new aircraft are designed for such lower cruise altitudes and lower speeds, a temperature damping of about 45 % seems to be possible.

It must be noted at this point, that further consequences like the capacity impact in air traffic have not yet been investigated in this study. This is focus of further research to confirm these trends and finally assess the realistic potential of this approach.

It is common understanding that the assessment of the aviation-related climate impact still holds too many uncertainties to draw conclusions (Fig. 10.14). Grewe (2008) has shown that although uncertainties are large, the approach is applicable to assess technologies with respect to their climate impact [27]. A detailed description how climate response models and associated uncertainties are included in the present approach is given by Dahlmann [28].

10.2.4 Measures for Emission Reductions

The previous sections have described how emissions from aviation affect the climate. In principle, three areas of measures are available to mitigate the emission

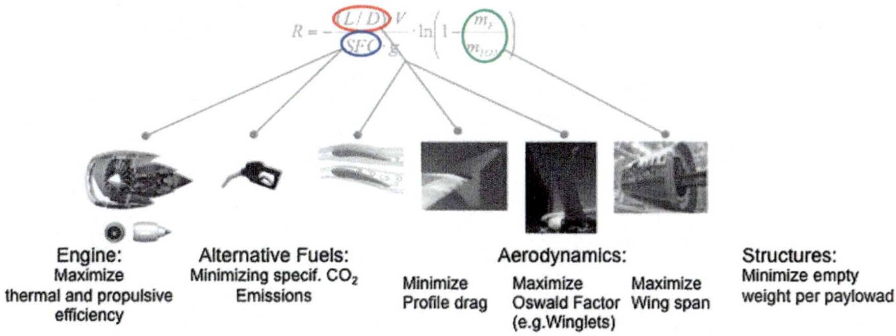

Fig. 10.18 Breguet-Formula showing technical potentials for emission efficiency improvements

impact. The first area addresses the overall air transport performance in terms of aircraft movements. In order to ensure people's mobility, aviation is needed to connect the world on a global basis. However, people´s mobility can be realized by frequently operated smaller aircraft or bigger aircraft flying not that often. To avoid the so-called "Rebound Effect", which describes, that saved energy, fuel or emissions of a given transport performance is used for increasing performance it is mandatory to limit the amount of aircraft movements worldwide, but to increase the capacity of aircraft. Second, there are some technical measures to improve the individual climate impact performance. Looking at the *Breguet*-formula as introduced in Chap. 5, in Fig. 10.18, the main technical areas become obvious.

The improvement of engine's overall efficiency offers a potential of about 25 % achieved by reaching the physical limits of thermal and propulsive efficiency. Recuperation and intercooling elements improve the thermodynamic efficiency of engines leading to a reduction in the specific fuel consumption (SFC). On the other hand, these new components itself increase the engine weight, which must be compensated by lighter materials and design.

Looking to the current high bypass engines, a decoupling of the fan speed and the core engine shaft speed using a additional gear box improves the efficiency of the engine in terms of fuel consumption and emissions as well.

Alternative fuels offer a range of 10–20 % of CO_2 overall balance reduction, heavily depending on the feedstock used. Aerodynamics in theory can provide indirectly reductions in CO_2 emissions of 20–30 % superposing all measures. However the efficiency of these measures, like laminarity, increase in wing span or winglets is strongly depending on the operational conditions, Sect. 11.4.1.3.

At last Fig. 10.19 shows structural weight reductions especially for primary structures offering an indirect potential of about further 15 % CO_2 reductions. However, practical experience has shown, that the empty weight of aircraft developments in the past did increase, especially due to the introduction of additional systems for the cabin.

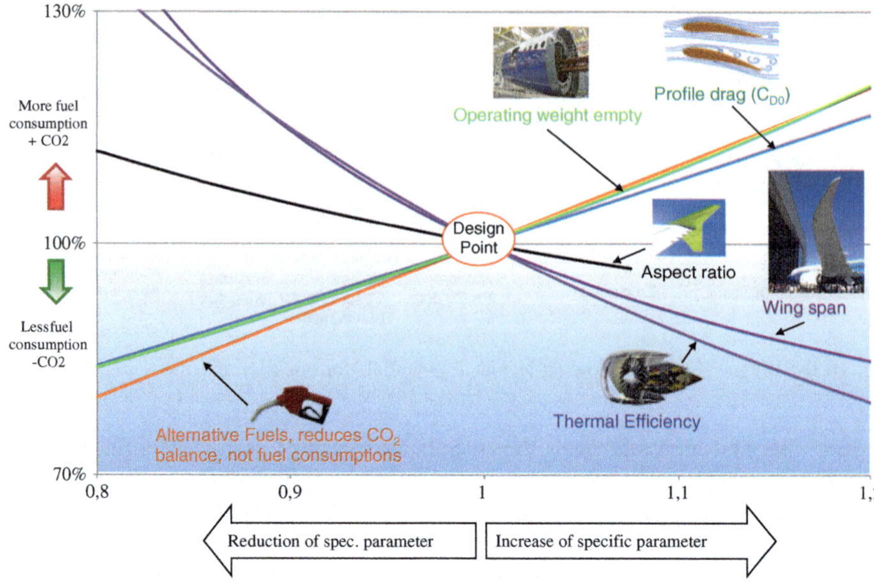

Fig. 10.19 Technology contributions to CO_2 efficiency of aircraft

The introduction of various emission trading schemes (ETS) in Europe enforces the air transport community to reduce the CO_2 emissions and represents an indirect regulatory measure to force the air transport industry to introduce emission reducing technologies. The actually stopped (2013) European ETS approach drives ICAO to develop a global approach to emission trading.

10.3 Noise and Sound of Air Transport

Noise and sound are one of the most sensitive issues in air transport. Because the impression of noise and sound is very subjective, a common assessment is very hard to achieve. This section provides principle knowledge about the physics of noise, the regulative environment of noise handling and some measures to reduce aircraft noise. Although the physical principles apply also to aircraft cabin noise, in the following only external aircraft noise aspects are considered.

10.3.1 Some Basics of Medical Noise Impacts

At a first look, air transport noise caused by aircraft operations seems to be much more relevant to people than emissions. This is due to the immediate recognition of noise. High power noise level influences mainly man's health but also has a significant but less obvious psychological aspect of annoyance. When people are continuously influenced by noise at day, this can cause hearing disorder, heart attack and disruption of communication, nervousness and psychological disruptions. If also continuous higher sound level overnight is given disruptions of sleep and wake ups are observed additionally. Further, it must be noticed, that not only noise intensity is relevant but also the related frequencies play a major role in noise annoyance.

Medical research has shown that noise impact on man is different whether it appears on day or night. During night, sudden wakeups or longer lasting disturbance during sleep may occur.

Comparing the acoustic properties of different transport systems *Basner* et al. have shown that the rising time of sound pressure level (SPL) and the duration of a noise event differ between rail, road and air traffic [29]. The rise time of sound pressure and the duration of noise impact show reciprocal behaviour for the different transport systems, Fig. 10.20.

While road transport noise impact is characterized by frequent short-term events air transport noise duration is much longer compared to road transport, but the rising time is much slower. The study previously mentioned also shows that the resulting awaking probability is significantly increasing with the SPL rise time. As a consequence, aircraft noise lead to remarkable lower awaking probabilities [29]. However, a subjective assessment of the noise impact leads to an inverse result the study has shown. It seems to be comprehensible that longer noise duration leads to a more intensive mental recognition, while a short-term event stimulates more the physiological reaction of the body. This excursion to medical and psychological impacts of noise instantly highlights how differently people recognize noise. Recent studies also have shown that at least the occurrence of an aircraft shadow at the sky let people associate noise impression although the physical sound level is fairly low. Understanding those basics is fundamental to derive right measures to improve the noise impact of the air transport system. These principle observations are added by lack of comfort, if the stay in an aircraft cabin is considered.

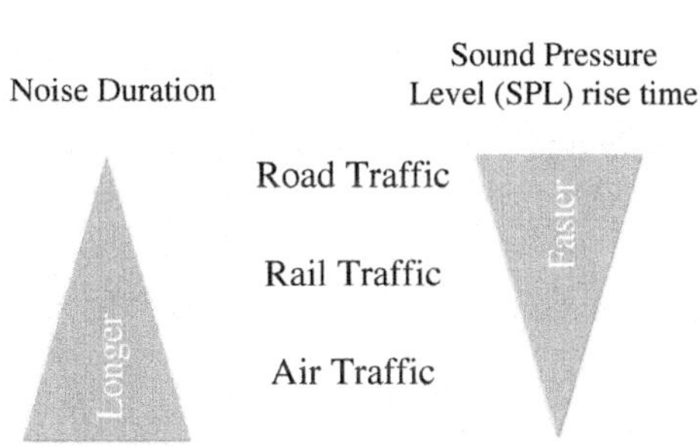

Fig. 10.20 Acoustic properties of different transport systems

10.3.2 Basics of Noise and Aeroacoustics

In order to provide a principle understanding of noise creation and measurement, the major but basic physical principles are described in the following.

Aeroacoustic noise results from mass, pulse and energy propagation in flowing air. Therefore, sound is a transient flow in a pressure field. In order to describe this behaviour, *sound pressure* $\acute{p}$ is the key parameter in acoustics:

$$\acute{p}(t) = p(t) - \bar{p} \tag{10.5}$$

The sound pressure is determined through a timely averaged constant pressure level $\bar{p}$ and time-dependent variations $p(t)$. Human hearing directly reacts to sound pressure and covers a frequency of 16–16,000 Hz.

The lowest level of sound pressure recognized by human beings is about 2×10^{-5} Pa. Sound begins to hurt at a pressure level of about 2×10^{2} Pa, which is around 10^{7} times more intensive than the lowest recognizable level of sound. In order to cope with this wide range of *sound pressure level resolution* (SPL) a logarithmic scale is used. The signal intensity of sound pressure is denoted by its *effective value.*

$$p_{eff} = \sqrt{\bar{\acute{p}}^2} \tag{10.6}$$

SPL is defined as the logarithmic relation of the measured sound pressure p to a reference pressure p_0:

$$\mathrm{SPL} = 20 \cdot \log\left(\frac{p_{\mathrm{eff}}}{p_0}\right) \qquad [\mathrm{dB}],\ \text{with } \mathrm{p_0} = 2 \times 10^{-5}\ \mathrm{Pa} \tag{10.7}$$

In order to address the frequency content of a sound signal, the autocorrelation $P(\tau)$ is used:

$$P(\tau) = \overline{\acute{p}(t) * \acute{p}(t+\tau)} \tag{10.8}$$

Using the Fourier transformation the corresponding power density $\hat{P}(f)$ provides information about the frequency content. Performing a transformation back from frequency to time domain, the frequency content of the effective pressure level becomes available for further analysis:

$$p_{\mathrm{eff}}^2 = P(0) = 2 \cdot \int_0^{\infty} \hat{P}(f)\mathrm{d}f \tag{10.9}$$

The analysis of the sound pressure signal is performed in two different ways using the third-octave band to emphasize the high frequency portion and the narrow band analysis to address the bandwidth of a signal. This calculation is only valid in

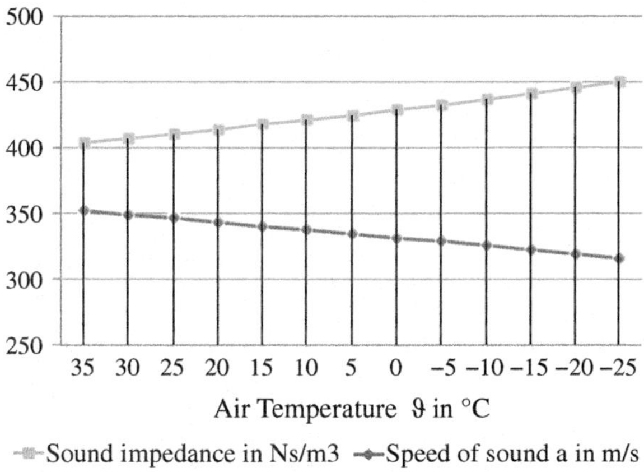

Fig. 10.21 Dependency of speed of sound, sound impedance, sound transmission and air density on air temperature

the fare field. Air density in conjunction with the speed of sound both depending on temperature can be understood as a form of impedance or damping of the air, inhibiting the sound pressure to transmit. Hot air increases the speed of sound, and therefore reduces the measured intensity, while high altitudes and low air density increase the intensity of the sound pressure, Fig. 10.21.

It is important to note, that an increase of the SPL by +10 dB (A) is recognized as a duplication of the perceived noise, whereas a duplication of the intensity of the noise source by +3 dB (A) doubles the noise level!

Further the perceived noise level (PNL) characterizes the sound pressure, which is recognized by people, while the emitted pressure level determines the pressure level of a source. In order to acquire the noise impact and assess its influence, the equivalent permanent SPL (L_{eq}) is the international measure.

There are different definitions or measures existing, which are used to quantify sound and noise. However, sound and noise are very subjective impressions. Therefore, the PNL, which represents the various noise frequencies in a weighted form is a well-known indicator for the momentary maximum noise level in air transport.

The effective perceived noise level (EPNL) which is used within the ICAO, Annex16 noise requirements for aircraft is a measure for the noise and sound impact taking also into account the duration of noise reception [30].

As shown in Fig. 10.22, considering the flyover of an aircraft, the equivalent perceived noise level (EPNL) covers the forward and the rearward radiated noise. For the calculation itself, the top 10 dB noise level is not considered because the timely intensity is more relevant than the peak level.

$$\mathrm{PNL} = \mathrm{PNL}_{\max} - 10\,\mathrm{dB}\ [\mathrm{PNLdB}] \tag{10.10}$$

Beside the pressure level itself also the transmission frequency is important to notice, because at lower frequencies a higher pressure level and therefore more

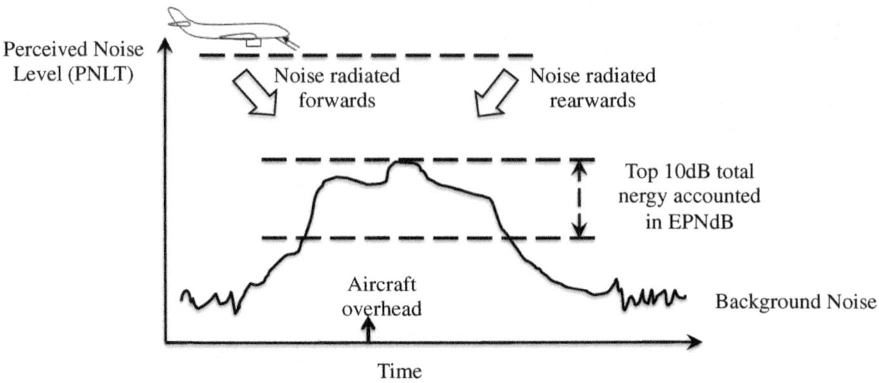

Fig. 10.22 Equivalent perceived noise level

energy is transmitted. For this reason, a frequency correction C is introduced to calculate the EPNL

$$\mathrm{EPNL} = \mathrm{PNL} + C\ [\mathrm{EPNdB}] \tag{10.11}$$

Generally, noise impact on people is different during day and night. While at day the permanent noise level is more important for men´s health, during night single events do have a more severe effect. For the assessment of noise events at night, similar to the equivalent noise level determination, the sound is measured but here the sound measurement (Day-Night-Level, DNL) is penalized by +10 dB.

$$\mathrm{DNL} = \mathrm{PNL}_{\mathrm{max}} + 10\,\mathrm{dB} \tag{10.12}$$

These findings are reflected in regulatory requirements concerning day and night noise limits, presented in Table 10.3.

Table 10.3 Noise impact on man's health and associated sound level

Time	Day	Night
Criterion	Permanent sound level (L_{eq}, DNL, …)	Permanent sound level and single events
Consequences	Disruption of hearing at $L_{eq} > 75$ dB(A) Increased risk of heart attack at $L_{eq} > 65$ dB(A) Disruption of communication $L_{eq} > 40$ dB(A)	Distortion of sleep or falling asleep Wake up events of 6 × 53 dB(A) inside or outside 6 × 63 dB(A) Permanent sound pressure level: disrupted sleep $L_{eq} > 32$ dB(A) inside
Regulatory limits (Germany)	$L_{eq} = 60$–65 dB(A): protection zone 2 $L_{eq} > 65$ dB(A): protection zone 1	Max allowed: Single events: 6 × 60 dB(A) inside or outside 6 × 75 dB(A) Permanent sound pressure level: $L_{eq} < 55$ dB(A) outside

In addition to the noise level definitions given before, Table 10.3 also incorporates another definition called Equivalent Noise Level, L_{eq}, which is used in Europe.

10.3.3 Noise Requirements for Aircraft

Since aircraft noise is immediately recognized by the people it is felt to be more important than emissions. In order to reduce aviation noise impact to a minimum, certification authorities as mentioned in Chap. 4 request for an aircraft noise certificate before entering a new aircraft into service. Therefore, engine noise regulations have been formulated in the ICAO Annex16, Volume I set measurement standards and maximum sound pressure limits, which have to be fulfilled [30].

Following ICAO Annex 16, Volume I, the noise certificate has to demonstrate the fulfillment of certain SPLs for approach, flyover and along the runway.

Figure 10.23 shows three measurement points before, aside and at the end of the runway, where noise measurements are to be performed. Each aircraft has to fulfil the limits at each measurement point but also a cumulated limit must be fulfilled. In the context of this section, all discussions about sound pressure limits refer to perceived noise. That means a potential reduction of the PNL by 50 % requests for a −10 dB decrease in the relevant value. The requested noise pressure level to be

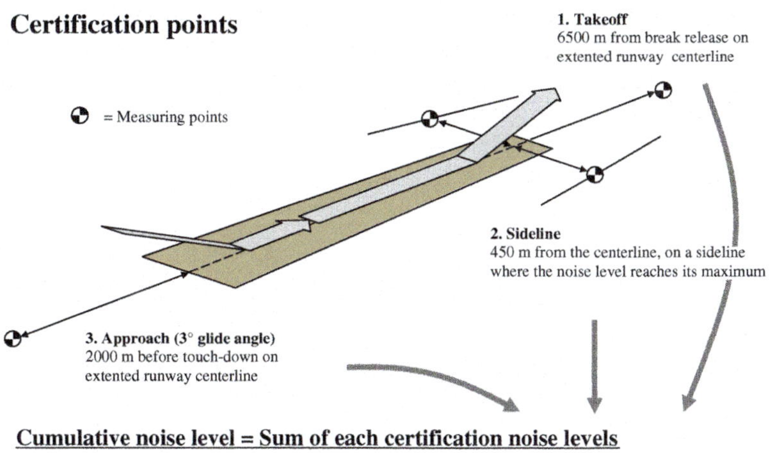

Fig. 10.23 ICAO Annex 16 noise certification measurement set up

Chapter 3

MTOW [t]	0	20,2	28,6	35	48,1	280	385	400
Sideline	94			80,87 + 8,51 log(MTOW)				103
Takeoff ≤ 2 TW	89				66,65 + 13,29 log (MTOW)		101	
Takeoff 3 TW	89		69,65 + 13,29 log (MTOW)				104	
Takeoff ≥ 4 TW	89	71,65 + 13,29 log (MTOW)					106	
Approach	98			86,03 + 7,75 log(MTOW)		105		

in [EPNdB]

Chapter 4: 10 EPNdB cumulative below Chapter 3
...but at each certification point at least 2 EPNdB

Chapter 14: 7 EPNdB cumulative below Chapter 4
...but at each certification point at least 1 EPNdB

Fig. 10.24 ICAO Chapter 3 noise certification requirements

fulfilled is depending on the age of the aircraft, the amount of engines and the maximum takeoff mass as shown in Fig. 10.24.

For example, according to Chapter 3 a twin engine A330 with 247t m_{TOM} requirements has to fulfil a sideline SPL of

$$\mathrm{EPNdB} = 66.65 + 13.29 \log(247) = 98.5 \; [\mathrm{dB}] \tag{10.12}$$

If such an aircraft would be designed and certified today, Chapter 4 is applying, and the relevant value has to be at least 2 dB lower for the sideline measurement point and 10 dB less cumulated. When the aircraft will be certified after 2017 Chapter 14 rules apply and the relevant limit will be another 1 dB lower, i.e. 95.45 EPNdB for the single measurement point and another 7 dB for cumulated limit.

For the cumulated limits, Fig. 10.25 shows the development of the overall noise levels from Chapter 3 to Chapter 14.

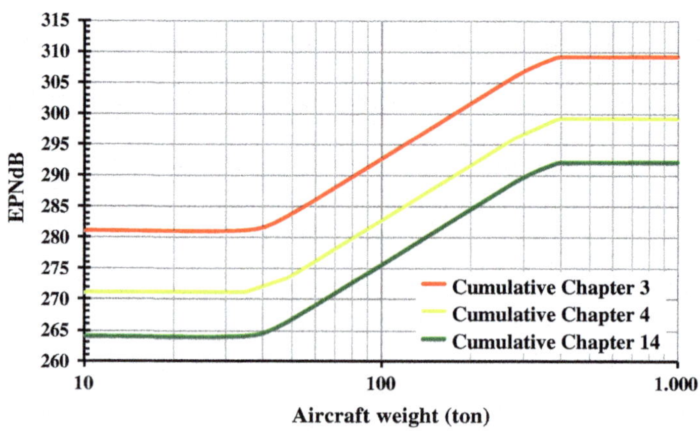

Fig. 10.25 ICAO Chapter 3–14 noise certification requirements development

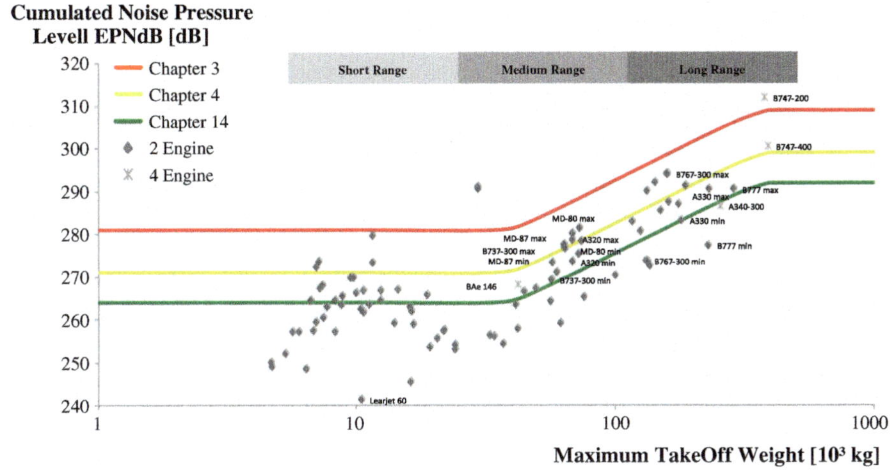

Fig. 10.26 Perceived noise level summary of real aircraft

The heavier an aircraft the higher the allowable noise level is defined. The more new and quiet aircraft enter into the market the more the allowable noise levels are strengthened. Figure 10.26 gives an overview about the overall noise characteristics of selected aircraft in the current world fleet. Most aircraft today fulfil the Chapter 3 noise requirements. Some are above Chapter 4 and also today many aircraft fulfil the 2017 Chapter 14 cumulated noise requirement.

The perceived sound pressure metrics and the noise exposure limits requested by the authorities lead to different noise limit levels around the runways of an airport. Those noise footprints of actual aircraft are given in Fig. 10.27 for one runway at London Heathrow, showing the 85 dB boundary.

As one can see actual aircraft developments provide much smaller noise footprint compared to older ones. This ranking is on the one hand driven by the amount of engines, where four engines create a larger footprint. On the other hand, actual aircraft and engine technologies offer some potential for noise reduction, which are described in the following section.

Fig. 10.27 Principle comparison of 70 dB noise footprints of different aircraft generations [31]

10.3.4 Aircraft Noise Sources and Potential for Reduction

Looking at an aircraft, various sources of noise are given, which are mainly associated with the engines, the landing gear and the wing flap and slat systems, Fig. 10.28.

When we are talking about technical options for aircraft noise reduction, the source of noise is addressed. Here as previously mentioned, −50 % reduction is achieved, if the SPL at its source is reduced by −3 dB.

Looking at the aircraft noise sources, the engine noise is predominant during takeoff and landing phases. At takeoff noise is created due to the full power setting by the fan outlet, the combustion and the turbine jet while the airframe contribution is significantly increasing during landing, when the landing gear and the high lift systems become more relevant. The engine is at a low power setting in this phase.

Despite the contribution to the overall SPL, also the frequency depending share of the different components is of interest as shown in Fig. 10.29.

High lift systems of aircraft provide the highest overall SPL with a maximum at about 100 Hz. Further, slats and overall wing contribute significantly while the landing gear shows a maximum contribution at 600 Hz.

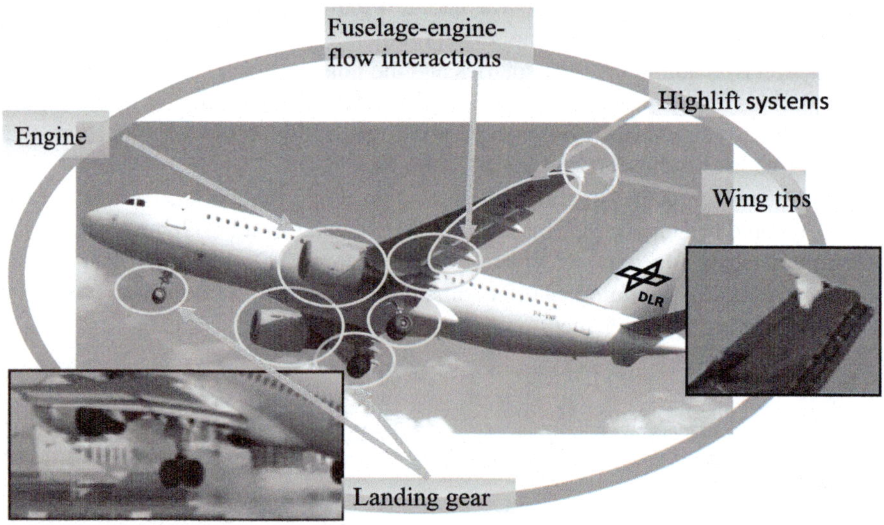

Fig. 10.28 Aircraft noise sources

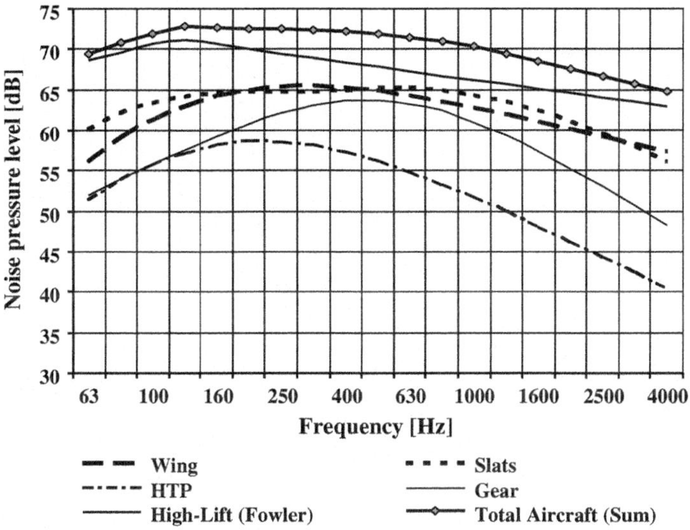

Fig. 10.29 Share of aircraft noise sources during takeoff and landing [32]

10.3.4.1 Approaches to reduce aircraft noise

There are various technical approaches to achieve lower noise creation levels at various sources. Although a single improvement may be not that much, the implementation of several solutions as a package can provide a significant reduction. In the following, some examples of potential noise reduction technologies are presented.

Since the high lift systems, consisting of various flaps and slats is contributing most to the overall SPL, filling cavities and slats can provide remarkable reductions in aircraft noise, Fig. 10.30.

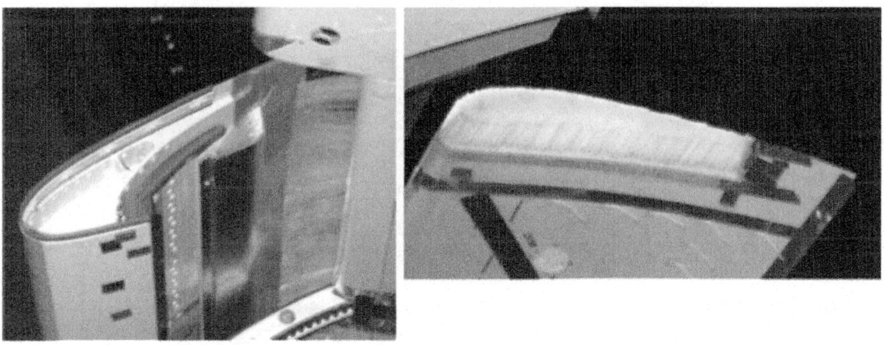

Fig. 10.30 Filling cavities and brushes to reduce noise, DLR [32]

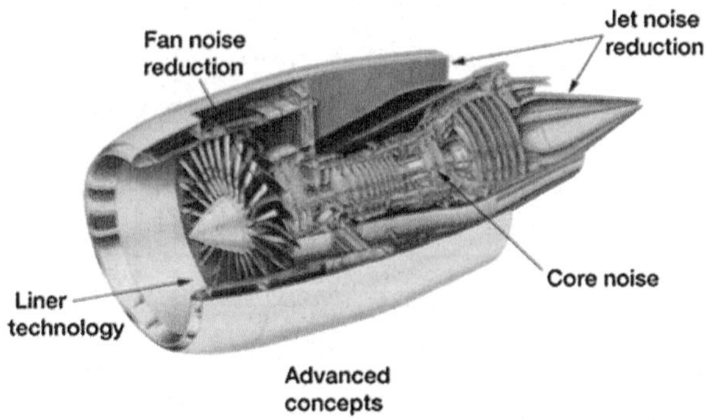

Fig. 10.31 Engine noise reduction potential

Research at the German Aerospace Center has shown that those simple measures, which close slots or cavities could reduce the source SPL by 3–5 dB, which means half of the original SPL. The principle is quite well known in bionics, where owls have a feathering, which closes slots as much as possible to fly quietly.

Looking at the landing gear fairings provide the potential for about 3–5 dB noise reduction of the SPL, which means a reduction about 50 % recognized noise level. Here again closing slots and filling cavities creating smooth flow surfaces is the key to reduce the noise level.

On engine side there are also pragmatic measures to reduce the SPL of the major components, Fig. 10.31.

Rolls Royce developed an actual solution, called chevron, which has been implemented into the actual B787 Dreamliner aircraft, Fig. 10.32.

Fig. 10.32 Chevron nozzle reducing exhaust jet noise, [Rolls Royce]

Such a fringed exhaust nozzle device reduces the far-field noise impact by about 1–3 dB EPNL by a smoother swirling transition and mixture between the hot high-speed engine exhaust air and the air flow of the surrounding.

In addition noise absorbing materials, like honey comb used as shielding provide also 1–3 dB EPNL reduction. Further, active noise suppression systems are under investigation, which enable a frequency specific sound suppression. However, such measures increase the engine weight, which must be considered as a drawback.

On the operational side, night curfews are used most especially in Europe to prevent people from air transport noise at night. There are different detailed procedures to define restrictions like general curfews, closing single runways or closing an airport at night after achieving a certain cumulated SPL.

Also special fees are set for airlines, which operate noisy aircraft. These measures are some of more, which are applied on a regulatory operational basis to limit the noise impact, Fig. 10.33.

Further continuous descent but also steep approaches have been developed as operational procedures to reduce noise intrusion by eliminating the thrust levelling phases (Fig. 10.34).

But also increasing the flight path angle from 3° to 6° is a measure to reduce the area of noise intrusion. While these procedures affect the vertical flight profiles horizontal approach trajectories defined as SID and STAR for each airport can be designed in order to avoid inhabited city areas as much as possible (Sect. 9.2).

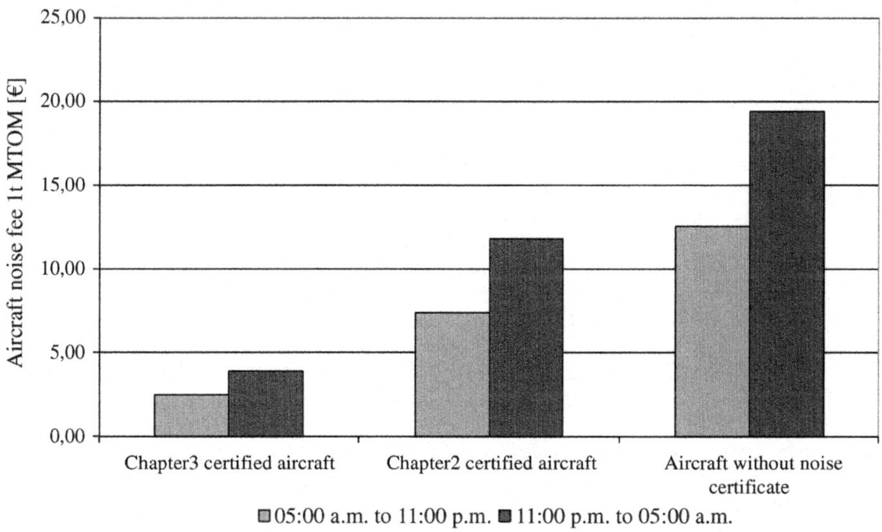

Fig. 10.33 Typical example of airport noise fees

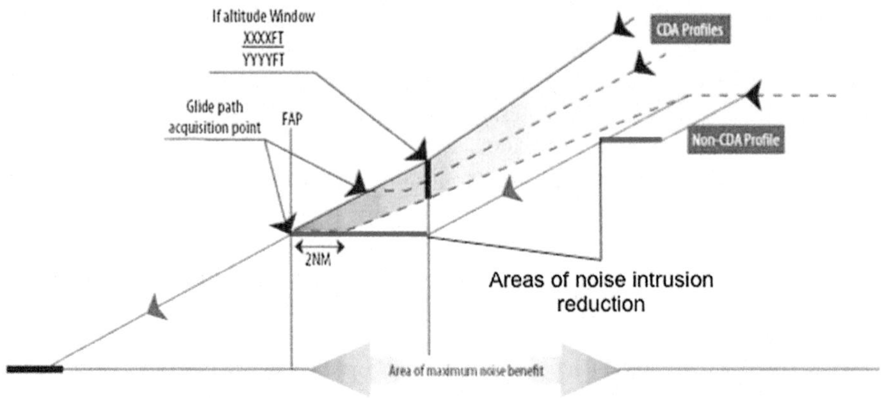

Fig. 10.34 Continuous descent approach to reduce noise intrusion, [Eurocontrol]

References

1. Gollnick, V.: Rotorcraft System Dynamics. Lecture at Technical University Munich, Institute for Aviation Technologies, German (2006)
2. ACARE: European aeronautics: a vision for 2020—meeting society's needs and winning global leadership, Advisory Council of Aeronautical Research Europe, European Community 2001. http://www.acare4europe.org/docs/Vision%202020.pdf (2012). Accessed 26 April 2012
3. IPCC: First assessment report. http://www.ipcc.ch/publications_data/publications_and_data_reports.shtml
4. N.N: Next generation air transport system. United States Government Accountability Office, August 2010. http://www.gao.gov/assets/310/308608.pdf (2012). Accessed 26 April 2012
5. DGTransport.: Flightpath 2050 Europe's vision for aviation. Directorate General for Mobility and Transport, European Commission (2013). http://ec.europa.eu/transport/modes/air/doc/flightpath2050.pdf. Accessed 21 Nov 2013
6. N.N: Air travel—greener by design—mitigating the environmental impact of aviation—opportunities and priorities. Royal Aeronautical Society, July 2005. http://www.greenerbydesign.org.uk (2012). Accessed 26 April 2012
7. Janic, M.: Sustainability of Air Transport. Ashgate, Aldershot (2009)
8. IPCC: Climate Change 2013—the Physical Science Basis Working Group I Contribution to the Fifth Assessment Report of the Intergovernmental Panel on Climate Change, 1st edn. Cambridge University Press, Cambridge, New York, Melbourne, Madrid, Cape Town, Singapore, São Paolo, Delhi, Mexico City (2013). www.cambridge.org/9781107661820
9. Schumann, U., Jeßberger, P., Voigt, C.: Contrail ice particles and their climate importance. Geophys. Res. Lett. **40**, 1–6 (2013). doi:10.1002/grl50539 (American Geophysical Union)
10. Vahrenholt, F., Lüning, S.: The Cold Sun, 2nd edn. Hoffmann & Campe Publishing, German (2012). ISBN:978-3-455-50250-3
11. Roedel, W.: PHYSIK unserer Umwelt – Die Atmosphäre, 3 Auflage. Springer, German (2000)
12. Fuglestvedt, J., et al.: Climate forcing from the transport sectors. Proc. Natl. Acad. Sci. **105**, 454–458 (2008)
13. Lee, D.S., et al.: Transport impacts on atmosphere and climate: aviation. J. Atmos. Environ. **44** (37) (2010). ISSN:1352-2310 (Elsevier Publishing)
14. Gmelin, T.: Comprehensive analysis of aircraft efficiency potentials, especially considering actual engine technology developments. Diploma thesis at Institute of Aeronautics and Astronautics, Technical University Berlin, German, Mar 2008

15. Gollnick V.: The Air Transport System, 6th edn. Lecture at the Technical University of Hamburg-Harburg (2011)
16. Schady, A.: Global NO_x distribution. Personal note (2014). http://wiki.bildungsserver.de/klimawandel/upload/Luftfahrt_ziv_NOx.png. Accessed 26 Feb 2014
17. Schumann, U., Graf, K.: Aviation induced cirrus and radiation changes at diurnal timescales. J. Geophys. Res. **118–5**, 2404–2421 (2013)
18. Lee, D.S., et al.: Aviation and global climate change D.W. in the 21st century. Atmos. Environ. (2009). doi:10.1016/j.atmosenv.2009.04.024
19. IPCC: Climate change 2007—synthesis report, published by the intergovernmental panel on climate change (2008). http://www.ipcc.ch
20. Koch, A.: Climate impact mitigation potential given by flight profile and aircraft optimization. PhD-thesis, Institute for Air Transport Systems, German Aerospace Center and Technical University Hamburg-Harburg (2013)
21. Pachauri, R.K., Reisinger, A., et al.: IPCC Forth Assessment Report: Climate Change 2007. IPCC, Geneva, Switzerland (2007)
22. ICAO: Environmental Protection, 3rd edn. ICAO Annex 16, vol. 2, International Civil Aviation Organization, Montreal (2008)
23. Sausen, R., Schumann, U.: Estimates of the climate response to CO_2 and NO_x emission scenarios. Clim. Change **44**, 27–58 (2000)
24. Koch, A., et al.: Climate impact assessment of varying cruise flight altitudes applying the CATS simulation approach. In: 3rd CEAS Air and Space Conference, The International Conference of the European Aerospace Societies, Venice (2011)
25. Wuebbles, D.J., Yang, H., Herman, R.: Climate Metrics And Aviation: Analysis of Current Understanding and Uncertainties. Aviation-Climate Change Research Initiative (2008)
26. Nolte, P., Gollnick, V.: Aviation of the future—vision 2020, a half time resume. In: Symposium Report, Symposium at Technical University Hamburg-Harburg, Hamburg, German, Sep 2011. http://www.luftverkehr-der-zukunft.de/Archiv/2011/
27. Grewe, V., Stenke, A.: AirClim: an efficient toll for climate evaluation of aircraft technology. Atmos. Chem. Phys. **8**, 4621–4639 (2008)
28. Dahlmann, K.: A method for efficient assessment of aviation climate impact mitigation. PhD-thesis, Ludwig-Maximilian-University, Munich, Germany (2011)
29. Basner, M., et al.: Single and combined effects of air, road, and rail traffic noise on sleep and recuperation. SLEEP **34**(1) (2011). www.journalsleep.org
30. ICAO: Environmental Protection, 3rd edn. ICAO, Annex 16, vol. 1, International Civil Aviation Organization, Montreal (2008)
31. Airinsight: Noisy neighbors? (2014). http://airinsight.com/2013/05/22/noisy-neighbors/. Accessed 26 Feb 2014
32. Gollnick, V., Weiss, M.: Applying Aircraft Noise Reduction Technologies at its Source—Progress in Technological Development
33. Zdunkowski, W., Bott, A.: Dynamics of the Atmosphere—A Course in Theoretical Meteorology. Cambridge University Press, Cambridge (2003)

Chapter 11
Challenges and Competition of Air Transport

Abstract This chapter gives a global view of the challenges of future air transport starting first with a reflection of the ACARE Vision 2020 and Flightpath 2050 goals. The situation and perspective of energy demand and provision for air transport addresses especially the current developments in alternative fuels. A deeper look is provided at the structural and competitive situation of multimodal transport in various regions of the world. This provides a basis to assess market potentials of air transport. Some perspectives in different aviation technologies are discussed to provide a basis to assess future opportunities. A further section describes an integrated systems and technology approach to optimize the introduction of technologies across different stakeholders and substructures in the air transport system. At the end the chapter concludes with some changes and measures which should appear to realize a more efficient and competitive air transport system.

In the previous chapters the air transport system has been introduced and developed, as it is up to date.

Air transport will be a major pillar of future mobility. Nevertheless, the environment of air transport of the past decades has fundamentally started to change. Like in the ACARE *Vision 2020* the air transport system has been understood as a fast and intensively growing system. Logically, the Vision 2020 requests for capacity growth in airspace since a three times increase in passenger movements and a two times increase in aircraft movements is expected in Europe until 2020 [1].

At the same time the environmental compliance of air transport has become an undisputable objective of future developments. Also the society's sensitivity in environmental noise impact has reached a threshold, which requires strong attention.

Further the availability and cost development of energy and fuel, as a major representation of energy, is crucial for the future development of air transport.

At last it is out of question that the classical aerospace technologies like aerodynamics, lightweight design, and turbo fan engines have reached a very high level of maturity, which allows only small incremental improvements of physical efficiency at high research and development cost.

Particular attention will be paid further to future energy provision to air transport. The competitive situation of air transport within transport gives an additional

D. Schmitt and V. Gollnick, *Air Transport System*,
DOI 10.1007/978-3-7091-1880-1_11

outlook to future research. Some major technology trends are addressed, which may offer some potential for improvements. Further an approach for future research and development is described, which gains holistic complex system level research, as air transport is. Taking this methodology into account, future concepts for air transport in a competitive but also multimodal transport system are described. They offer a way for a balanced approach to reach the Flightpath 2050 targets.

11.1 Global Challenges for Air Transport 2050

Summarizing the global developments described previously, air transport grew tremendously in terms of passenger and aircraft movements, like it was shown in Fig. 1.5. The later is based on a significant increase in the amount of aircraft. The amount of aircraft causes limitations in airport and airspace capacities especially in Europe and Northern America, [2–4]. In the growing regions like Asia and Southern America those capacity limits are partially not yet reached, but need to be considered for future developments.

Responding to this in 2001 the Advisory Council of Aeronautical Research in Europe (ACARE) has defined high-level targets, for future improvements to make the global air transport system competitive and attractive for the twenty-first century. It has to be pointed out that all these requirements are referring to a single aircraft flight. This implicitly leads to the request to replace the worldwide aircraft fleet until 2020, which is neither feasible nor realistic. These high-level political targets as introduced in Chap. 1 are picked up in Table 11.1.

In addition, Table 11.1 also provides the evolution of these targets to the European *Flightpath 2050* vision and also a comparison to the American vision to the future air transport, [5].

Also in the United States targets for the future air transport have been formulated [3]. Here, in the operational field the *NextGen* program defines especially objectives for more efficiency in air transport flow. The American *N+3* project driven by NASA additionally sets requirements on improved aircraft performance [2].

Comparing both approaches the American *NextGen* Air Transport System can be understood as more holistic, while the European Vision 2020 addresses more aircraft-related technologies to improve the air transport system. Here a significant difference between Europe and the United States becomes obvious in the understanding of air transport industry. While the European view is very much focusing at the aircraft manufacturing industry, the American view is much more transport system oriented. This difference becomes visible, if one is looking at the detailed requirements. While the Vision 2020 requests for a percentage reduction, e.g. in CO_2 emissions by 75 % until 2050, the *NextGen* requirement asks for a countrywide absolute reduction of 14 Mio. tonnes/year.

These goals of the *Vision 2020* are set to be achieved until 2020 and refer to the ATS performance of 2000 as the reference. As mentioned in Chap. 1, Fig. 1.1 different stakeholders shall contribute with different technologies.

Table 11.1 European and American high-level targets for future air transport systems

Objective	EU (ACARE, 2000)	EU (Flightpath 2050, 2000)	US (N + 3, user def. reference)	US (NextGen)
CO2 Emissions	−50 %	−75 %	Better than −70 % fuel burn	−14 million tonnes cumulative
NOx Emissions	−80 %	−90 %	Better than −75 %	Improved environmental performance
Noise	−50 %	−65 %	−71 dB (below Stage 4)	
Punctuality	99 % within 15 min	within 1 min regardless of weather conditions	N/A	20–35 % delay reduction
Security	Zero Hijack	Seamless security for global travel; resilient air vehicles; secure data network	N/A	Provide air transp. security
Safety	Fivefold reduction in average accident rate	<1 accident per 1 billion commercial flights	N/A	Improved safety operations
Quality and Affordability	<15 min in airport for short-haul <30 min for long haul	90 % of travellers within Europe able to complete door-to-door journey within 4 h	N/A	Integrated NextGen information
Throughput	16 million flights/year	25 million flights/year	N/A	Collaborative capacity management
				Collaborative flow contingency management
				Flexible separation management
				Efficient trajectory management
Others	Cut time to market in half	Emission-free taxiing	Exploit metroplex concepts	Flexible airport facility and ramp operations
		All air vehicles recyclable		

A midterm resume, however, indicated in 2011 that not all of these goals could be achieved until 2020 [6]. While the environmental goals concerning CO_2 and NO_x emissions are achievable by more than 50 % an extension of the airport and airspace capacity as well as the improvement of punctuality are hard to reach until 2020. Further actual research on climate impact of aviation has raised the question whether the percentage requirements on reductions of emissions are the right one, because the impact on global warming in terms of contribution to ΔT seems to be more appropriate. This metric covers interdepending effects in a better way as *Schumann* demonstrated [6].

Although some goals of the *Vision 2020* will be not achieved until 2020, the European aviation community developed a Flighpath 2050 as a vision for the next four decades.

Looking at this vision and its challenges as listed in the second column of Table 11.1 one can see, that the requested reductions in emissions are again formulated in percentages referring to 2000 for a single aircraft mission. The grade of reduction is again about 50 % of the 2020 targets or 75 % of CO_2 reduction compared to 2000. Also for NO_x the amount of reduction is again 50 % of the 2020 target, i.e. 20–10 % related to the 2000 reference figures.

Such a requirement is as questionable as the request for 75 % CO_2 emission reduction, while the temperature increase in the atmosphere is the more relevant criterion, which is depending on more than CO_2 emissions only. The request for 4 h *door-to-door travelling* time for the performance area "quality and affordability" is indicating the beginning of a fundamental change in aviation. On the one hand this requirement implies the multimodal aspect, since 4 h door-to-door cannot be realized by air travel only, on the other hand prime focus is put on the "quality and service aspect" of air transport. In this context also the request for *1 min punctuality* under all weather conditions must be discussed.

A minute punctuality is practically unfeasible. Along a transport chain a lot of unexpected disturbances may occur, and corrective measures also need some reaction time. Further to react for 1 min punctuality in case of any disturbance will make the overall system very sensitive and unstable. Small delays will occur very often, and robustness of the system will disappear as far as incremental punctuality is requested. To ensure predictability of the entire transport process, also including delays with adequate accuracy is in fact what is meant by this requirement.

Looking again at the requirements about environmental impact it must be noted, that the dynamic development of the atmospheric warming is the key parameter to be addressed. As mentioned in Chap. 10 for some years CO_2 was considered to be the most critical greenhouse gas. Today science knows, that NO_x and contrails as well as water vapour have a similar impact on the global warming trend, but have a quite different behaviour in time, as shown in Fig. 10.4.

Therefore, future research shall be spent on finding the adequate requirement formulation to reach the relevant targets.

Further requirements for air transport in 2050 address "seamless security" and an increase in aviation safety by a factor of 10. At this point it is to be noted how far "seamless security" is compatible with the legal data privacy. Therefore, this issue is more a legal rather than a technical one. Concerning safety improvement it is to be questioned where the real benefit of such an effort might be, since today aviation provides by fare the safest way of travelling.

Associated to these targets for aviation in 2050 some other aspects must be addressed, which have been mentioned in the previous chapters, too.

As described in Chap. 1, Fig. 1.11 the cost of crude oil as the basis for kerosene increased over the years from about \$30 up to \$100 between the late 1980s and 2010. This trend will go on for the next decades leading to significantly increasing operating cost.

Further air mobility reaches a certain saturation, if the GDP reaches a level of about 25.000 Billion \$, as shown in Fig. 1.4. Consequently, no more quantitative growth can be expected in those regions. Since the growing countries started to develop their own aviation industry, and e.g. Chinese COMAC intends to cover at least 50 % of its home market, the potential for Airbus and Boeing aircraft sales is questionable, [7].

In the 1970s public awareness of the environmental impact of man-made emissions and mobility worldwide growth developed, one conclusion of this awareness, but also of the economical development resulted in a statement that quantitative growth is limited as described by the *Club of Rome* [8]. Although these trends did not affect the air transport system until the early years of the twenty-first century, they influence further progress of air transport.

11.2 Future Energy Provision and Alternative Fuels for Air Transport

A prerequisite to enable air transport is energy to drive the vehicles. Over the last decades turbo fan engines fired by kerosene dominated civil transport aircraft. Although new oil and gas reserves are discovered every year, it became common sense that the peak oil and gas production has been passed and the worldwide production will decrease to about 50 % of the production in 2050.

This development is associated with growing increase in the oil and gas price, as it is shown in Fig. 11.1.

Referring to the International Energy Agency (IEA), in 2012 transport as a whole took nearly 50 % of the worldwide oil consumption, while air transport covered 10 EJ or 5.7 % of the global consumption [10]. This picture will change until 2050, when air transport will share 7.7 EJ of 56 EJ overall oil consumption. On the other side biofuels will become a remarkable energy supply with 27 EJ for overall transport, where aviation will consume about 2.9 EJ. This development correlated with a global increase of transport energy consumption from 92.6 to 100.7 EJ between 2009 and 2050.

Consequently, alternative energy carrier needs to be discovered and developed. There are various alternatives currently known, like synthetic fuels formed from coal (Coal to Liquid, CtL), gas (Gas to Liquid, GtL) or biomass (Biomass to Liquid,

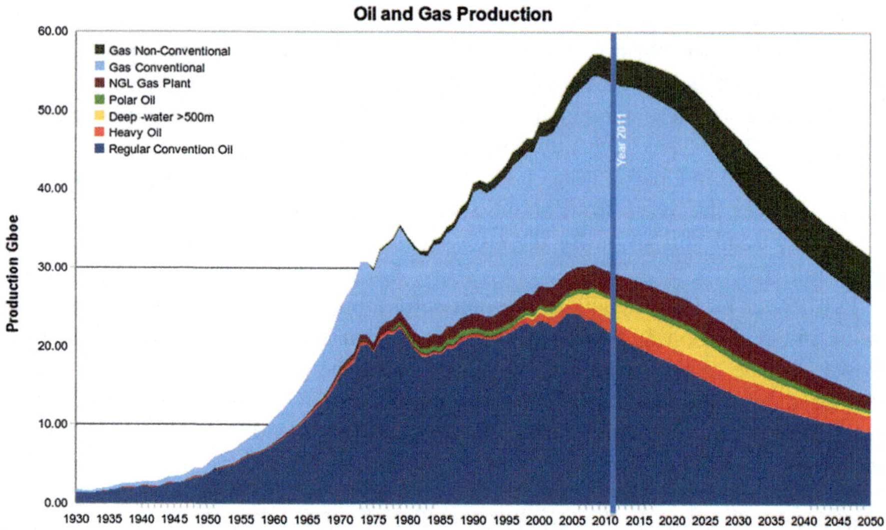

Fig. 11.1 Global oil and Gas production and estimated forecast [9, 10]

BtL) using the *Fischer–Tropsch* process. Further hydrogen, ethanol or methane have been investigated in the past. In the recent years biological fuels based on plants like jathropha and soja have been developed, but also methyl ester derived from fatty acids (FAME) or rape (RME) have been used. At last algae became of interest.

Looking at this different kind of fuel for aviation the production process is to be considered from different perspectives as shown in Fig. 11.2. The *Fischer–Tropsch* Process requires for a higher energy effort and also produces higher CO_2 emissions. On the other hand the quality of the fuel received is higher because the energy content is increasing through different process steps. Enzymatic conversion using formation or catalysis offers the opportunity to receive high-quality fuel with less energy effort.

Fig. 11.2 Fuel quality and process energy effort [11]

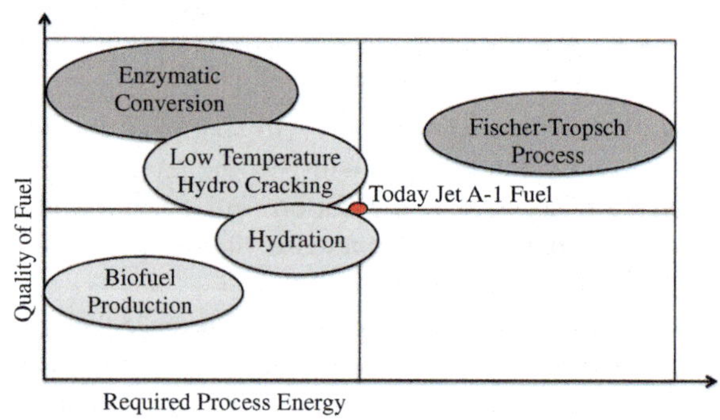

The greenhouse gas impact, however, of ethanol as the final product, which is got, e.g. from woody or agricultural biomass is heavily depending on the basic biomass material being used. Also the required land use is of crucial relevance. If the raw material crop is in competition with food production and woody areas, essential for CO_2 reduction this process becomes very critical for ethic reasons.

For Low Temperature Hydrocracking applied, e.g. to ethane, propane or butane more energy is required and the quality of the resulting fuel is lower in terms of energy content. Hydrocracking describes the chemical process to crack complex molecules into more simple ones. Since this process requires some more energy the overall balance is not as good as for enzymatic conversion.

Figure 11.3 presents the principle chain of fuel production. The upper chain shows the process of crude extraction, which is representative for raw oil, coal, uranium and other fossil or natural primary energy carrier. The lower chain represents the production chain for biomass cultivation.

The major difference in both processes is the fact, that for biomass cultivation water and CO_2 are used to let the plants grow. Here atmospheric CO_2 is used for the growth of the plants, which reduces the overall CO_2 balance. The raw material

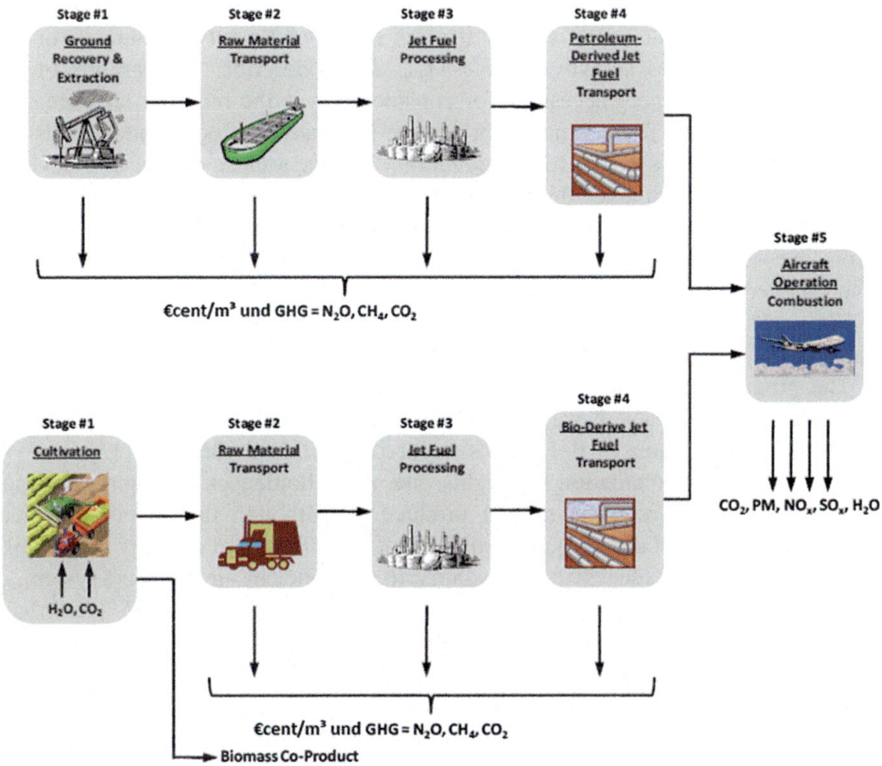

Fig. 11.3 Principle chain of biomass and crude conversion to kerosene [12]

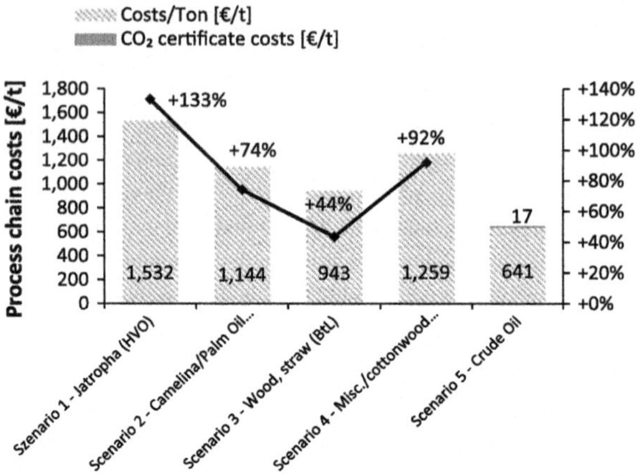

Fig. 11.4 Cost comparison of different fuel production chains [12]

harvest and extraction require different amount of energy for the machines used, while the transport can be seen similar. To compare the different processes the following Fig. 11.4 provides a comparison of the cost of the different fuel production chains.

It becomes obvious, that all biomass-based fuels are still more expensive than crude based fuel, although a crude oil price of 110 *\$/Barrel* is assumed for the forecast year 2020 as shown in Table 11.2, which describes the baseline of the analysis. Another aspect necessary for consideration is the required land area for biomass cultivation. There is no doubt that the land used for food production is not available for biomass-based fuel production. This is a clear ethical and societal issue. Therefore, biomasses derived from drop or plants like jatropha, which can grow on non-arable areas, are suitable resources for future alternative fuels. Further also algae become a more and more interesting raw resource, since they offer a much higher areal density compared to other raw materials.

Figure 11.5 provides a rough comparison of the required land use for soybeans and algae.

The productivity of algae earning 94.000 litre biofuel per hectare is 168 times higher than for soybeans. Nevertheless to ensure the worldwide airline fleet need of 322 billion litre an area of Belgium is required to fulfil this demand [13].

This exemplary calculation highlights one major bottleneck for alternative biofuels, which is the required rate of production. Although this is not a typical aviation issue in itself the adequate provision of fuel in terms of quantity and quality is a crucial factor of success for a sustainable and efficient air transport system.

Regarding fuel quality the international standard ASTM D1655 (American Society for Testing and Materials) sets the quality and performance requirements for aviation jet fuel. Since June 2011 the standard also includes the certification of 50 % biofuel blend within crude oil kerosene.

Here the energy density, the ignition temperature and the freezing temperature of the fuel are of paramount importance affecting directly the aircraft performance.

Table 11.2 Baseline assumptions for fuel cost analysis [12]

Assumptions				
Time reference	2020			
Interest rate	8 %			
Crude oil price [US $/ Barrel]	110			
Exchange rate [US $/€]	1.35			
CO_2 certficate cost [€/t]	12			
Share of free certificates	56 %			
Scenario	**Feedstock (share)**	**Transport Means**	**Transport Range (Phase 2 + 4)**	**Conversion Efficiency (Phase 3)**
1 (pract. Example)	Jatropha (100 %)	Sea Vessel + Truck	11,600 +220 km	85 % (HVO)
2	Camelina (50 %) Palm oil (50 %)	Sea Vessel + Truck	16,500 + 220 km	85 % (HVO)
3	Wood (50 %) Straw (50 %)	Truck	220 + 220 km	45 % (BtL)
4	Micanthus (50 %) Cottonwood (50 %)	Truck	220 + 220 km	45 % (BtL)
5	Cruide Oil (100 %)	Sea Vessel + Pipeline	8800 + 450 km	90 %

The following Fig. 11.6 provides an overview about the principle characteristics of synthetic and biofuels.

Compared to the ASTM D1655 specification most alternative fuels offer lower freezing temperatures, which provides better operational safety when flying at high altitudes. On the other hand the ignition temperature is higher and can adversely affect the reignition performance. Further alternative fuels offer a lower density as shown at the right figure leading to a better flight performance of the aircraft if concurrently considered with the higher energy density, shown on the left side.

Summarizing these effects with respect to the aircraft flight performance some improvements can be achieved for the payload range of an aircraft as shown in Fig. 11.7.

The benefit of lower density can be used, if the overall range of the aircraft shall be extended, as shown at the right figure. For a given maximum fuel volume of the

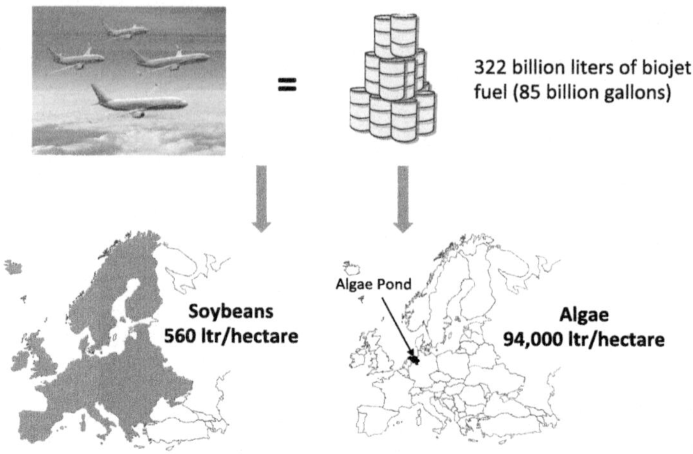

Fig. 11.5 Land use of algae and soybeans for worldwide fleet demand 2004 [13]

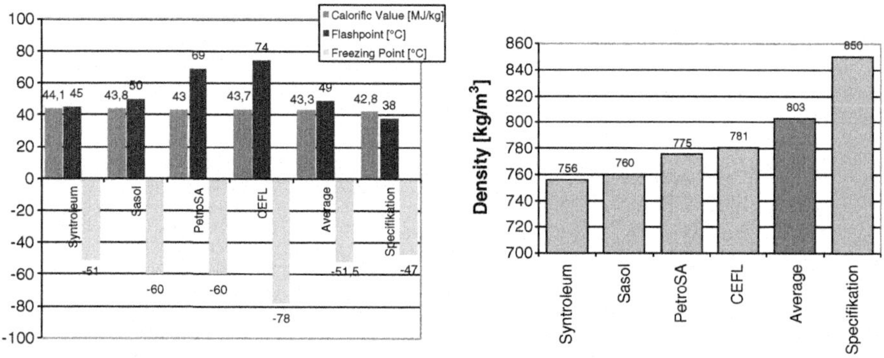

Fig. 11.6 Alternative fuel characteristics [14]

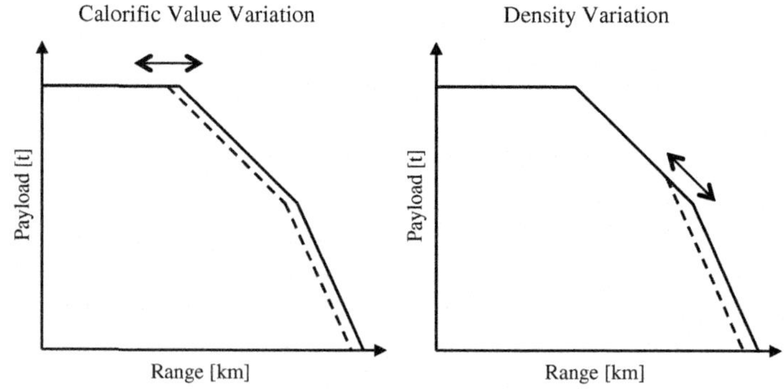

Fig. 11.7 Alternative fuel impact on payload—range [15]

Table 11.3 Commercial biofuel demonstration flights [16]

Date	Operator	Platform	Biofuel	Notes
June 2011	KLM	Boeing 737-800	Used cooking oil	World's first commercial biofuel flight, 171 passengers from Amsterdam to Paris
Juy 2011	Lufthansa	Airbus A321	Jatropha, camelina plants and animal fats	6 month regular series of flights from Hamburg to Frankfurt with one engine using biofuel
Juy 2011	Finnair	Airbus A319	Used cooking oil 50/50 blend	1,500 km journey between Amsterdam and Helsinki
Juy 2011	Interjet	Airbus A320	Jatropha	27 % jatropha between Mexico City and Tuxtla Gutierrez
August 2011	AeroMexico	Boeing 777-200	Jatropha	World's first trans-Atlantic revenue flight, from Mexico City to Madrid with passengers
October 2011	Thomson Airways	Boeing 757-200	Used cooking oil	
November 2011	Continental/United Airlines	Boeing 737-800	Algae	Houston to Chicago

aircraft the lower density leads to less aircraft mass and therefore fuel burn as a snowball effect. Since more than 90 % of all continental and also intercontinental flights are performed at much shorter ranges at point 2 of the payload–range diagram, the advantage of a slightly better energy density becomes more relevant [15].

Especially in 2011 various commercial flights had been performed to investigate and demonstrate the operational applicability of alternative fuels (Table 11.3).

Looking at the practical tests mainly jatropha and cooking oil have been tested. Most of the flights were performed on short ranges around the world. In addition various test flights have been performed and are performing to investigate further the operational relevance of different biofuel designs.

Concluding this section alternative fuels cover a wide range of types. They are capable to fulfil the technical requirements and offer in addition slight advantages concerning flight performance. However, the availability of sufficient quantities will be the crucial factor of success, since the required land use and potential competition with food production is an indisputable restriction.

There will be only a short time, when the production cost of biofuels will become competitive to fossil kerosene, as shown in Fig. 11.8.

Due to the decrease of crude oil reserves its price will increase and alternative energy resources must be developed for aviation right now on a broad level.

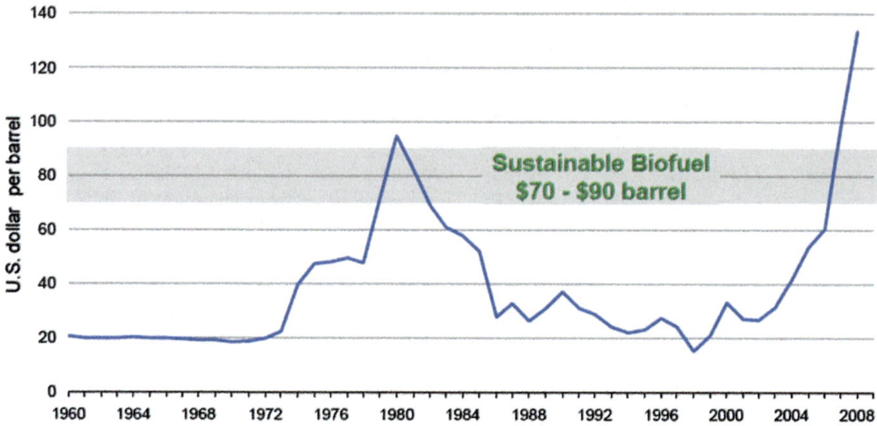

Fig. 11.8 Alternative fuel break even price [14]

11.3 Competitive and Multimodal Air Transport

Air Transport is one major pillar of transport. Therefore, its role has to be analysed to find out its major strength and future perspectives. Looking at the competitive situation between various transport systems, as shown in Fig. 11.9, high-speed trains, automotive and civil air transport compete mainly at ranges of about 600–750 km today in Europe. Due to the development of high-speed trains the relevant range of competition will develop up to 1000 km approximately in the next decades. Nevertheless, there will be an upper limit in this range for the competitiveness of railway systems, because they cannot increase speed much more above 300–350 km due to physical reasons. In addition, the energy consumption of trains

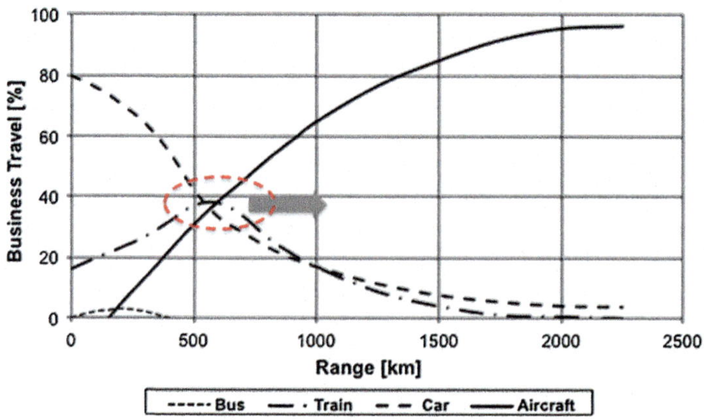

Fig. 11.9 Competition between transport systems

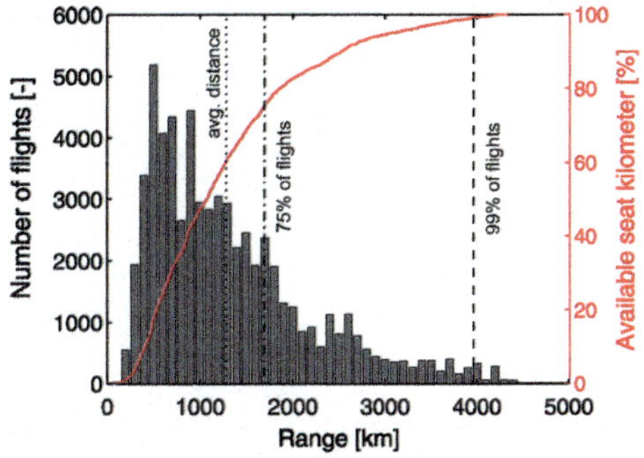

Fig. 11.10 Real worldwide operational ranges of short-range aircraft [18]

increases by the factor of nearly 4 when the speed doubles (E $\sim$ V^2). When the train is running at more than 150 km/h the aerodynamic drag of the train is becoming the dominant drag part compared to the roll-resistance drag of the wheel–rail system, as explained by *Niedzballa*, [17].

Also automotive transport will not extend its competitiveness in range and speed beyond 750 km because of the infrastructure and traffic density. Busses are not a competitor of civil air transport today and seem to have no real potential to become competitive in the future.

On the other hand it has to be noticed, that about 75 % of the worldwide distances flown by so-called short-range aircraft of Airbus A320 and Boeing B737 type are below 2000 km (1000 nm) [18].

However, the typical design range of short-to-medium range aircraft is about 6000 km (3240 nm), which may indicate an oversize from operational point of view (Fig. 11.10).

Here an antagonism becomes visible, because the airlines on the one hand request for highly flexible aircraft, which implies at first range flexibility. On the other hand, those flexible aircraft are mostly operated far away from the optimized design point, which makes them less efficient. Although during those off-design missions less fuel is carried on-board the aircraft structure and mass is dimensioned to the maximum fuel load. This additional mass is pure ballast, which causes additional fuel consumption.

Looking at the operational competitive situation and at the European railway network structure it becomes obvious, that its density and connectivity among the bigger cities in Western Europe will make high-speed trains more competitive. Figure 11.11 provides an overview about the 2013 high-speed railway network structure. The development in Eastern Europe is actually behind this due to its history. Here improvements are mandatory. Alternatively, due to overall investment cost and land use air transport can be considered as an alternative, as proposed by *Schmitt*, [19]. A major advantage of high-speed railway transport is that it brings people directly to the centre of the cities, while most airports are located outside the cities.

Fig. 11.11 European high-speed railway system network [20]

Central city airports like *Berlin Tegel* today or *Munich Riem* in the past will be closed and replaced by airports located outside. This is due to the infrastructural limitations for future expansions and the increasing recognition of air transport noise impact. At this point airport connectivity becomes a crucial success factor for the future development of air transport. Therefore, future air transport must be understood as an intermodal system including alternative transport systems like public transport and it is not the aircraft, which is the key.

Although this situation seems to be very specific for Europe similar situations could be observed also in the United Stated and China [21–24]. However, in these regions the density and network size are limited to some areas because of the size of the entire country. In the United States separated West Coast and East Coast networks can be identified, while the central country is not densely covered by railway systems except the Chicago and South Central regions.

In a different way the Chinese railway system is stepwise expanding from the coast to the inner country [21]. While a railway network densely covers the coastal regions network only a very view tracks are going to central and westerns regions. However, looking at the development perspectives more and more railway tracks will be built.

It can be concluded from both countries that high-speed railway systems are discovering a country from highly populated coastal regions. This is complementary to the competitiveness analysis discussed in Fig. 11.9, where the advantages of rail systems are limited to approximately 750–1000 km. A similar development is visible, if the modal split in China is considered.

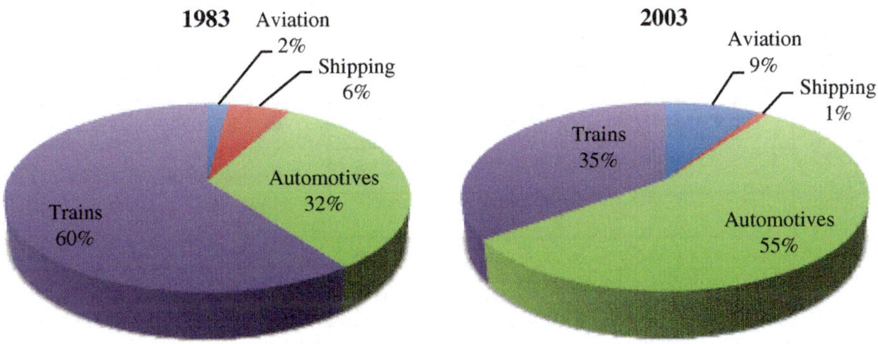

Fig. 11.12 Transport modal split China 1983 and 2003 [23]

A clear trend is visible in Fig. 11.12 from railway transport towards road transport. Also the percentage of air transport is significantly increasing from 2 to 9 % indicating the growing relevance of air transport in China to explore and develop the country.

Further air transport is of advantage in regions with less developed infrastructure like in central USA, China or Brazil, where jungle areas do not allow for a fast development of road and railway infrastructures. Those regions can only be developed by air transport, which is the reason why, for example in China about 150 new airports are intended to be built in the next 20 years.

Beside these topological considerations also the efficiency of the entire transport process needs to be considered. Figure 11.13 demonstrates the relation of the side course time (SCT) to the overall course time (OCT) for different door-to-door multimodal passenger travel in Germany [25].

Starting at different locations in the centre and vicinity of Munich as a representative example the timeshare of side course time to overall course time is analysed. It becomes obvious, that for multimodal air transport about 70 % of the overall course time is spent for the approach and departure to and from the airport,

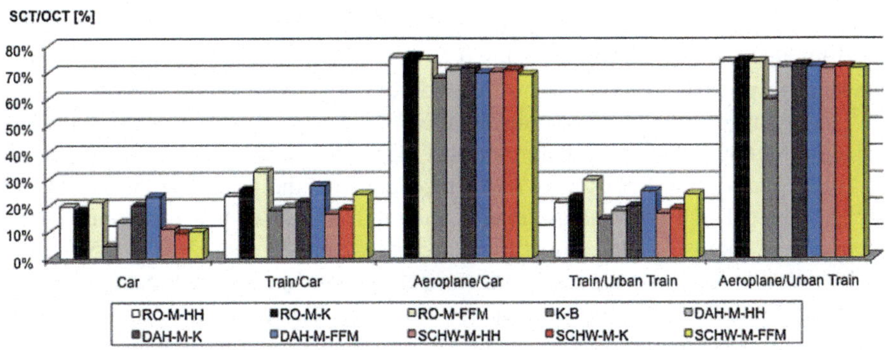

Fig. 11.13 Time share of different intermodal transport modes [25]

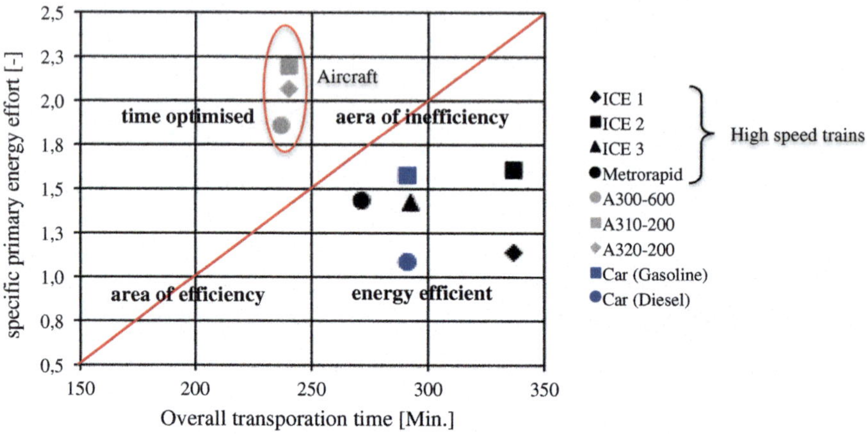

Fig. 11.14 Time and energy efficiency of various transport systems on the O–D pair Rosenheim-Frankfurt [25]

including the time spent at the airport. All other transport systems like automotive and train show a side course timeshare of 20–30 % at maximum. That means in fact, that air transport considered from door-to-door is not as time efficient as it should be if distances between 350 and 800 km are investigated. If side course time will share about not more than 50 %, air transport becomes much more competitive.

Further, if the associated overall time and energy efficiency of the different multimodal transport solutions is compared, air transport shows the best time efficiency, but also the highest primary energy effort, when looking at Fig. 11.14. Here time and energy effort of different transport systems is described on the origin-destination (O-D) pair Rosenheim-Frankfurt, where Rosenheim is a small city in the southern vicinity of Munich. As the transport distance and therefore the mission is the same for all systems the performance is comparable. The maximum range capability of a transport system is not relevant because it is not used in most cases.

On the other hand especially high-speed trains, like the German Inter City Express (ICEx) types are much more energy efficient compared to air transport but need longer overall travel times. This principle situation does not change, whether the starting point is in the center of Munich or somewhere in the vicinity.

Taking this observation into consideration together with the timeshare of the side courses of air travel, the overall air transport seems to be less efficient, because during the flight phase time losses have to be recovered for the price of a higher overall energy effort. The time—energy relation shown in this analysis is representative for all transport chains on the competitive range between 350 and 750 km [25].

In order to improve air transport competitiveness the entire transport process must be set into the focus, to balance time and energy effort of the different contributing systems and processes along the chain [26]. That means in practice from time perspective, all processes and systems contributing to approach to and

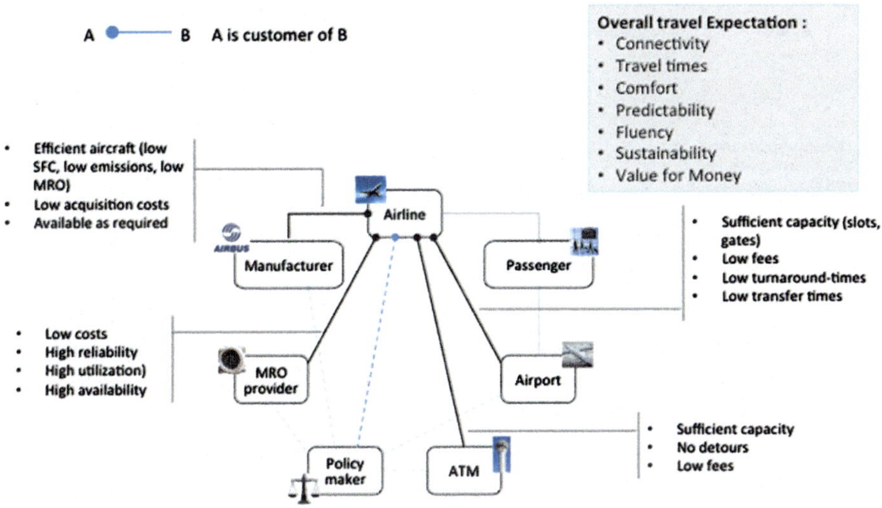

Fig. 11.15 Airline expectations in air transport [26]

stay at the airport must be accelerated in order allow the aircraft to fly slower at lower energy level. The airline as a customer also requests for those functional process performances in terms of no detours during flight to be offered by the ANSP or short turnaround times provided by the airport ground services, see Fig. 11.15.

Those requirements are very closely linked to the airport and airspace capacities, which are more and more reaching the limits. Such a development becomes more severe, if the forecasted increase in passenger and aircraft movements is taken into account.

In addition also physical performances like aircraft range and seat capacities are demanded from the aircraft manufacturer in conjunction with low-cost flight performance like low fuel burn or reduced crew operations.

Therefore, these stakeholders have to collaborate more in the future in order to balance their expectations and capabilities in a more process and operation oriented way [26].

In Fig. 11.16 looking at passenger expectations, the airline and the airport are the direct service provider, where the passenger especially expects seamless travel functions like connectivity and predictability but also services like shopping, lounges and comfort in flight like inflight entertainment or catering. Those services imply less physical performances like speed or range but more functional process performances.

There is no doubt that each customer being, e.g. an airline or a passenger requests for best service at lowest price, but this principle request is not key for a change in mindset. Moreover, overall process orientation door-to-door and increased common process improvement rather than single stakeholder and

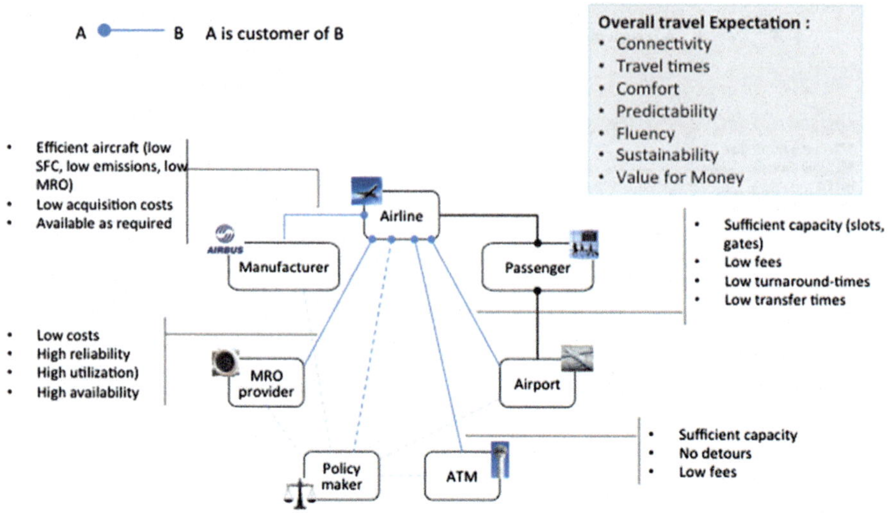

Fig. 11.16 Passenger expectations in air transport [26]

technical optimization will become the major challenges for future intermodal air transport. This change in mindset can be understood as a paradigm shift compared to the past, where mainly single stakeholder interest had been addressed.

Further the climate impact of air transport emissions must be taken into account as well as air transport noise, as introduced in Chap. 10.

Both issues are of high societal relevance and must be addressed as in *Flightpath 2050* to keep societal acceptance of attractive air transport [5].

Concluding, this situation will lead to a stronger competition between civil air transport and high-speed railway systems in the future due to less efficient aircraft operations on the relevant extending transport ranges.

To cope with these differentiated aspects an approach for a balanced holistic view on air transport is required, which is presented in Sect. 11.5. Before that, some technology trends and challenges for future air transport will be discussed in the next section.

11.4 Technology Trends

Chapters 1 and 5 showed that aircraft technologies have reached a very high level of maturity. Further some trends are shown, indicating some kind of saturation in global mobility can be envisaged in the next decades, as shown in Fig. 1.4. In Chap. 9 the operational needs to improve Air Traffic Management efficiency by 4D flight trajectories and reduced amount of airspace sectors in Europe have been described. This section provides some perspectives in technologies, which will drive the future development.

11.4.1 Technology Perspectives in Aircraft Design

Typically, in the past the design of an aircraft has been characterized by the aerodynamic performance in terms of minimum drag or high glide number (Lift to drag ratio) and the structural lightweight design. Also low thrust-related specific fuel consumption was always a major design target. The classical tube-wing configuration has been proven as the most efficient layout over the decades.

Figure 11.17 provides an overview about actual and envisaged aircraft programmes until 2050. New aircraft platforms for new technologies will not occur until the 2030s, because new relevant aircraft like the A320NEO and B737Max for short range as well as B787 Dreamliner and A350XWB are under development or started to enter the market.

They will be produced for at least 20 years. Also in the field of macro- and ultra-wide body aircraft the A380 and B747-8 have just entered production and need some decades until they reach the breakeven point to become profitable for the manufacturer. So the next decades will be characterized by improving production efficiency and slight product upgrades. This is also documented in the vision of the European research establishments [28, 29].

Nevertheless, we are experiencing actually a change in relevant technologies. While in the past aerodynamics and light weight structures drove the aircraft development in the future communication and information technologies will more and more determine the value of a new aircraft. The aircraft itself can be seen as an

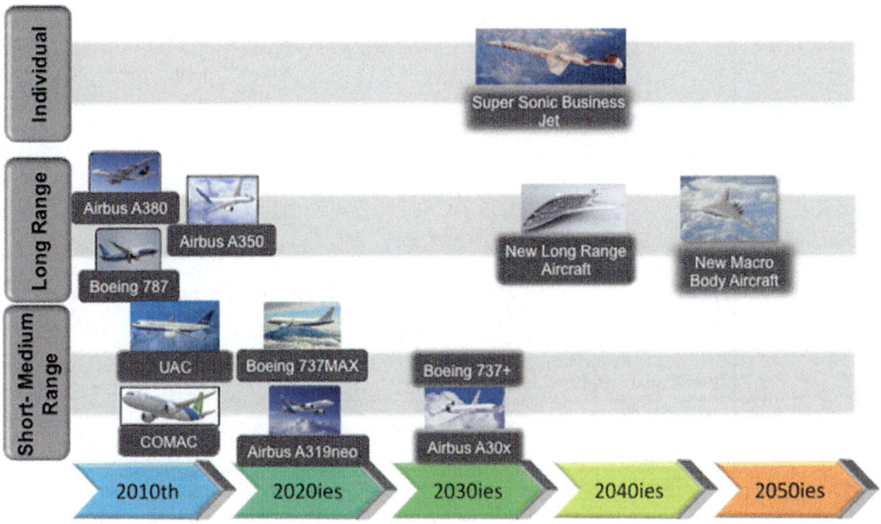

Fig. 11.17 Long-term perspectives of aircraft programmes [27]

investment product, which has a long product life cycle. On the other hand the aircraft cabin and also systems and avionics are more consumable products with much shorter lifetimes. Consequently, due to these principle characteristics of aircraft and equipment but also due to general technology trends the future aircraft innovations are made on functional level rather than in classical aeronautical disciplines.

11.4.1.1 Lightweight Structures

Regarding the development of lightweight structures composites became more and more relevant during the last programmes as shown in Fig. 11.18. The actual experiences with Airbus A350 and Boeing B787 Dreamliner indicate, that a portion of about 50 % composites on aircraft structures is at the moment the upper limit and this will only change if at least 20 years of operation will identify further potential for weight reduction. The trend line of Fig. 11.18 also highlights that A350 and B787 incorporate more composite share than it might be in a pure evolutionary trend. This is in line with the observation, that composites were a significant design driver forced by the market.

Especially, the risk of damage and damage detection at the fuselage may become a critical issue. During turnaround the fuselage is the major interface for boarding, loading and servicing, which includes the higher risk for potential damages. On the other hand high-quality smooth surfaces, which can be realized by composites, are of paramount interest for low drag wing designs.

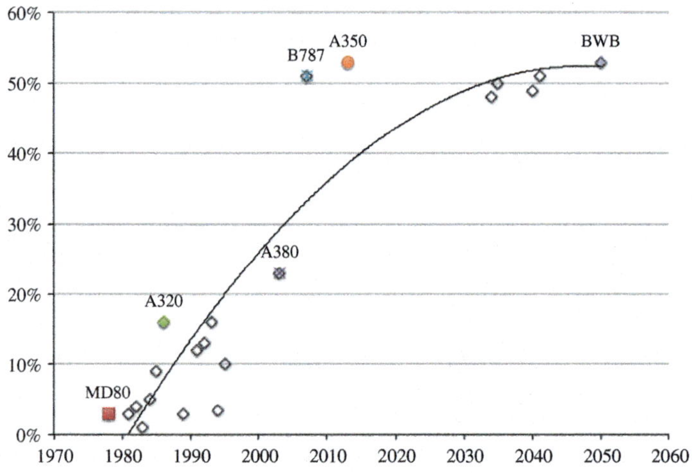

Fig. 11.18 Perspectives of composite structures

11.4.1.2 Aeroelastic Tailoring

Aeroelastic Tailoring describes the integrated technology, which allows for structural weight reduction by active suppression of gust loads using high dynamic flaps. This approach might offer indeed some potential at the main wing box. The secondary structures will not contribute because they carry the control surfaces. Further it must be considered that high dynamic control requires higher forces and therefore energy. Future research has to show the real potential of this approach.

11.4.1.3 Aerodynamic Drag Reduction

In the field of aerodynamics laminarity of the airflow is again in the focus of actual research. Laminarity describes the behaviour that the airflow is streaming around the wing profile without wake turbulences in the boundary layer resulting in lower drag than with turbulent boundary layers as today. Former studies have shown a potential of about 10 % drag count reduction using this technology [18]. It is out of question, that laminarity will provide its best performance under steady state long distance cruise conditions, which are typically given for long-range flights. Recent studies investigated how far the potential of natural laminar flow can be made available, also under real operational conditions of short-range aircraft [18]. The following figure shows the spread of natural laminar flow benefit for a complete European flag carrier fleet.

Looking at Fig. 11.19 short-range aircraft operated on long legs of about 3000–4000 km contribute only minor block fuel savings to the entire fleet, because those leg length are served only rarely in a real fleet network. The more real operational legs are considered, which are shorter, the more the graph is approaching to the left, the more block fuel savings can be achieved on fleet level. However, under best, i.e. full laminar flow conditions, an overall benefit of up to 5.5 % on fleet level can be

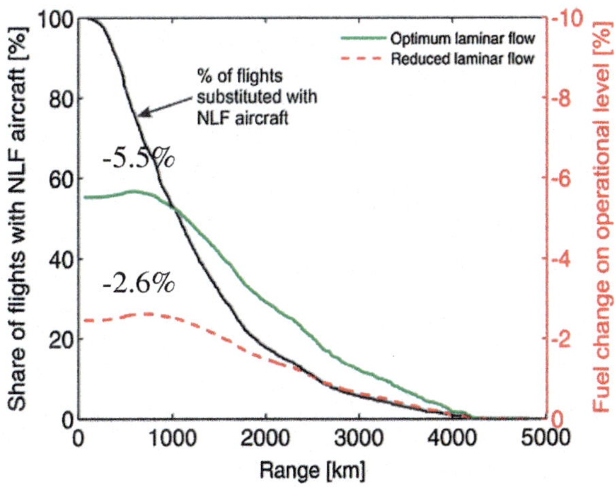

Fig. 11.19 Operational benefit of laminar flow technology on short-range aircraft [18]

reached. It has to be taken into account, that also minor contaminations will reduce the laminar performance, which reduces block fuel savings to 2.6 % in this case. Those contaminations are caused by insects at low level flight during take off and landing or ice crystals, soot or aerosols at high cruise altitudes. Further intensive cleaning must be considered during each turnaround. Therefore, the benefit of laminar flow technology is questionable for short-range aircraft, which offer only very limited cruise conditions. The aerodynamic performance of winglets with about 3.5 % block fuel saving seems to be less costly at the same efficiency.

Alternatively, hybrid laminar flow technologies using boundary layer absorption or air injection could be used to establish more stable laminar conditions. Those technologies have to be applied at the leading edge of the wing (up to 25 % chord) to ensure laminar flow up to more than 60 % of wing chord. Therefore, hybrid laminar flow technology might be applied for long-range aircraft only. Big efforts are made at the moment in the European research community, including the European Commission to investigate these technologies and identify all repercussions like additional weight, system complexity and cost, and production of high quality surfaces, which have to be understood, before introduction on production aircraft can be considered. It is not yet clear whether these complex technologies will ever appear at the market.

11.4.1.4 Future Engine Development

Over the last decades aeroengines achieved a very high level of efficiency as shown in Chap. 5, Fig. 5.26. The current state of the art engines like the A380 GE90 or the actual geared turbo fan engine PW1000G offer more than 60 % reductions in fuel burn, CO_2 and NO_x emissions compared to the JT3 engine, which was operated on the B707.

Like the classical aircraft technologies also further improvements in engine technologies will require much effort in terms of money and time for limited steps forward. Nevertheless, further reductions in engine/aircraft noise creation and engine emissions remain major tasks. Looking at Fig. 5.25 there is saturation in further emission and fuel burn visible in the region of 35 % efficiency compared to the JT3 engine. Two principle engine architectures, the geared turbo fan and the open rotor concept offer the highest potentials. While the geared turbo fan will be about 15 % more efficient compared to the A380 GE90 engine it also provides a very good noise shielding due to the encapsulated fan. On the other hand the open rotor concept indicates a further 15–20 % fuel burn reduction benefit. This remarkable improvement is associated to less reduction in engine noise creation compared to the geared turbo fan as shown in Fig. 11.20.

These engine technologies are also part of the European "Clean Sky Program" and a lot of additional research is required before final conclusion can be drawn! The open rotor concept but also an extended geared turbo fan fit well to aircraft configurations like the Blended Wing Body aircraft, which offers a good noise shielding due to its integrated fuselage wing shape [30]. Combined with its drag reduction

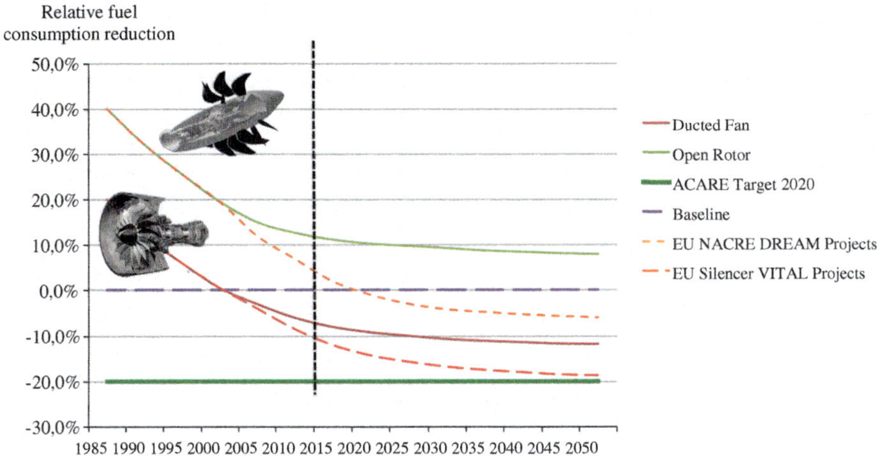

Fig. 11.20 Impact of engine architecture on engine noise potential, [Rolls Royce Plc.]

potential of about 20 % the BWB shows a realistic potential to reach the *Flightpath 2050* goals. Another trend in future engine technologies also associated with challenging integration efforts is described by distributed propulsion concepts. Those concepts aim on one hand to a multiplication of several complete engines or on the other hand, to have one or two core engines driving several fans integrated on the aircraft. While the first approach will increase the accumulated fan diameter of the engines to improve efficiency, also the entire weight of the entire engine system will increase significantly. Alternatively, the distributed approach with one or two main drives offers the chance to limit the increase of overall engine system weight.

Other future concepts look at electrical engines. The main challenge of these approaches is associated to the energy conversion and weight of the energy storage. Also the development of solar panel technologies may find a way of application on the aircraft, as shown by the first solar power gliders and the Solar Impulse concept, prepared to show the capability of a man flight around the world only by solar energy.

11.4.1.5 More Electric Aircraft

Having shown in Chap. 9 electronics and avionics represent today about 30 % of the aircraft value. This is the same order as engine and the rest of the aircraft are representing each. As in consumer or automotive industries electronics and software systems are taking over more and more functions, which have been realized by mechanical systems in the past. In addition the aircraft cabin incorporates an increasing amount of new electronic passenger services like entertainment systems and cabin management systems.

Since the beginning of the twenty-first century "More Electric Aircraft" has been established as a major technology trend to investigate the replacement of mechanical systems by electronics. Further there is a significant increase in electrical power consumption on-board, which must be handled. Up to now electrical power is generated be engine shaft power off take to drive electrical generators. For the future it is envisaged to use fuel cells as electrical power generators to discharge the engine. A potential of about 6–10 % fuel savings depending on the flight phases might be possible, if more electrical systems will be integrated in future aircraft. Decoupling the air condition system from the engine, which makes it bleedless, may save further 10 % of fuel in cruise and flight idle. Concluding more electric aircraft technologies are much promising technologies to design energy efficient aircraft in the future.

11.4.1.6 New Business Areas in Aircraft production

The overall lifetime of an aircraft is about 40 years. Consequently, the aircraft being developed today will determine the future of air transport for the next 60 years considering 20 years of production and sorting out the last aircraft 40 years later.

On the other hand today the value of an aircraft from operator's and passenger's point of view is mainly defined by the attraction of the cabin and avionics functions.

Those mainly software-based technologies are also characterized by a short life time cycle, which requires for permanent new development. Therefore as in other industries like consumer and automotive industry software-based systems are the key technology for improved more efficient and sustainable air transport, which also generates interesting profit, rather than former mechanical components.

11.4.2 Perspectives in Air Traffic Management

Air Traffic Management based on airspaces, communication, navigation and surveillance systems for flight guidance and management system offers different opportunities for further improvements. Since airport and airspace capacity are crucial for increased efficiency, a new set-up of the European airspace structure is the key. While the European airspace is roughly of the same size as the US airspace, 47 ATC service providers operate it, while in the US only one provider is responsible. In addition air traffic control is done by 58 centres, serving about 9 million flights per year instead of 18 million flights in the US.

Figure 11.21 shows the actual segmentation of the European airspace, which shall be revised by the European SESAR program. This essential issue is much more a political one than a technical, which makes realization much more difficult. At the end European countries have to give up national authority on their airspaces, which is sensitive from national security point of view. However, most of the technologies to improve ATM efficiency are available today.

In order to cope with an expected increase of up to 30 billion aircraft movements worldwide until 2030, 4D aircraft trajectories planning, tracking and management

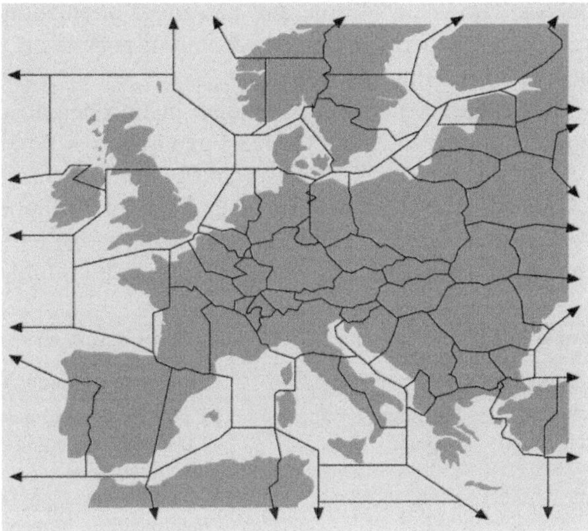

Fig. 11.21 European airspace segmentation

will become one crucial technology. Further aircraft self-separation will contribute to more efficient tactical flight management to use the airspace more efficiently. However, most ATM technologies are still available and the mandatory task is to perform a block wise worldwide implementation [31]. The realization of those technologies is a prerequisite to achieve more seamless and efficient aircraft movements with reduced and at least predictable delays. Because this activity requires simultaneous investments of ANSP, airports and airlines the realization is that challenging.

11.4.3 Perspectives in Airport Operations

Today, more than 40 % of the airport business is based on non-aviation turnover. Asset management covering parking areas, shops, restaurants and travel agencies is the main pillar, while the original airport functions to serve the aircraft during turnaround and to perform boarding and deboarding are cost-intensive. More and more check-in is done via internet and only baggage delivery and security check remain as major process elements in the terminal.

Based on these developments airport operator and airline interests are going into different directions. Airport operators prefer to have the passengers as long as possible in the terminal to motivate them for shopping. Airlines like to move passengers, especially business traveller, as quick as possible through the terminal.

In the future individual passenger guidance systems allowing for bidirectional communication will enable leading the passenger through the airport considering the individual customer needs. The major challenge is to predict potential delays and to alert the passenger early for departure but also to inform him early about delays. The value of time and passenger orientation is addressed so that the passenger can experience the stay on an airport as an attraction or while being on a business trip without haste.

Regarding airport infrastructures in Europe and the U.S. not a lot newly built airports are expected but the existing airports will be rebuilt within the given areas. Such recreations will direct to more seamless passenger flows. In Asia and the Middle East new airports are built considering the most efficient passenger flows. A particular challenge is associated with so called Mega Airports of 4 and more runways. Although those airports provide a huge airside capacity the large

Table 11.4 Key technologies for future airport developments [28]

Airport component and functions		Promising technologies	Drivers
Airport Network	Passenger-oriented airport	Two-way communication devices: - Datalink technologies and wireless communications - Cloud computing, computer identification, human–machine interfaces - Robotics: assistance robots. Terminal transport systems	User Comfort Capacity
	Operations	Data fusion techniques, field sensor network Vision techniques, remote sensors and acquisition devices for security check Information and communication technology, dynamic multi-risk management New aircraft configurations New aircraft procedures and new procedures management	Safety Security Environment Capacity
	Information handling and collaborative decision making	Optimisation analysis based on multi-actor, multi-objective, risk monitoring, management system, sensors and ambient intelligence Near-field communication technologies, RFID, Bluetooth and mobile devices, LED bar codes Development of multi-scanning devices Voice and data link between base-station and vehicles, various transmission technologies GPS, Galileo, high resolution radar	Capacity Safety Security
Airside	Ground-based noise measurements	Develop noise-preferential approaches Noise monitoring and modelling Noise walls or anti-noise interferences systems	Environment
	All weather operations	The aircraft as a meteorological sensor Improved instant forecasting (i.e. for strong winds) Information sharing with other air vehicles and the ground	Capacity Safety

dimensions might cause significant time to move on the apron, which reduces the attraction. Summarizing the future perspectives on airports the following table provides some key issues, which will drive the future developments (Table 11.4).

In addition also new business concepts will be considered. Since neither airport nor airline business is really profitable both stakeholder will consider whether they will develop from a pure infrastructure or flight service provider to a mobility provider.

The airport hosts a lot of travel agencies and aircraft from different airlines. The airport while offering a full service a door-to-door travel service can use these resources. Further an air trip can be offered independent from an individual airline schedule, which allows for much more flexibility in travelling.

The airlines on the other hand also can offer total travel services from door to door, when cooperating with railway, bus and taxi companies. At many points, those cooperation are still existing, e.g. in Vienna, but further evolution is expected. Also in this business modern communication technologies based on smart phones and software will drive the development.

11.5 Integrated Approaches Towards Future Air Transport

In order to cope with the different challenges of the air transport substructures and the need to improve the entire system a change towards comprehensive research is needed to cover all relevant stakeholders. Such an approach will lead to interdisciplinary and integrating research and development. This will not mean, that disciplines are no longer needed, but their relevance will change.

To assess the global impacts of potential future technologies an overall system concept is required [32]. The selection of technologies to compose new systems is driven by knowledge, experience and gut feeling. Here a process-oriented approach following functional chains is the fundamental strategy to select and combine the right technologies.

The key focus will be at the interfaces and interferences between the different air transport subsystems to achieve balanced and efficient overall solutions.

One approach, which incorporates such an approach is called "Virtual Integration Platform (VIP)", developed by *Szodruch* and *Gollnick*, [33]. A VIP is a description of a future air transport system concept consisting of aircraft, airport, airline and air traffic infrastructure concepts, which are composed of potential new or/and existing technology solutions ensuring interface compliances. The intention is to define a set of appropriate technologies in terms of physical and procedural solutions, which will be commonly developed and optimized across all stakeholders in an integrated approach. Figure 11.22 provides an example for a future long-range air transport concept, where different stakeholder-related technologies are brought together for a balanced and optimized composition.

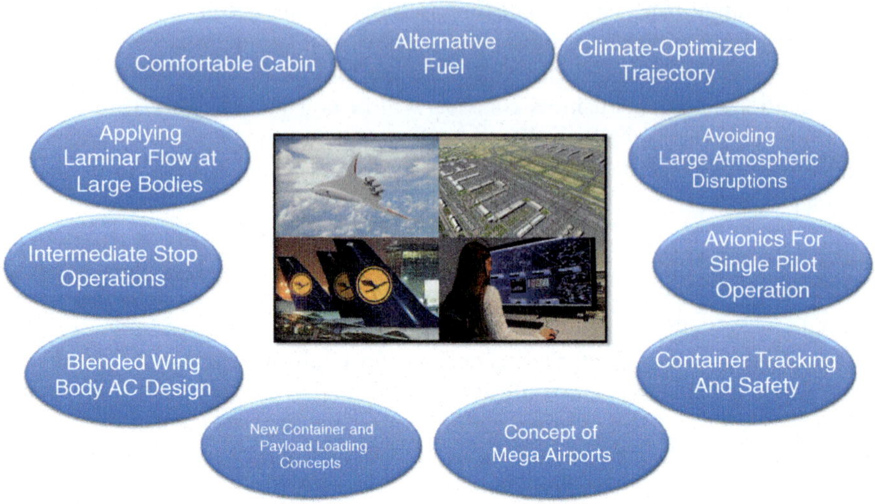

Fig. 11.22 Virtual integration platform—long range [25]

Future sustainable and efficient long-range transport will be driven by high-altitude climate compatibility and intensive comfort expectations of the passengers, despite the permanent request for cost efficiency. Following these drivers, the future long-range concept called "Comfortable And Clean" addresses future technologies, which are of particular but not exclusive interest for this kind of transport. For example, the blended wind body aircraft configuration provides special advantages for mass transport, which is relevant for long-range flights. Single pilot operations may enable mass and cost reduction for long-range aircraft, but can be also applied die short-range aircraft. Large atmospheric disruptions like volcanic eruptions or large adverse whether conditions affect the global traffic flow but not only the regional traffic. These are some examples of the technology ensemble, which explain the philosophy. The realization of such a VIP concept requires the development of appropriate design methods in terms of tools but much more in interdisciplinary ways of working. The integrated layout of aircraft and airport and the integration of the aircraft in an individual airline fleet and also the global fleet is necessary to assess at the end the impact of such a technology ensemble.

11.6 Compliance Achievement with Flightpath 2050

Summarizing the different aspects of air transport, which have been addressed throughout the book it can be concluded, that the challenges for air transport in 2050 politically defined in the *Flightpath 2050* can be achieved only through paradigm shifts in thinking and working in the air transport industry.

Because single classical disciplinary areas like aerodynamics, structures and have reached a very high level of maturity they will not enable significant changes in the future alone.

Further also the aircraft market will reach certain saturation in the next 20–30 years globally, while such saturation is still visible in Europe and U.S.

Increasing energy cost, either crude oil price increase or shift to alternative energy carrier, will damp the people's willingness to move.

Environmental responsibility directing to emissions but also noise drives to additional fees and flight restrictions.

Limitations in maximum airport and airspace capacities require for new operational concepts to cover the people's mobility demand.

Some necessary paradigm shifts derived from these main trends require:

- More process orientation and process technologies in production and operation of aircraft resulting in less stakeholder orientation to make the transport process more efficient and attractive;
- More orientation to inter- and multimodality to understand air transport as part of the bigger transport system;
- Change from disciplinary research and development to interdisciplinary and integrative work as the future main driver for innovation and improvement;
- Communication and software technologies as the key enablers for efficient operation and production;
- Focus short lifetime products like cabin interiors and systems, communication and software systems, which need to be updated in short sequences, instead of long living aircraft fuselages.

At the end the vision of "Air Transport 2050" is better flying instead of more flights to enable peoples mobility in the future.

References

1. ACARE: European Aeronautics: Vision for 2020. www.acare4europe.org/docs/Vision%202020.pdf (2001). Accessed 27 Feb 2011
2. D'Angelo, M., Gallman, J., Johnson, V., et al.: N+3 Small commercial efficient and quiet transport for year 2030–2035, NASA/CR-2010-216691 final report for contract NNC08CA85C, 1 May 2010
3. NASA: NASA & the next generation air transport system (NEXTGEN). http://lp.ncdownloader.com/eb2/?q=nextgen%20whitepaper%2006%2026%2007%20pdf. Accessed 26 July 2006
4. Plath, F.: Analysis and summary of various market forecast of civil avaition und development of a data base for individual data analysis. DLR Institute of Air Transport Systems, Technical University Hamburg, LK-BA-02/2008, Hamburg (2008) (in German)
5. European Commission: Flightpath 2050 Europe's vision for aviation report of the high level group on aviation research, Brussels. http://ec.europa.eu/transport/modes/air/doc/flightpath2050.pdf (2011). Accessed 31 Aug 2014

6. Nolte, P., Gollnick, V.: Future of Aviation—Vision 2020—a midterm resumee. Symposium of the Institute of Air Transport Systems, Hamburg, 30–31 Aug 2011
7. Toh, M.: Comac's sights on single –aisle sector, Flight International, 20–26 Nov 2012
8. Meadows, D.-H., Meadows, D.-L., Randers, J., Behrens, W.: The limits to growth, Universe Books (1972). ISBN 0-87663-165-0
9. ASPO: World oil and gas production, Ireland Association for the study of peak oil & gas. http://aspoireland.files.wordpress.com/2011/05/aspo-oil-and-gas-production-profiles1.pngaspoireland.org (2013). Accessed 04 Jan 2013
10. IEA: Energy technology perspectives, International Energy Agency. http://www.iea.org/etp/explore/. Accessed 04 Jan 2013
11. Kuhn, H., Falter, C., Sizmann, A.: Renewable energy perspectives for aviation. In: Proceedings of the 3rd CEAS Air & Space Conference, Venice, Italy (2011)
12. Nittinger, N.: Value anaylsis of HVL and BtL bio fuel production chains for aviation. Master Thesis, IB-328-2011.25, Institute for Air Transport Systems, Technical University Hamburg, Harburg (in German)
13. Dagett, D.: Alternate fuels for commercial aviation, Boeing Company, Greenfuel Workshop, Cleveland, 18 Oct 2007
14. Gollnick, V.: Air Transport Systems, Lecture Series. Institute for Air Transport, Technical University Hamburg, Hamburg (2013)
15. Isfort, J., Nittinger, N., Gollnick, V.: Imapct of HVO-Fuel Characteristics on the Payload Range Performance, Conference Paper, German Aerospace Congress 2012, Berlin, 12–14 Sept 2012 (in German)
16. Wikipedia: Commercial biofuel flight demonstrations. http://en.wikipedia.org/wiki/Aviation_biofuel#Commercial_flights. Accessed 3 April 2013
17. Niedzballa, H., Schmitt, D.: Comparison of the specific energy demand of aeroplanes and other vehicle systems, Aircraft Design 4/2001, Elsevier, pp. 163–178
18. Wicke, K., Wunderlich, T.: Mission and economic analysis of aircraft with natural laminar flow technology. In: AIAA 11th Aviation Technology Integration and Operation Conference, Virgina Beach, 20–23 Sept 2011
19. Schmitt, D.: Air transport 2025 chances and challenges for Eastern Europe. In: Proceedings of READ Conference in Brno (2012)
20. Wikipedia: The high-speed railroad map Europe. http://upload.wikimedia.org/wikipedia/commons/thumb/e/ea/High_Speed_Railroad_Map_Europe_2011.svg/2000pxHigh_Speed_Railroad_Map_Europe_2011.svg.png,Wikipedia.org. Accessed 21 Mar 2013
21. Wikipedia: The Chinese railway network system. http://upload.wikimedia.org/wikipedia/commons/9/97/4%2B4_PDL_network_in_China_%28English_version%29.png,Wikipedia.org. Accessed 21 Mar 2013
22. Wikipedia: United States highspeed railway system http://upload.wikimedia.org/wikipedia/commons/6/6f/High_Speed_Rail_07-09-2009.JPG,Wikipedia.org. Accessed 21 Mar 2013
23. Knie, A. Marz, L., Wagner, H.-C., Wolf, J.: People Republic of China: Perspectives for Mobility in the 21 Century. In: Int. Verkehrswesen, vol. 60. Deutsche Verkehrswissenschaftliche Gesellschaft (2008) (in German)
24. Siemens: Faster to the future. J. Complete Mobility Siemens 5. www.siemens.com/mobility (2010)
25. Gollnick, V.: Potential for transport efficiency improvements of aviation transport systems. In: 25th International Council of Aeronautical Sciences, (ICAS), Hamburg, 11–13 Sept 2006
26. Gollnick, V.: Customer perspectives in aviation—a process oriented view. In: Hamburg Aviation Conference on Conference Key Note, 22–24 Feb 2012
27. Gollnick, V.: The aircraft of the future—Flying 2050. Guest Lecture University Zurich, 30 Oct 2012 (in German)
28. Brunet, M., deBoer, A., Gollnick, V., et. al.: The EREA vision on high priority research axes towards Air transport System 2050, Paper 229. In: 28th International Congress of the Aeronautical Sciences, Brisbane, 23–27 Sept 2012

29. EREA: From air transport system 2050 vision to planning for research and innovation, European research establishment association, Brussels. www.erea.org, http://s144928.gridserver.com/wordpress/wp-content/uploads/2012/07/EREA-ATS2-Booklet-red.pdf (2012). Accessed 11 Apr 2013
30. Gollnick, V.: The blended wing body—a green future air transport concept, Greener Skies Ahead, Berlin, 13 Sept 2012
31. Learmount, D.: ICAO warns of ATM paralysis, Flight International, 27 Nov–3 Dec 2012, flightglobal.com
32. Griffin, M.D.: System engineering and the two cultures of engineering, Boeing Lecture, Purdue University. http://www.spacecom/news/viewsr.html?pid=23775 (2007). Accessed 18 Aug 2011
33. Gollnick, V., Stumpf, E., Lehner, S., Szodruch, J.: Virtual Integration Platforms (VIP)—a concept for integrated interdisciplinary air transport research and assessment. In: AIAA 11th Aviation Technology Integration and Operation Conference, Virgina Beach, 20–23 Sept 2011
34. N.N.: Growing in significance: biofuel flight testing, aviation and the environment, Issue Feb.-March 01/2009. www.aviationandenvironment.com. ISSN 1755-9421
35. Hirschel, E., Prem, H., Madelung, G.: Aviation research in Germany, Bernhard und Graefe Publishing (2001) (in German)
36. La Rocca, G., van Tooren, M.: Knowledge based engineering to support aircraft interdisciplinary design and optimization. J. Aerosp. Eng. **224**, 1041–1055 (2010)
37. Schumann, U.: Atmospheric Physics, Springer (2012). ISBN 978-3-642-30182-7

Printed by Printforce, the Netherlands